Saas-Fee Advanced Course 37

For further volumes:
http://www.springer.com/series/4284

Joss Bland-Hawthorn · Kenneth Freeman
Francesca Matteucci

The Origin of the Galaxy and Local Group

Saas-Fee Advanced Course 37

Swiss Society for Astrophysics and Astronomy
Edited by B. Moore

Joss Bland-Hawthorn
Sydney Institute for Astronomy
University of Sydney
Sydney, NSW
Australia

Kenneth Freeman
Australian National University
Australian Capital Territory, ACT
Australia

Francesca Matteucci
Astronomy Department
Trieste University Osservatorio
    Astronomico (INAF)
Trieste
Italy

Volume Editor
Ben Moore
Institute for Theoretical Physics
University of Zürich
Zürich
Switzerland

This Series is edited on behalf of the Swiss Society for Astrophysics and Astronomy: Société Suisse d'Astrophysique et d'Astronomie Observatoire de Genève, ch. des Maillettes 51, CH-1290 Sauverny, Switzerland

Cover Illustration: The cover image shows the Andromeda galaxy superimposed over the Mönch, 4107 m (not to scale). This was the view from the Saas Fee school in Mürren, photographed by Ben Moore. The image of Andromeda is from http://commons.wikimedia.org/wiki/File%3A Andromeda_Galaxy_(with_h-alpha).jpg, by Adam Evans [CC-BY-2.0 (http://creativecommons.org/licenses/by/2.0)], via Wikimedia Commons from Wikimedia Commons.

ISSN 1861-7980   ISSN 1861-8227 (electronic)
ISBN 978-3-642-41719-1 ISBN 978-3-642-41720-7 (eBook)
DOI 10.1007/978-3-642-41720-7
Springer Heidelberg New York Dordrecht London

Library of Congress Control Number: 2014931389

© Springer-Verlag Berlin Heidelberg 2014
This work is subject to copyright. All rights are reserved by the Publisher, whether the whole or part of the material is concerned, specifically the rights of translation, reprinting, reuse of illustrations, recitation, broadcasting, reproduction on microfilms or in any other physical way, and transmission or information storage and retrieval, electronic adaptation, computer software, or by similar or dissimilar methodology now known or hereafter developed. Exempted from this legal reservation are brief excerpts in connection with reviews or scholarly analysis or material supplied specifically for the purpose of being entered and executed on a computer system, for exclusive use by the purchaser of the work. Duplication of this publication or parts thereof is permitted only under the provisions of the Copyright Law of the Publisher's location, in its current version, and permission for use must always be obtained from Springer. Permissions for use may be obtained through RightsLink at the Copyright Clearance Center. Violations are liable to prosecution under the respective Copyright Law.
The use of general descriptive names, registered names, trademarks, service marks, etc. in this publication does not imply, even in the absence of a specific statement, that such names are exempt from the relevant protective laws and regulations and therefore free for general use.
While the advice and information in this book are believed to be true and accurate at the date of publication, neither the authors nor the editors nor the publisher can accept any legal responsibility for any errors or omissions that may be made. The publisher makes no warranty, express or implied, with respect to the material contained herein.

Printed on acid-free paper

Springer is part of Springer Science+Business Media (www.springer.com)

Preface

Understanding the origin of galaxies is one of the major research goals of astrophysicists. Our own Milky Way and its neighbouring galaxies provide the ideal laboratory to facilitate a deeper understanding of how galaxies form. We can look in detail at the different components of our Milky Way and try to reconstruct events in the distant past through present-day clues—in essence we are carrying out archaeology on a galactic scale.

Our Galaxy resides in The Local Group, an overdense region out to about ten megaparsecs from the Milky Way. This region includes Andromeda (M31) a similar galaxy as our own, as well as a few dozen smaller galaxies and satellites. This environment is typical of most galaxies in our Universe. In a few billion years, the Local Group will have evolved into a single large elliptical galaxy as its most massive members merge together.

The origin of the Galaxy and Local Group is placed within the framework of the standard LCDM big bang cosmology. The Milky Way and its satellites continue to provide tests and constraints on theories of galaxy formation and on the standard cosmological model—namely a hierarchical universe in which structure formation is driven by an underlying dominant component of cold dark matter.

The following chapters contain a broad and detailed overview of our current understanding of the origin of our Milky Way Galaxy and the Local Group. They represent the current state of the art in the exciting topic of Near Field Cosmology.

These up-to-date reviews are based on lectures given at the inspiring 37th Saas Fee School held in Muerren, Switzerland. Muerren is a car-free mountain village high above the spectacular Lauterbrunnen valley in the Bernese Oberland. The school was entitled "The Origin of the Galaxy and Local Group" and it was attended by over 100 young astronomy students. The School was organised by Prof. Ben Moore (University of Zurich) and Prof. Eva Grebel (University of Heidelberg) and the lectures were given by world experts on these topics: Prof. Joss Bland-Hawthorn (Sydney Institute for Astronomy), Prof. Kenneth Freeman (Mount Stromlo Observatory) and Prof. Francesca Matteucci (Trieste University).

Zürich, June 2013   Ben Moore

The Origin Of The Galaxy and Local Group
37th Saas Fee Advanced Course

Contents

1 Near Field Cosmology: The Origin of the Galaxy and the Local Group 1
Joss Bland-Hawthorn and Kenneth Freeman
- 1.1 Prologue 1
- 1.2 Far Field Cosmology 2
  - 1.2.1 The Cosmic Microwave Background 4
  - 1.2.2 The First Stars 7
  - 1.2.3 The First Black Holes 10
  - 1.2.4 The First Dark Haloes 12
  - 1.2.5 Reionization and the First Galaxies 15
- 1.3 Lessons from Galaxy Redshift Surveys 16
  - 1.3.1 Evolution and Environment 19
  - 1.3.2 Accretion and Feedback 20
  - 1.3.3 Baryon Inventory and Metal Enrichment 22
  - 1.3.4 Chemical Evolution in Galaxies 25
  - 1.3.5 Milky Way and Local Group Analogues in the Real Universe 25
  - 1.3.6 Milky Way and Local Group Analogues in Simulated Universes 26
- 1.4 Gas Accretion onto Galaxies 27
  - 1.4.1 Introduction 27
  - 1.4.2 Earliest Epoch of Gas Accretion 28
  - 1.4.3 Early Ideas on Galaxy Accretion 30
  - 1.4.4 Accretion Shocks 32
  - 1.4.5 Cooling Flows 34
  - 1.4.6 Cold Flows 35
  - 1.4.7 Warm Flows 38
  - 1.4.8 Accretion via Major and Minor Mergers 38
  - 1.4.9 Accretion of High Velocity Clouds 39
- 1.5 Near Field Cosmology 45
  - 1.5.1 Introduction 45
  - 1.5.2 A Working Model of How the Galaxy Formed 47

		1.5.3	Timescales and Fossils	49
		1.5.4	Stellar Age Dating	51
		1.5.5	Goals of Near Field Cosmology	54
	1.6	Structure of the Galaxy		55
		1.6.1	The Bulge	55
		1.6.2	The Disk	59
		1.6.3	The Stellar Halo	60
		1.6.4	The Dark Halo	62
	1.7	Signatures of Galaxy Formation		63
		1.7.1	Zero Order Signatures: Information Preserved Since Dark Matter Virialized	63
		1.7.2	First Order Signatures: Information Preserved Since the Main Epoch of Baryon Dissipation	68
		1.7.3	Second Order Signatures: Major Processes Involved in Subsequent Evolution	74
	1.8	Reconstructing the Past Through Chemical Tagging		86
		1.8.1	Unravelling a Dissipative Process	86
		1.8.2	How Many Star Clusters?	88
		1.8.3	Cluster Chemistry	89
		1.8.4	Chemical Homogeneity	90
		1.8.5	Unique Chemical Signatures	92
		1.8.6	Primary Requirements of Chemical Tagging	92
		1.8.7	Candidates for Chemical Tagging	97
		1.8.8	Short-Term Goal: Size and Structure in a Multi-Dimensional C-Space	100
		1.8.9	Long-Term Goal: Reconstructing Ancient Star Groups from Unique Chemical Signatures	101
	1.9	Epilogue: Challenges for the Future		104
		1.9.1	The Limitations of Near Field Cosmology: Are We Really Putting ΛCDM to the Test?	104
		1.9.2	Future Surveys	106
	Appendix A: The Discovery of Dark Matter in Galaxies			109
	Appendix B: Stellar Data: Sources and Techniques			112
		B.1	Data Needed for Galactic Archaeology	112
		B.2	Sources of Data	118
		B.3	Sources of Models	125
	References			128
2	**Chemical Evolution of the Milky Way and Its Satellites**			**145**
	Francesca Matteucci			
	2.1	How to Model Galactic Chemical Evolution		145
		2.1.1	The Initial Conditions	146

		2.1.2	Birthrate Function	146
		2.1.3	Stellar Yields	150
		2.1.4	Gas Flows	158
	2.2	Basic Equations for Chemical Evolution		158
		2.2.1	Yields per Stellar Generation	158
		2.2.2	Analytical Models	159
		2.2.3	Detailed Numerical Models	162
	2.3	The Milky Way		164
		2.3.1	The Formation of the Milky Way	164
		2.3.2	The Two-Infall Model	166
		2.3.3	Detailed Recipes for the Two-Infall Model	167
		2.3.4	The Chemical Enrichment History of the Solar Vicinity	170
		2.3.5	The Galactic Disk	181
		2.3.6	The Galactic Bulge	186
	2.4	What We Have Learned About the Milky Way		194
	2.5	The Time-Delay Model and the Hubble Sequence		194
		2.5.1	Star Formation and Hubble Sequence	195
	2.6	Dwarf Spheroidals of the Local Group		198
		2.6.1	How do dSphs Form?	198
		2.6.2	Observations of dSphs	200
		2.6.3	Chemical Evolution of dSphs	200
		2.6.4	What Have we Learned About dSphs?	208
	2.7	Ultra-Faint Dwarfs in the Local Group		212
	2.8	Other Spirals		214
		2.8.1	Chemical Models for External Spirals	215
	2.9	Cosmic Chemical Evolution		217
	References			222
Index				229

Chapter 1
Near Field Cosmology: The Origin of the Galaxy and the Local Group

Joss Bland-Hawthorn and Kenneth Freeman

The Galaxy has built up through a process of accretion and merging over billions of years which continues to this day. Astronomers are now embarking on a new era of massive stellar surveys over the coming decade. These campaigns will derive three-dimensional space motions and heavy element abundances for millions of stars throughout the Galaxy and its neighbours. The new observations will reveal signatures of the formation and early evolution of the Local Group; this is what we mean by 'near field cosmology.' We set this new course of study within the context of fossil signatures from galaxy surveys and the high redshift universe. We discuss the complex relationship between baryons and dark matter over cosmic time, and introduce a synthetic framework that will allow both numerical simulations and the impending data deluge to be compared. We also include relevant source materials for the young near-field cosmologist and some historical perspectives.

1.1 Prologue

In an earlier review, we outlined the concept of near field cosmology to emphasize the fact that there are ancient signatures all around us today providing evidence of the formation processes that led to the Galaxy and the Local Group (Freeman and Bland-Hawthorn 2002). We see ancient stars around us in the old thin disk, the thick disk, the stellar halo, the inner bulge, and in satellite dwarf galaxies. About half of all stars in the Galaxy today formed before a redshift $z \sim 1$. The Hubble Space Telescope (HST) probes these early formation epochs directly during a time when

---

J. Bland-Hawthorn (✉)
Sydney Institute for Astronomy, University of Sydney, Sydney, NSW 2006, Australia
e-mail: jbh@physics.usyd.edu.au

K. Freeman
Mount Stromlo Observatory, Australia National University, Weston Creek, ACT 2611, Australia
e-mail: kcf@mso.anu.edu.au

gas accretion, star formation and feedback processes were greatly enhanced. But how do we connect the past with the present? There are no easy answers. Here we revisit our earlier review, extend our discussion to observable fossils from the Big Bang to the present day, with an updated discussion of the complete inventory of baryons.

The desire to connect present-day galaxies and the distant Universe has a long history that began long before it was possible to compare both epochs directly. We now speak of galactic and stellar archaeology. Eggen et al. (1962) were the first to show that it is possible to study galactic archaeology using stellar abundances and stellar dynamics; this is probably the most influential paper on the subject of galaxy formation. In 1966, Fred Hoyle wrote "It is not too much to say that the understanding of why there are these different kinds of galaxy, of how galaxies originate, constitutes the biggest problem in present day astronomy. The properties of individual stars that make up the galaxies form the classical study of astrophysics, while the phenomena of galaxy formation touches on cosmology. In fact, the study of galaxies forms a bridge between conventional astronomy and astrophysics on the one hand, and cosmology on the other." (Hoyle 1966).

In the Local Group, we can see the imprint of processes that date back to at least a redshift $z \sim 5$ and maybe earlier still. Reliable ages do not exist for individual stars beyond the Sun but at least some of the most metal-poor stars discovered so far may date to early generations immediately following the first stars. We are coming into a new era of galactic investigation, in which one can study the fossil remnants of the early days of the Galaxy in a broader and more focussed way, not only in the halo but throughout the major luminous components of the Galaxy. The goal of these studies is to reconstruct as much as possible of the early galactic history. We will review what has been achieved so far, and point to some of the ways forward.

But first we review important fossils from the high-redshift universe ($z \gtrsim 10$) and provide a summary of some key findings from galaxy redshift surveys ($0 \lesssim z \lesssim 10$). These provide a context for our later discussions on the formation and evolution of the Local Group.

1.2 Far Field Cosmology

It is now firmly established that the Universe underwent a period of rapid exponential expansion (acceleration) about a picosecond after the Big Bang. This was the birth of space-time, the forces of nature, and the Higgs field[1] that imparted mass to energetic particles and dark matter. The Universe today is isotropic, homogeneous and flat on the largest scales. A mathematical description of the expanding Universe is given by the Friedmann-Robertson-Walker (FRW) metric of general relativity although a useful description is provided by Newtonian analysis (Wright 2003). Using either

---

[1] It is presently believed that the Higgs field is distinct from the inflaton field that drove inflation (Guth 1997).

approach, for an appropriate choice of density and pressure, it is possible to arrive at Hubble's law , $v = H_o d$, where v is the recession velocity[2] and d is the proper distance. This simple formula describes the linear expansion of the Universe (Lemaître 1927; Hubble and Humason 1931) for which modern estimates of Hubble's constant fall within $H_o = 70 \pm 4$ km s^{-1} (e.g. Freedman et al. 2001).

It therefore came as a surprise to (most of[3]) the astronomical community that distant supernovae appear to be accelerating from us (Riess et al. 1998; Perlmutter et al. 1998), providing clear evidence for a second period of inflation for the Universe as a whole. The acceleration is powered by a mysterious source of 'dark energy' that has dominated the energetics of the Universe only relatively recently ($z \lesssim 0.5$; Peacock 1999). Dark energy is the name we give to Einstein's cosmological constant Λ or something that acts like it. As we shall see, this secondary inflationary phase is already affecting the way in which structure forms throughout the Universe. The evidence for the existence of the dark sector is strong, but its properties are only loosely understood.

The first inflationary phase diluted all previous structure except for quantum fluctuations which cannot be smoothed out. These primordial fluctuations grew in time to become the seeds of all structure formation (Liddle and Lyth 2000). As the Universe expanded, dark matter across the hierarchy began to collapse to form larger structures (Fig. 1.1), and the cosmic radiation began to cool (Davis et al. 1985). By a redshift of $z \sim 3000$, the Universe cooled enough to allow electrons and protons to form atoms. This was the the 'recombination era' when a fog of atomic hydrogen and helium permeated the Universe for the first time.

In the 1940s, Gamow, Alpher and Herman began to consider the consequences of a hot expanding Universe and realised that the Universe must be filled with cool radiation today, a relic of these early times. This ancient signal was first detected in the 2.3 K rotational temperature of cyanogen (McKellar 1940; cf. Roth et al. 1993) although was not recognised as such until Penzias and Wilson (1965) detected the microwave signal at a temperature of 2.7 K using a horn antenna. This relic radiation is known as the cosmic microwave background (CMB) and provides us with our most powerful probe of the high-redshift universe.[4]

It is instructive to ask what is the source of the CMB photons? During the 'recombination era,' any excess energy from the recombination of electrons and protons radiates away as photons. But these photons made a negligible contribution to the CMB. A viable source of photons must have occurred at an earlier epoch. In the very early Universe, during the so-called 'baryogenesis era,' electrons and positrons

---

[2] We note with interest that, exactly 100 years ago, the first galaxy radial velocity was measured by Slipher (1913). Two years later, the first 'rotation curve' was measured for NGC 4594 (Slipher 1914), long before it was understood what these observations were telling us. We refer the reader to an excellent review by Peacock (2013) on the remarkable achievements of V. Slipher that have been largely overlooked.

[3] Few can claim to have foreseen $\Lambda > 0$ from their published work (e.g. Efstathiou et al. 1990; White et al. 1993; Yoshii and Peterson 1995).

[4] Gravity wave (Abbott et al. 2009) and neutrino (An et al. 2012) experiments can also provide constraints on the post-inflationary epoch but these endeavours are in their infancy.

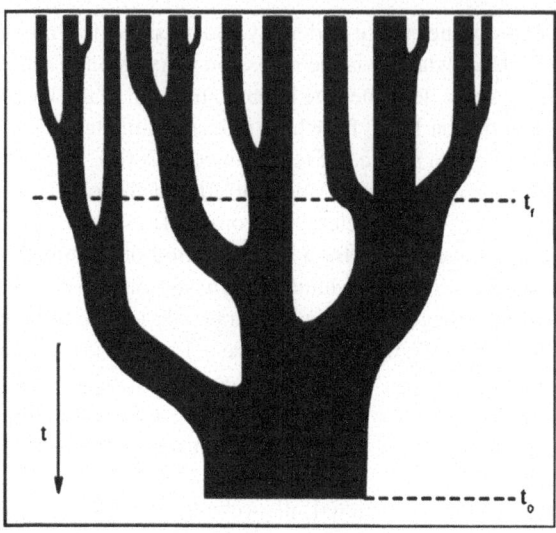

Fig. 1.1 A merger tree depicting the hierarchical merging of dark matter structures over cosmic time t (Lacey and Cole 1993); the time t_o corresponds to a dark matter halo today. We can identify a time in the past t_f when the halo first formed

annihilated to produce 511 keV photons. Interestingly, the neutrinos were not heated by this process such that their temperature today is predicted to be 1.9 K rather than 2.7 K characteristic of the CMB.

The baryogenesis era also accounts for one of the most extraordinary fossils from the inflationary epoch—matter-antimatter asymmetry. The Universe began with equal amounts of matter and antimatter, and yet today matter dominates over antimatter by up to ten orders of magnitude, the same ratio by which photons dominate over baryons today. The reason for the spectacular asymmetry is not known, but it may involve a charge conjugate-parity (CP) violation in quark/anti-quark interactions, a complex process that will soon be explored directly with neutrino/anti-neutrino experiments (An et al. 2012).

1.2.1 The Cosmic Microwave Background

The CMB is a redshifted relic of our hot beginnings taking us back to a time within 99.99 % of the age of the Universe (Fig. 1.2). See Lineweaver (1999) for an exceptionally intuitive discussion of the CMB. The CMB is of immense importance in our understanding of the early universe because it allows us to see the imprint of dark matter fluctuations within 370,000 years of the Big Bang, i.e. a redshift of $z \approx 1100$. As we travel back in time, the Universe was smaller, hotter and denser. Anyone who has tried to see through mist with car headlights will understand what is meant by

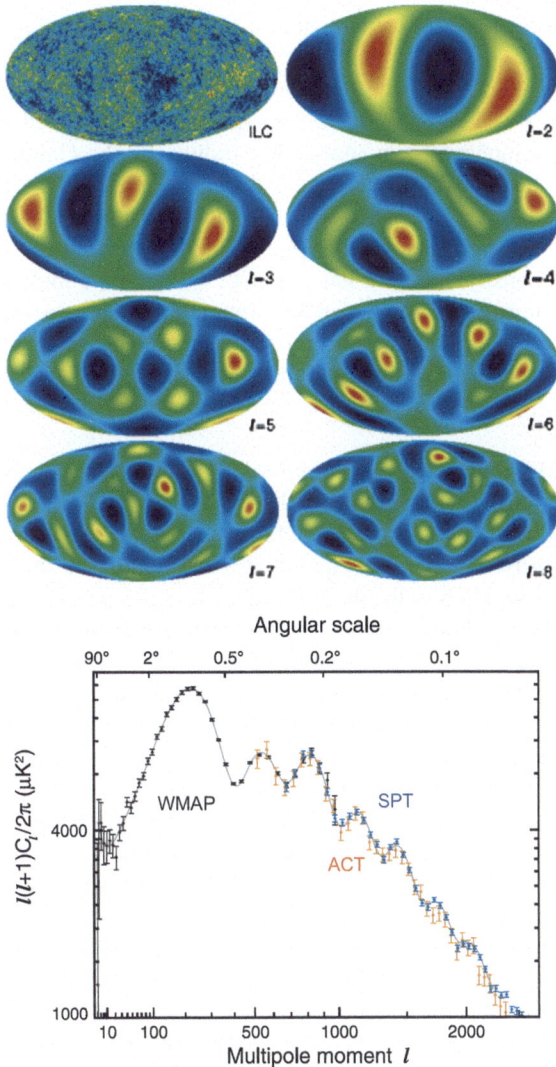

Fig. 1.2 *Upper* Cosmic microwave background (CMB) observed by the WMAP experiment separated into its fundamental and higher harmonics; the dipole contribution due to the Sun's motion has been subtracted. *Lower* 9 year analysis of the angular power spectrum where the fundamental harmonic is clearly visible. The declining tail—due to 'Silk damping' in the early Universe (Sect. 1.2.1)—is compiled from observations by the South Pole Telescope (SPT) and the Atacama Cosmology Telescope (ACT) (we acknowledge the WMAP team for this figure)

the 'surface of last scattering'—the light does not penetrate very far into the mist. And so it is with the CMB which arises from the 'surface of last scattering' that we can penetrate with microwave photons. Our view back to this time is optically thin and so we are probing large-scale structures when the Universe was a billion times smaller by volume.

The CMB is highly thermalized and therefore a near-perfect isotropic black body at a temperature of 2.725 ± 0.002 K. The spectrum is described by the Planck function as was demonstrated in spectacular fashion by the Cosmic Background Explorer (COBE) satellite. When the Principal Investigator John Mather first showed the spectrum at the American Astronomical Society in January 1990, he received a standing ovation. This remains the most precise Planck spectrum across all fields of applied physics. Consistent with a flat universe, wherever we look across the sky, the mean temperature is the same to within roughly 0.1 %. But stamped onto this uniform signature is the effect of our motion through the Universe: CMB radiation is slightly warmer in the direction of our motion, and slightly cooler in the opposite direction. From this signal, we measure our motion to be 552 ± 6 km s^{-1} in the direction of Hydra.

When the CMB dipole signal is removed, small temperature fluctuations are observed across the sky with amplitudes less than 0.01 %. These are the imprints of early structure formation. Before recombination, the free electrons had scattered and trapped the thermal radiation in a photon-baryon fluid; the abrupt elimination of these free electrons allowed the thermal radiation to move nearly without scattering, and so the tiny variations in temperature trace variations in the local potential at $z \approx 1100$. When we form a power spectrum of the CMB variations, a set of 'acoustic peaks' can be seen on angular scales below $2°$, the projected horizon scale. These are due to sound waves within the photon-baryon fluid as it reacts to the presence of dark matter fluctuations. Lineweaver (1999) provides a marvellous series of illustrations to describe how the acoustic peaks came about. Structure on angular scales greater than $2°$ dates back to the first picosecond after the Big Bang (inflation), and therefore constitute our most ancient fossils from the early Universe at the present time.

It is often overlooked that the dark matter fluctuations are essential to our very existence (unlike dark energy)! Before $z \sim 1100$, the photon-baryon fluid was subject to strong diffusion (Silk damping) due to viscosity and thermal conductivity. This allowed photons to diffuse from hot regions into cold regions wiping out any fluctuations on scales less than about 10 Mpc as observed today. But fortunately for us, the dark matter hierarchy continued to evolve and thereby preserved a semblance of structure. At a later time, this mass spectrum of fluctuations was to drive the formation of the first stars and first galaxies. Precision measurements of the distribution of CMB radiation, most recently by the Wilkinson Microwave Anisotropy Probe (WMAP) satellite (Komatsu et al. 2011) are in beautiful agreement with the relativistic theory of how the present concentrations of mass in the Universe grew out of the distribution of dark matter at the time of radiation breaking free from the photon-baryon fluid. The baryon acoustic oscillations (BAO) detected in the CMB have now been detected in galaxy redshift surveys at low redshift (Percival et al. 2001; Eisenstein et al. 2005; Cole et al. 2005; Blake et al. 2011).

1.2.2 The First Stars

How did the first stars form? Since Big Bang nucleosynthesis—which takes us back to the first minute of cosmic time—only produced elements up to boron, the existence of metals (C, N, O...) is proof enough that something dramatic took place after the recombination era. The first stars were unique to their time—very massive, free of elements heavier than boron (metal free), and very short-lived (Yoshida et al. 2004; Gao and Theuns 2007). Such a star has never been observed directly but we live in hope of identifying its chemical signatures in low-mass stars that formed in the second or subsequent generations.

A useful summary of how the first stars formed is given in a highly readable, short text by Loeb (2010); a more extensive review is given by Bromm and Yoshida (2011) and Karlsson et al. (2013). By a redshift of $z \sim 200$, gas began to accrete through gravitational instability towards local dark matter concentrations (Barkana and Loeb 2007). Whether a forming gas cloud is able to collapse is a competition between gravity and internal pressure. The increase in gas density towards the centre of the cloud generates a pressure (sound) wave that propagates outwards. The pressure wave attempts to counteract the pull of gravity but if the wave does not have sufficient time to cross the cloud, the cloud will collapse. This situation is usefully described by the Jeans mass given by

$$M_J = \frac{4}{3}\pi \langle \rho \rangle r_J^3 \quad (1.1)$$

where $\langle \rho \rangle$ is the mean density within a radius r_J. If the gas is collapsing onto a dark matter fluctuation, which is assumed in most contemporary models, the mass density includes the combined contribution of the gas and dark matter. If the gas temperature is roughly the temperature of the CMB, $M_J \sim 10^5$ $M_\odot$ independent of redshift (Peebles 1993; Haiman et al. 1996).

The first stars—so-called Pop III—are likely to have formed roughly 200 Myr ($z \sim 20$) after the Big Bang (Bromm et al. 1999; Abel et al. 2002) when the primordial gas was first able to cool and collapse onto small dark matter 'mini haloes' with masses of about 10^6 $M_\odot$. Their large masses (>30 $M_\odot$) arose from the limited cooling channels available to primordial gas, more specifically, cooling through radiation from simple molecules. It is now thought that Pop III star formation consisted of two distinct modes: one where the primordial gas collapses into a dark matter minihalo via H_2 cooling (Pop III.1), and one where the metal-free gas becomes significantly ionized prior to the onset of gravitational runaway collapse via HD cooling (Pop III.2).

Pop III.1 mode. The formation of the first stars is much simpler than the corresponding star formation process in the local Universe where the environment of giant molecular clouds is extremely complex (Bromm and Yoshida 2011). The Universe had no heavy elements, and therefore no dust to complicate the physics of cooling and opacity. The early post-recombination Universe was devoid of ionizing radiation and strong drivers of turbulence, such that the collapse into the minihalos assisted by H_2 cooling proceeded in a rather quiescent fashion.

The formation of the first stars inside minihalos provides us with a well-posed problem which can be tackled through rigorous computational studies. Since the late 1990s, several groups have simulated the formation of the first stars with sophisticated numerical algorithms with either smoothed-particle hydrodynamics (SPH) or adaptive-mesh refinement (AMR) techniques (Abel et al. 2000, 2002; Bromm et al. 1999, 2002; Yoshida et al. 2006; O'Shea and Norman 2007, 2008). These calculations have converged on key parameters describing the formation of the first stars.

The most important result, where there is general agreement, is that Pop III.1 stars were predominantly massive. To first order, this can be understood as the consequence of a large Jeans mass in gas that cools only via H_2. Prior to undergoing runaway collapse, the primordial gas settles into what is sometimes termed a quasi-hydrostatic 'loitering' state. This state is characterized by temperature and number density $T_{ch} \simeq 200$ K and $n_{ch} \simeq 10^4$ cm^{-3}. Recent numerical simulations indicate that Pop III.1 stars formed with characteristic masses of $M_\star \gtrsim 30$ M$_\odot$.

It is possible that the first stars formed with a range of masses. The initial mass function (IMF) gives the number of stars formed per unit mass, $N_\star$, where present-day star formation is often described with a power-law with index β: $dN_\star/dm_\star \propto m_\star^{-\beta}$, or a sequence of such power-laws (Kroupa 2001). For Pop III, we do not know the complete functional form of the underlying IMF with any certainty. A rough estimate of the outcome of the star formation process can be made by focusing on the average mass: $\langle m_\star \rangle \propto \int m_\star dN_\star$. We refer to this as the 'characteristic mass' as it describes the typical result of star formation, with the understanding that some stars form with lower, and some with higher masses. It is not yet feasible to realistically simulate the assembly of an entire Pop III star, starting from realistic cosmological initial conditions. Recently, *ab initio* calculations have traced the evolution up to the point where a small protostellar core has formed at the center of a minihalo (Yoshida et al. 2008).

This initial hydrostatic core has a mass $m_{ch} \sim 10^{-2}$ M$_\odot$ very similar to present-day Pop I protostellar seeds. The subsequent growth of the protostar through accretion, however, is believed to proceed in a markedly different way. In the early Universe, protostellar accretion rates are believed to have been more enhanced, due to the higher temperatures in the star forming clouds, which in turn is a consequence of the limited ability of the primordial gas to cool below the ~100 K accessible to H_2 cooling. The higher accretion rates, together with the absence of dust grains, and the reduced radiation pressure that could in principle shut off the accretion, conspire to yield heavier final stars in the early Universe. Estimates for the masses thus built up are somewhat uncertain, but a rough range is $m_\star \sim 30$–100 M$_\odot$ although a higher mass cut-off is possible.

Recent simulations have shown that Pop III.1 stars may have formed as members of a binary or small multiple system (Turk et al. 2009; Stacy et al. 2010). Here the evolution is followed beyond the initial collapse of the first high density peak where one sees the emergence of a compact disk around the first protostar. This disk is gravitationally unstable and fragments into a dominant binary, possibly with a few more lower-mass companions. These simulations have not yet been able to reach the asymptotic end state, where all available mass is either accreted or permanently expelled. The reason is that radiative feedback processes in partially optically thick material cannot be neglected once a protostar has grown to 10 $M_\odot$, when full radiation-hydro simulations are required (Hosokawa et al. 2011; Stacy et al. 2012). Currently, the binary fraction of Pop III stars is a major unknown.

Pop III.2 mode. While H_2 cooling drives the formation of Pop III.1, the HD molecule can also play an important role in cooling primordial gas in situations where the gas experiences strong ionization. This is largely due to the high abundance of electrons which serve as a catalyst for molecule formation (e.g. HD) in the early Universe (Shapiro and Kang 1987). The temperature can then drop well below 200 K. In turn, this efficient cooling may lead to the formation of primordial stars with characteristic masses of the order of $m_\star \sim 10\ M_\odot$, so-called Pop III.2 stars.

Efficient HD cooling can be triggered within the relic H II regions that surround Pop III.1 stars at the end of their brief lives owing to the high electron fraction that persists in the gas as it cools and recombines (Nomoto et al. 2005; Yoshida et al. 2007). The efficient formation of HD can also take place when the primordial gas is collisionally ionized, such as behind the shocks driven by the first SNe or in the virialization of massive DM halos. There is a critical HD fraction necessary to allow the primordial gas to cool to the temperature floor set by the CMB at high redshift: $X_{HD} = n_{HD}/n \sim 10^{-8}$ where n_{HD} is the number density of HD molecules and n that of all particles. Except for the gas collapsing into neutral minihalos, the fraction of HD typically increases quickly enough to play an important role in the cooling of the gas, facilitating the formation of Pop III.2 stars. Other environments conducive for Pop III.2 formation are the so-called 'atomic cooling halos' which have been suggested as candidates for the first galaxies (Bromm et al. 2009).

In summary, there is an emerging consensus on the main processes involved in forming the first stellar generation because the hydrodynamical processes appear to be relatively well defined. But there are many questions. In particular, what is the role of turbulence in shaping the primordial mass function? If sufficient turbulence were present, a broad range of fragment masses could result, similar to the power-law extension towards high masses observed in the present-day IMF (Clark et al. 2011). In comparison, the details of the hydrodynamic processes leading to subsequent stellar generations are poorly understood (Clark et al. 2011; Greif et al. 2011; O'Shea and Norman 2007; Norman 2011).

One of the most important themes of near field cosmology (Sect. 1.5) is the use of chemical signatures in the oldest stars today to interpret the sequence of events in the Galaxy's formation. Unfortunately, the chemical yields of the first stars, and therefore their chemical signatures, are not well constrained at the present time (Karlsson et al. 2012). The signatures are complicated by a number of factors, in particular, the unknown combination of metals that falls back onto any black hole that forms in the subsequent supernova/hypernova explosion (Podsiadlowski et al. 2002; Umeda and Nomoto 2003).

1.2.3 The First Black Holes

Stars in the mass range $9 \lesssim m_\star/M_\odot \lesssim 40$ are believed to explode as core collapse SNe. At the end of hydrostatic burning, stars above $\sim 10 \, M_\odot$ contain an electron degenerate iron core. As the Fe core continues to grow by silicon shell burning, the electron pressure can no longer counterbalance the increasing gravitational pull and the core collapses. An explosive instability develops when the infalling outer layers bounce off the collapsed core (proto-neutron star) and are ejected into the ISM.[5] The details of the explosion mechanism is still a topic of debate (Burrows et al. 2006; Janka et al. 2007). Stars around 9–10 $M_\odot$ instead form a degenerate $O + Ne + Mg$ core which may collapse due to rapid electron captures on ^{20}Ne and ^{24}Mg, prior to the ignition of Ne (Nomoto 1987). If so, an electron-capture, or ONeMg SN is formed. Alternatively, the star loses its outer layers and forms a white dwarf (Garcia-Berro and Iben 1994).

At the high mass end of the core collapse SN regime (25–40 $M_\odot$), the potential energy of the stellar envelope is comparable to the kinetic energy of the explosion and the innermost layers of the star fall back onto the central neutron star which eventually collapses to a black hole (Fryer 1999). A weak or faint SN is formed if the black hole is non-rotating. For rotating black holes, however, a much stronger explosion may instead be expected in the form of a hypernova (HN; Fryer and Heger 2000). Stars with masses above 40 $M_\odot$ collapse directly to a black hole.

For Pop. III stars with masses $\gtrsim 100 \, M_\odot$, the pair instability kicks in after central carbon burning, driven by the creation of electron-positron pairs. Below $\sim 140 \, M_\odot$, this instability causes the star to pulsate violently. While the outer layers are lost in SN-like explosions, the star settles down to form a massive Fe core. Eventually the star collapses quietly to form a black hole. On the other hand, if the star is in the mass range $140 \lesssim m/M_\odot \lesssim 260$ it will face complete disruption as a result of the pair instability, and no remnant is formed (Heger and Woosley 2002). In still more massive stars, the energy released from explosive oxygen and silicon burning, caused by the pair instability, is instead used to photodisintegrate the nuclei in the central core. The explosion is halted and a black hole is formed (Fryer et al. 2001).

Thus some of the first stars may well have seeded the first black holes in their cores. The existence of high-redshift quasars provides direct fossil evidence of these first seeds. To see why, consider a black hole that converts rest-mass energy with an efficiency ϵ into radiation. Its luminosity is given by

$$L_A = \epsilon \dot{m} c^2 \qquad (1.2)$$

$$= 7 \times 10^{11} \left(\frac{\epsilon}{0.05}\right) \left(\frac{\dot{m}}{M_\odot \, yr^{-1}}\right) L_\odot \qquad (1.3)$$

---

[5] An elegant visual aid of a related process is to drop a basketball onto the ground with a tennis ball rested on it. When the basketball hits the ground, it bounces a little while the tennis ball shoots up to a great height.

for which $\dot{m}$ is the mass accretion rate. In reality, the accretion disk luminosity can limit the accretion rate through radiation pressure. The so-called Eddington limit is given by

$$L_E = \frac{4\pi G m_{BH} m_p c}{\sigma_T} \quad (1.4)$$

$$= 2 \times 10^{13} \left(\frac{m_{BH}}{10^9 \, M_\odot}\right) L_\odot \quad (1.5)$$

where m_{BH} is the black-hole mass, m_p is the proton mass and σ_T is the Thomson cross-section for electron scattering. This equation allows us to derive a maximum accretion rate $\dot{m}$, and therefore a maximum luminosity, for a given black-hole mass. (The 1990 Saas Fee lectures provide an excellent introduction to the physics of black holes.)

The critical Eddington accretion rate is proportional to the mass of the accreting black hole ($\dot{m} \propto m_{BH}$) which provides us with an exponential timescale for a black hole growing at its maximum rate. The so-called Salpeter timescale T_S is given by

$$T_S = \frac{\epsilon \sigma_T c}{4\pi G m_p} \approx 4.5 \times 10^7 \text{ yrs} \quad (1.6)$$

Consider the case of the newly discovered $z \approx 7.085$ quasar ULAS J1120+0641 (Mortlock et al. 2011). A crude estimate of its mass ($m_{BH} \sim 1\text{--}3 \times 10^9 \, M_\odot$) can be obtained from its Mg II line width and luminosity assuming that it is shining close to its Eddington luminosity. We observe this object at a time when the Universe was only 770 Myr old.[6] If our current picture is correct, that the first stars could *not* have formed before $\sim$100 Myr, then only $\sim$15 Salpeter e-folding times had elapsed before this time. Thus an initial black hole seed of $\sim$30 $M_\odot$, say, could have grown at the maximum rate to a mass of about $m_{BH} \sim 10^8 \, M_\odot$ in the time available. This is low compared to the estimated mass and therefore, given the uncertainties, only barely adequate to explain the quasar. Sources like ULAS J1120+0641 clearly occupy special environments (e.g. high pressure regions in multiple mergers) where sustainable long-term accretion was possible. These may have been the same environments that produced the oldest globular clusters ($\gtrsim$13 Gyr) observed in the Galactic halo today. Similar QSO sources at even higher redshift would be a major challenge for our 'standard model' for the first stars, or tell us something entirely new about how black holes are formed (in merging groups?) or evolve in the early universe.

---

[6] See Ned Wright's cosmology calculator at http://www.astro.ucla.edu/~wright/ACC.html (Wright 2006).

1.2.4 The First Dark Haloes

Most of the structure in the Universe came together in the redshift range $z \sim 3000$ to about $z \sim 0.5$ when the Universe was flat and matter dominated. Excellent discussions are provided by Peacock (1999), Kaiser (2002) and Binney and Tremaine (2008). From the FRW metric, the universal scale factor a obeys $a(t) \propto t^{2/3}$ and the mean density is given by $\rho_o(t) = (6\pi G t^2)^{-1}$. Consider a spherically symmetric density fluctuation δ_i within the expanding Universe at a time t_i which, when averaged over a volume with radius r_i, is much less than unity. The total mass in this sphere is then

$$M = \frac{4\pi}{3}(1+\delta_i)\rho_s(t_i)r_i^3. \tag{1.7}$$

If there is no flow of material across the surface defined by r_i, and no pressure within the collapsing region, then a small lump of matter experiences an acceleration given by

$$\ddot{r} = -\frac{GM}{r^2(t)}. \tag{1.8}$$

The parametric (cycloidal) solution to this equation is

$$r(\eta) = r_o(1-\cos\eta) \tag{1.9}$$

$$t(\eta) = \sqrt{r_o^3/GM}(\eta-\sin\eta) \tag{1.10}$$

where we have assumed $r = 0$ at the Big Bang. (These same equations can be used to describe the expanding Universe where $\dot{a} = (da/d\eta)/(dt/d\eta)$ and Hubble's scale factor is given by $H_o = \dot{a}/a$.) The 'turnaround radius' is defined as $r_{\text{turn}} = 2r_o$ where the expansion halts and the collapse begins ($\eta = \pi$).

If the average density inside the sphere is $\rho_s(t)$, then the average density compared to the universal mean is

$$\delta(t) = \frac{\rho_s(t)}{\rho_o(t)} \tag{1.11}$$

$$= \frac{9}{2}\frac{(\eta-\sin\eta)^2}{(1-\cos\eta)^3} \tag{1.12}$$

which at turnaround is equal to $\delta_{\text{turn}} = \frac{9\pi^2}{16} = 5.55$. If we now expand both Eqs. 1.10 and 1.12 to eliminate r_i from Eq. 1.7, we arrive at equations for the turnaround radius and turnaround time:

$$r_{\text{turn}} \approx \frac{(GMt_i^2)^{1/3}}{\delta_i} \tag{1.13}$$

1 Near Field Cosmology

$$t_{\text{turn}} \approx 1.1 \frac{t_i}{\delta_i^{3/2}} \tag{1.14}$$

This last equation is particularly interesting in the context of ΛCDM because it says that the larger the initial fluctuation δ_i, the sooner the collapse process gets going.

In practice, the 'spherical top hat' model for collapse within ΛCDM is overly idealised compared to what is seen in simulations (see Sheth and Tormen 2001 for a treatment of ellipsoidal collapse). The collapse cannot be purely radial otherwise the Universe would be the exclusive domain of black holes. In reality, the collapsing dark matter forms a halo which becomes virialized (stabilised) within some radius (see Fig. 1.3), the virial radius r_{vir}. After virialization, all of the energy is conserved and the potential energy of the system is $W = -\frac{3}{5} GM^2/r_{\text{turn}}$ such that the half-mass radius of the relaxed system is $r_h \approx 0.375 r_{\text{turn}}$, or about a third of the turnaround radius. The virialized halo density ρ_h can be compared to the mean universal density after the halo has virialized, $\rho_o(2t_{\text{turn}})$, such that

$$\frac{\rho_h}{\rho_o(2t_{\text{turn}})} = \frac{9\pi^2}{8} \left(\frac{r_{\text{turn}}}{r_h} \right)^3 \tag{1.15}$$

$$\approx 200. \tag{1.16}$$

Thus we speak of r_{200}, i.e. the radius within which the mean density is 200 times the universal density at a given epoch, as the virial radius of a halo, and the mass within is taken to be the mass of the dark matter halo. In words, the factor of ~ 200 is made up of three parts: (i) a factor of 5.55 is needed to ensure that matter freezes out of the Hubble flow; (ii) a factor of 8 to describe the increase in density after the twofold collapse in radius; and (iii) a factor of 4 to reflect the drop in the mean background density during the time of the collapse, i.e. $t_{\text{turn}} < t < 2t_{\text{turn}}$. The precise value of the virial radius depends on the underlying cosmology, but $r_{\text{vir}} = r_{200}$ is a useful measure for most relevant cosmologies.

It is possible to go beyond this description of the virialized halo and derive some properties of the internal structure (Gunn and Gott 1972). An isolated point mass M_o generates a divergence-free (pressure free) flow since the infall velocity increases as $\delta v \approx GM_o/(H_o r^2)$ which compensates for the decreasing area of the advecting surface as one approaches the point mass. The amount of mass across the surface is given by $\delta M \approx 4\pi r^2 \rho_o \, \delta v/H_o$ or equivalently $\delta M \sim M_o$. In an Einstein-de Sitter universe, the perturbation $\delta \rho/\rho = \delta M/M$ grows linearly with the scale factor a, such that the turnover radius is $r_{\text{turn}} \sim (\delta M/\rho_o)^{1/3} \propto a^{4/3}$. If we assume that the specific binding energy $\delta \phi \propto M(r)^{-1/3}$ of the infalling shell is conserved at some final settling radius r_{fin}, then it follows that $M(r_{\text{fin}}) \propto r_{\text{fin}}^{3/4}$ and $\rho(r_{\text{fin}}) \propto r_{\text{fin}}^{-9/4}$. This profile is *almost* equivalent to an isothermal halo, i.e. the expected form giving rise to flat rotation curves seen in galaxies. If we consider the timing (phase) of infalling material, it is possible to learn about distinct core-halo structures in galaxies even before considering the role of baryons (e.g. Binney and Tremaine 2008, Sect. 9.2.4), but we leave this discussion to a later review.

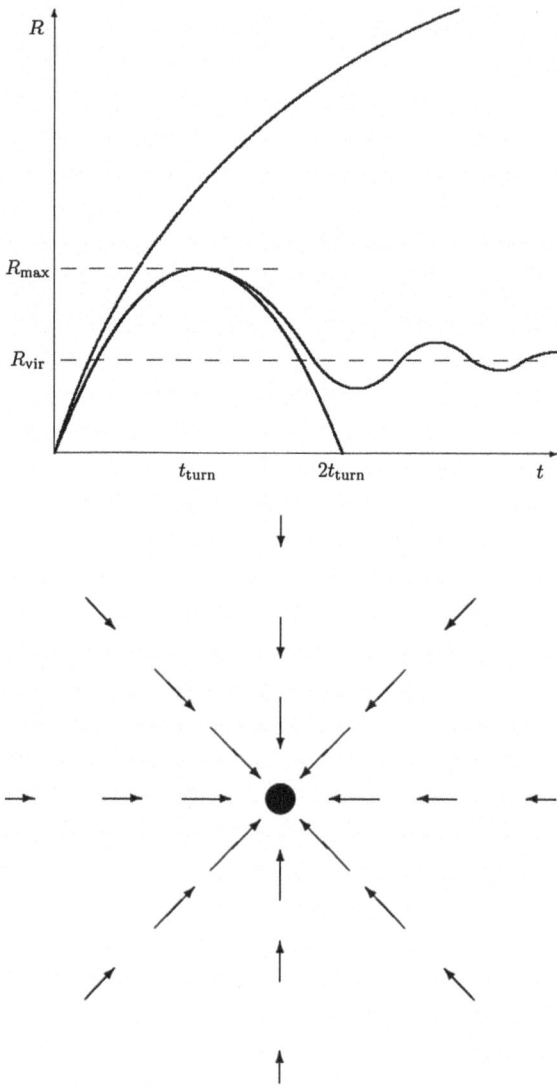

Fig. 1.3 *Upper* Evolution of a 'spherical top hat' perturbation (after Kaiser 2002). The *top curve* describes the expansion of material external to the perturbation. The *bottom curve* describes the formation of a black hole in the unlikely event that the mass fluctuation collapses perfectly radially. The *middle curve* describes the formation of a virialized halo: here the collapsing material essentially bounces before achieving some form of equilibrium. *Lower* Divergence-free flow pattern due to the presence of a point mass. Mass accumulates at the origin at a constant rate because the infall velocity increases as r^{-2} (shown by the vectors) while the surface decreases by the same factor

1.2.5 Reionization and the First Galaxies

In Sect. 1.2.2, we attempted a consensus across a wide literature on the likely properties of the first stars. We know they must have existed but at present there are many unknowns. Did the first stars form in isolation or in groups? Were relatively few massive stars responsible for reionization or was it triggered by the collective effect of massive star clusters?

One of the most important questions is what constituted the earliest stable baryonic *systems* and whether relics of these survive to the present day. The first stellar generations changed the Universe in many ways, in particular, the chemical properties and the equation of state of the intergalactic medium. Most of the diffuse atomic hydrogen and even gas collapsed within small halos became fully ionised. This is known as the 'epoch of reionization.' Is this exclusively the domain of the most massive stars, or can substantial intermediate and low mass stars form (Tsuribe and Omukai 2006, 2008; Clark et al. 2008)? In other words, did stellar populations observable today exist before reionization (Tumlinson 2010; Okrochov and Tumlinson 2010)?

Peebles and Dicke (1968) considered the possibility that the first structures were fully baryonic, more specifically globular clusters that formed before the first galaxies came together (see also Ferraro et al. 2009). These early baryonic systems may have been more massive than the typical cluster masses observed today in order to account for their enhanced metallicities (Conroy and Spergel 2011; Conroy 2012). How globular clusters fit into the early history of galaxy formation, and what they have to tell us about the first stellar generations (e.g. Caloi and d'Antona 2011), remains a mystery. But the contemporary view is that the earliest baryonic systems must have contained some dark matter to allow baryon cooling to proceed in order to form any stars at all, even the first generation of stars.

During reionization, the cosmic gas was heated above 10^4 K which raised the critical Jeans mass to roughly $m_{\rm crit} \sim 10^8$ $M_\odot$ (with roughly 10 % in baryons). In order for the local gas density to survive heating through photoionization (which suppresses star formation), the confining dark-halo mass must have exceeded $m_{\rm crit}$ (Rees 1986; Ikeuchi 1986; Efstathiou 1992; Barkana and Loeb 1999). These arguments based on ionization balance and heating are supported by 3D numerical simulations (Quinn et al. 1996; Navarro et al. 1997; Gnedin 2000; Okamoto et al. 2008). Naively, this means that galaxies (baryons + dark matter) less massive than the $m_{\rm crit}$ threshold could not have formed.

We stress that these earlier simulations do *not* completely exclude ongoing star formation for masses below $m_{\rm crit}$ during the reionization epoch (e.g. Dijkstra et al. 2004). The calculations do not consider the time it takes for the gas to evaporate from the low mass halos. For a cosmic ionizing UV intensity $J_0 = 10^{-21}$ erg cm^{-2} s^{-1} Hz^{-1} sr^{-1}, with a moderate power–law form of $f_\nu \propto \nu^{-2}$, we obtain an ionisation flux of $\sim 2 \times 10^5$ photons s^{-1} cm^{-2}. The potential rate of evaporation of hydrogen atoms off the surface of a confined gas cloud, of radius 50 pc ($\sim 3 \times 10^{41}$ cm^2), assuming one ionising photon evaporates one hydrogen atom, is then given by

$$\dot{m}_{\text{evap}} \sim 1.5 \times 10^{-3} (n_{\text{r}} + 1)^{-1} \, M_\odot \, \text{yr}^{-1}, \tag{1.17}$$

where the n_{r} term allows for the number of recombinations in the escaping wind. Thus, the timescale for evaporation for the 10^4–10^5 $M_\odot$ of gas in the models shown here ($\tau_{\text{evap}} \sim f_b M_{\text{min}}/\dot{m}_{\text{evap}}$) can exceed 10^8 yr, which is long enough for star formation to proceed in the interior while the outer regions are evaporating. Others have also argued that star formation was diminished but not fully suppressed, such that galaxies with masses of order 10^8 $M_\odot$ form stars during and immediately after reionization (Gnedin and Kravtsov 2006; Salvadori and Ferrara 2009). An alternative route to suppressing star formation in low mass systems has been to consider the effect of starburst-driven winds (Mac Low and Ferrara 1999).

So what were the first stellar systems? There are few if any reliable observational constraints thus far. Primordial objects are important to identify at any redshift because these may retain chemical signatures of the first and second generations of stars (Karlsson et al. 2012). Whether long-lived, low mass stars emerged depends in large part on whether dust formed from the ejected material of the first stars (Clark et al. 2008). If dust did not form in appreciable amounts, the CMB inhibits gas cooling, in which case low mass stars may not have emerged until $z \lesssim 10$ when mean metallicities reached $\sim 10^{-4} Z_\odot$ (Schneider and Omukai 2010).

1.3 Lessons from Galaxy Redshift Surveys

Over the past 20 years, multi-object spectrographs around the world have revolutionised our understanding of galaxies out to redshifts $z \sim 1$ and beyond (see Fig. 1.4). These surveys have been made possible by high quality u, g, r, i, z photometry from various sources, in particular, the HST and the Sloan Digital Sky Survey (www.sdss.org). A complete list of all surveys to date is maintained by I.K. Baldry (www.astro.ljmu.ac.uk/~ikb/research). The highest redshift sources confirmed spectroscopically are a gamma ray burst at $z \approx 8.2$ (Tanvir et al. 2009), a QSO (Mortlock et al. 2011) and two Lyman break galaxies at $z \sim 7$ (Vanzella et al. 2011), and a host of higher redshift 'candidates' with less accurate photometric redshifts (e.g. Robertson et al. 2010).

A comprehensive review of galaxy surveys requires an entire book (e.g. Mo et al. 2010); we can only give a flavour of the excitement and progress that has been possible in an era of HST and 8–10m telescopes to give some context to the importance of the Local Group. Useful recent reviews include Glazebrook (2012), Silk and Mamon (2012) and Driver et al. (2013). We encourage the reader to consider the ramifications of Fig. 1.5: understanding galaxies is an extremely challenging proposition requiring as much information as we can gather at all cosmic epochs.

1 Near Field Cosmology

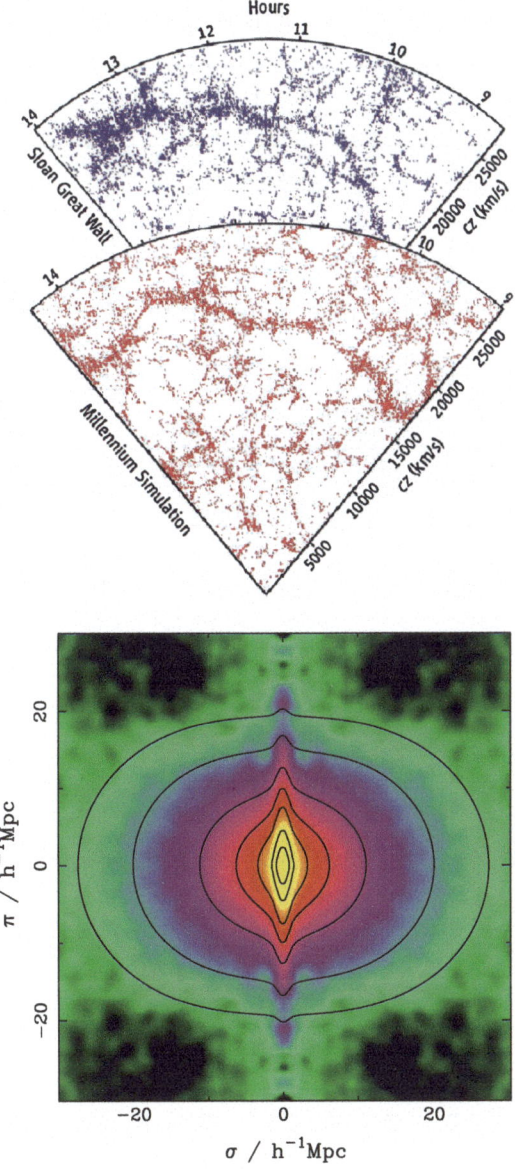

Fig. 1.4 *Upper* The "Great Wall of Galaxies" that emerged from the SDSS and 2dFGRS galaxy redshift surveys and a 2D slice from the Millennium ΛCDM simulation which shows that such structures, while relatively rare, are not unexpected (Springel et al. 2005). The Great Wall contains in excess of 10,000 galaxies and stretches over 75° (400 Mpc at $z \approx 0.07$). *Lower* The redshift-space correlation function $\xi(\sigma_{12}, \pi_{12})$ for the 2dFGRS galaxy redshift survey plotted as a function of the transverse (σ_{12}) and radial (π_{12}) pair separation (see Peacock et al. 2001). This figure shows clear redshift distortions with 'fingers of god' on small scales and a pronounced flattening on large scales due to infall onto filaments and sheets (see text)

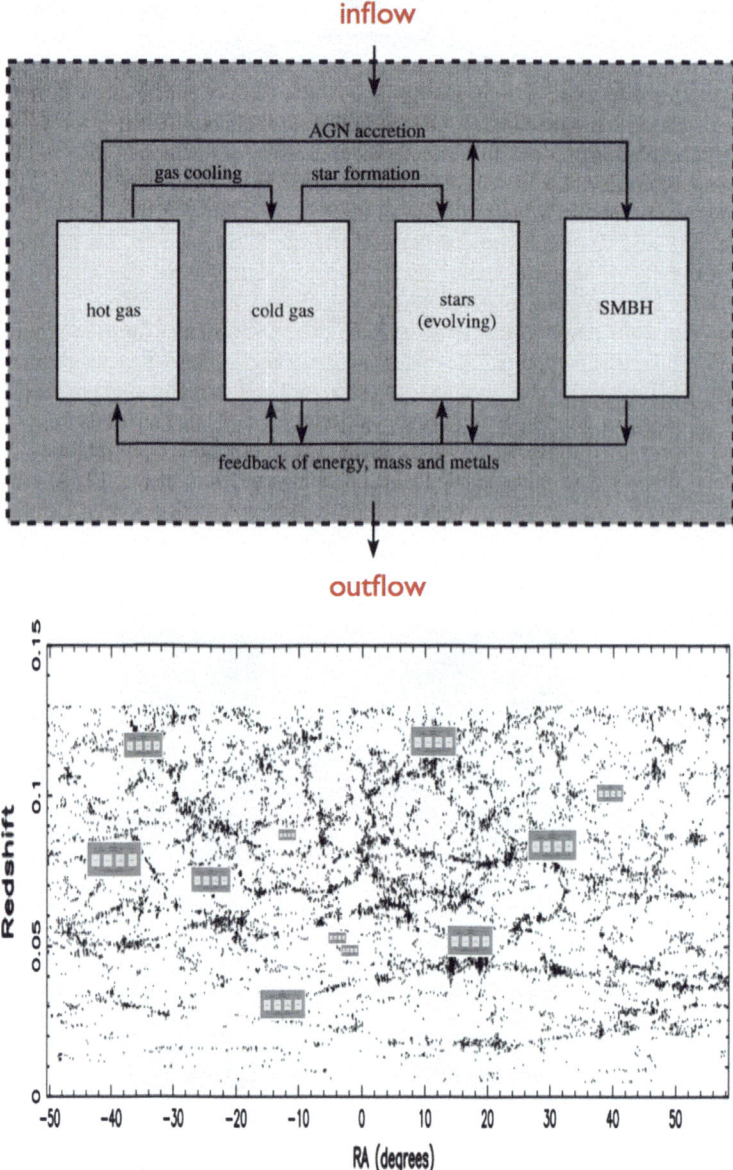

Fig. 1.5 *Upper* A schematic of the inner workings of a galaxy (adapted from Mo et al. 2010). *Lower* An illustration of the manifest complexities involved in galaxy studies. The simple model is distributed across large-scale structure (RA, redshift) identified in Stripe 82 from the Sloan Galaxy Redshift survey. The size of the schematic is proportional to a galaxy's halo mass. Model galaxies are distributed across a wide range of density contrast. Gas may be entering or leaving the galaxy, or a combination of both, or there may be no inflow or outflow (e.g. ram pressure stripping). Our goal is to understand the inner workings of galaxies across all environments and through cosmic time

1.3.1 Evolution and Environment

Most forms of galaxy activity (star formation, AGN, winds, mergers, etc.) appear to have peaked in the redshift range of $z = 1$–5 with a steady decline thereafter (Lilly et al. 1996; Hopkins and Beacom 2006). Within the ΛCDM paradigm of an evolving hierarchy, galaxies must have been smaller in the past, and this appears to be supported by extremely deep imaging from the HST (Bouwens et al. 2007). Galaxies were certainly more complex in character, with higher levels of asymmetry, clumpiness and internal velocity dispersion measured in the gas (Genzel et al. 2006; Law et al. 2007; Förster-Schreiber et al. 2009). From a detailed study of the Hubble Ultra Deep Field, Elmegreen et al. (2005) identify four main classes of galaxy morphology: 'chains,' 'clump clusters,' 'doubles' and 'tadpoles.' Out of this mess emerged modern-day disk galaxies which account for roughly half of the total stellar mass today (Driver et al. 2007; Bower et al. 2006). While well formed disks are a relatively modern phenomenon, there are occasional discoveries of 'normal'-looking disk galaxies at early times, in particular, a grand-design spiral recently discovered at $z \approx 2.18$ (Law et al. 2012).

One of the most discernible aspects of evolution is the phenomenon of 'downsizing' over cosmic time (for an updated discussion, see Fontanot et al. 2009). Before 2000, it was assumed that, in a hierarchy, small things form first, and larger objects appear through a process of merging of smaller systems (e.g. Fig. 1.1). In fact, with the benefit of hindsight, it was clear even in the 1960s that something was amiss. A large fraction of the most massive galaxies today are very old, red and dead (de Vaucouleurs 1961), whereas most low-mass systems show evidence of ongoing star formation. We now know that ellipticals have higher metallicities (Faber 1973) and enhanced α/Fe ratios (Ziegler et al. 2005), consistent with the idea that massive red galaxies inhabit dense cluster regions where much of the activity took place at early times ($z \gtrsim 1$; Gonzalez et al. 2011; Robertson et al. 2010). Presumably all of the smaller building blocks with active star formation that likely filled the early Universe have mostly been swept up in dense regions.[7]

We can look at this evolution in another way. Galaxy distributions in colour space (e.g. $g - r$) show a bimodal distribution where the details are strongly dependent on environment (Strateva et al. 2001; Blanton et al. 2003; Baldry et al. 2004). The 'blue sequence' is dominated by disks with ongoing star formation; the 'red sequence' is more of a mixed bag of dusty spirals and spheroids. The role of environment in galaxy evolution has been reviewed comprehensively by Blanton and Moustakas (2009). A complicating factor is that galaxy surveys provide accurate positions on the sky but only rough distances can be deduced from a redshift. Thus a formal definition of environment is hard to come by. Many schemes have been explored but

---

[7] This is reminiscent of the early Solar System when proto-fragments left over from planet formation bombarded the Earth and Moon (Late Heavy Bombardment) giving rise to the familiar 'face in the moon' around 4 Gyr ago.

these amount to little more than statistical measures of crowding (Muldrew et al. 2012, Table 1).

In addition to environment, a galaxy's evolution is affected by its mass at a given epoch, although mass and environment are coupled in the sense that more massive galaxies appear to be highly clustered at all epochs (Mo et al. 2010). Galaxy masses are difficult to determine at the present time although 'photometric masses' using red or infrared bands are a useful proxy (Baldry et al. 2008). Whether evolutionary affects are dominated by a halo's mass (or its central black hole) or its association with clustered galaxies embedded in a hot cluster medium are major questions that remain unresolved.

Thus we see evolution in time and evolution across large-scale structure. Interestingly, we also see evolutionary differences between groups of comparable mass. For example, the M81 group today has a similar inventory to the Local Group by mass, but appears to have more neutral gas, with extended tidal arms of HI and a HVC population dominated by massive gas clouds. It seems likely that we are seeing comparable mass groups at different stages in their evolution (Nichols and Bland-Hawthorn 2009, 2013).

1.3.2 Accretion and Feedback

Just how gas and dark matter move out of voids and into filaments, sheets and clusters are topics of key interest in modern cosmology. The Two Degree Field Galaxy Redshift Survey (2dFGRS; $z \approx 0.05$) provides clear evidence that 80 % of galaxies are collapsing into sheets and filaments (Peacock et al. 2001), the so-called Kaiser effect (Kaiser 1987). From these data, it is possible to determine a 2-point correlation function $\zeta(\sigma_{12}, \pi_{12})$ from the transverse (σ_{12}) and radial (π_{12}) pair separation. The ζ distribution measures the excess probability over random of finding pairs; for an isotropic random distribution, ζ has circular (iso)likelihood contours. But the radial vector π_{12} is determined from the redshift and therefore suffers redshift-space distortion. Small scales are dominated by the internal motions of clusters which leads to the familiar 'finger of god' effect. At larger separations, the likelihood contours are clearly flattened in the transverse direction consistent with galaxies having fallen away from the general Hubble flow (Peacock et al. 2001). Inflow onto individual groups and clusters is also seen in limited redshift surveys (Ekholm and Teerikorpi 1994; Ceccarelli et al. 2005) and possibly even in samples at $z \sim 2.4$ (Rakic et al. 2011).

Sheet structures may already be evident at large redshift. High resolution spectroscopy of paired QSOs provides two closely spaced sight lines to explore all IGM material from the present day to $z \sim 4$. Rauch et al. (2005) find evidence of gas sheets at different orientations presumably associated with galaxies in large-scale structure (see also Penton et al. 2004), although large-scale expansion of the sheets cannot be ruled out. The distribution of Lyα galaxies at these epochs appear to show evidence of sheets and filaments (Matsuda et al. 2005).

1 Near Field Cosmology

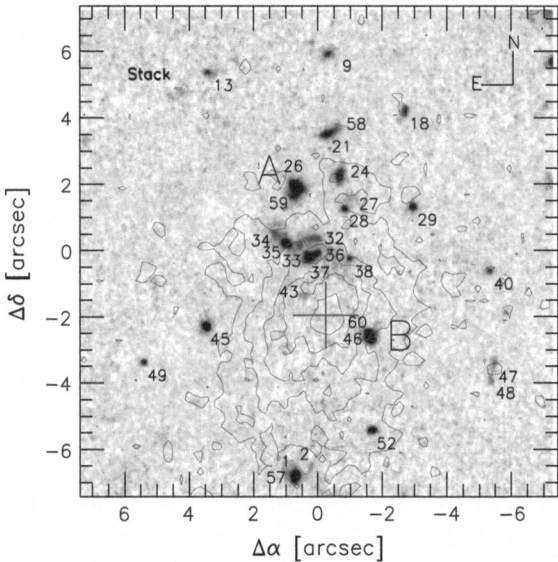

Fig. 1.6 Giant Lyα nebula (*blue* contours) in LABd05 ($z = 2.66$) where the *greyscale* image comes from HST broadband imaging. The centroid of the Lyα emission observed with Subaru is shown as a red cross. The extent of the gas ($\gtrsim 100$ kpc) is consistent with the early formation of a compact massive group (Prescott et al. 2012)

It is extremely difficult to probe gas beyond the nearby universe. HI 21 cm emission is only observed to $z \sim 0.1$ directly although $z \sim 0.4$ has been achieved through stacking observations. Reaching HI 21 cm beyond $z \sim 1$ is a major science goal of the Square Kilometre Array in the next decade. In the nearby universe, all HI clouds are associated with galaxies without exception (Sancisi et al. 2008). Early claims of accreting protoclouds on megaparsec scales (e.g. Blitz et al. 1999) have never been firmly established at any epoch. On smaller radial scales, evidence has been found for metal-rich gas accretion onto star-forming galaxies (Rubin et al. 2012). *Clear unambiguous evidence of gas accretion on galaxy scales is rare at any epoch.*

Lyα imaging has been used to detect diffuse gas in halos of high redshift $z \sim 4.4$ sources, most notably with the Grantecan Osiris tuneable filter (Kuiper et al. 2011). Another spectacular example is shown in Fig. 1.6. An impressive demonstration of the generality of Lyα halos at this epoch has come from stacking ~100 images of Lyman break galaxies (Steidel et al. 2011) where diffuse gas is seen out to 80 proper kiloparsecs. Galaxies must accrete gas in order to form stars, but whether this accretion is largely through infalling dark halos or a smooth flow associated with dark matter remains unclear. The specific processes involved with how gas gets into galaxies is discussed in Sect. 1.4.

Numerical simulations appear to show that there must be some form of 'feedback' within a forming galaxy to produce realistic disks and to match the details of the bimodal colour distribution. Starburst winds are often invoked to remove the low angular momentum build-up of material seen in synthetic simulations (Sharma et al. 2012; Brook et al. 2011). These are known to occur at all redshifts (Heckman et al. 1989; Veilleux et al. 2005), but whether the numerical outflows have any relationship to real winds is unclear (Springel and Hernquist 2003).

A stronger argument for feedback stems from the fact that the mass distribution of dark matter haloes predicted in ΛCDM simulations looks very different from the observed stellar mass distribution of galaxies (Fig. 1.7). Semi-analytic models (SAMs) that attempt to paint galaxies onto ΛCDM models through increasingly complex prescriptions fail to account for this behaviour without feedback. It seems increasingly likely that the complexity of baryonic physics has been underestimated and not adequately resolved nor understood in existing numerical simulations. In particular, star formation seems to be too complex to be simply gravity-induced, which has been the driving assumption in many numerical models.

In a recent review, Silk and Mamon (2012) appeal to processes that are strongly dependent on the dark halo mass. Supernova-driven feedback in starbursts may provide an acceptable fit to the turnover in the low mass end of the galaxy luminosity problem, but there are numerous problems. Low-mass galaxies have cores rather than cusps that arise in ΛCDM simulations so there are clearly other processes at work. It is possible that outflows can flatten cusps to cores (Governato et al. 2012) if most of the activity occurred in the early universe, but these models fail to account for the rapid dwarf evolution at low redshift (Weinmann et al. 2012), even accounting for the suppression of gas accretion onto dwarfs during reionization.

At the high mass limit, AGN feedback appears to account for the sharp cut-off in the bright end of the galaxy luminosity function (Croton et al. 2006; Panter et al. 2007). In a recent model, the cold flows onto the high mass haloes that drive the star formation are eventually interrupted by AGN-driven superwinds (Dubois et al. 2012). But if the wind or jet momentum is sufficient to stop gas inflow, it may even *induce* in situ star formation (Silk and Mamon 2012).

1.3.3 Baryon Inventory and Metal Enrichment

The universal baryon fraction of the total matter content of the Universe $f_b^o \approx 0.17$ is well determined from the angular power spectrum of the CMB (Komatsu et al. 2011). There is a well established trend in the baryon fraction observed in dark matter haloes as a function of halo mass (Fig. 1.7). The most massive systems, i.e. clusters with total masses of order 10^{15} M$_\odot$, have baryon fractions that approach the universal value (Dai et al. 2010). For $L_\star$ galaxies like the Milky Way, $f_b \lesssim 5\%$; in the limit of ultrafaint dwarf galaxies, f_b approaches zero. The explanations for this phenomenon are many and varied: either specific accretion rates decline with halo mass, or the lower mass haloes are interacting with their environment, or starburst-driven winds remove baryons more easily, or the baryons are less visible in lower mass systems.

Shull et al. (2012) have recently provided a useful baryon census of the local universe. In Fig. 1.8, the baryon census is given as 25 % in a warm-hot intergalactic

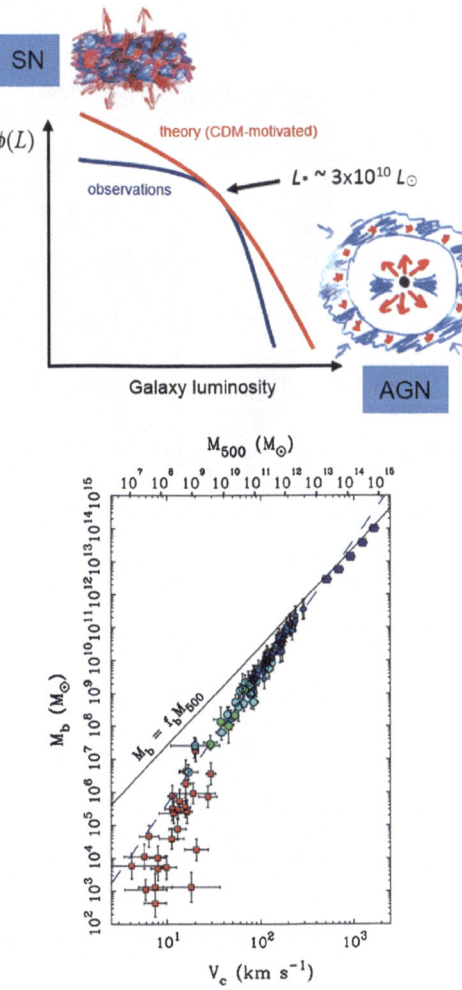

Fig. 1.7 *Upper* Schematic drawing of the ΛCDM "luminosity function" versus the stellar luminosity function $\phi(L)$ illustrating the possible role of feedback in modifying the observed distribution. The discrepancy at the low mass end may arise from starburst activity, and at the high mass extreme from AGN activity (from Silk and Mamon 2012). *Lower* Baryon mass M_b in clusters, groups (*purple*), galaxies (*blue, green*), Local Group dwarfs and ultra-faint dwarfs (*red*). The halo mass scale is shown at the top. The *solid line* is the derived baryon mass from assuming a universal baryon fraction $f_b^o \approx 0.17$ (McGaugh et al. 2010)

medium, 28 % in the Lyα forest, 4 % in the intracluster medium, 9 % in galaxies and a further 5 % in the circumgalactic medium (i.e. on scales of the halo virial radius); a further 29 % is listed as 'missing in action.' Circumgalactic media (CGM) are in essence multiphase gaseous haloes that extend to the halo virial radius. Spectacular evidence for the widespread occurrence of the CGM has come from OVI absorption observed through the haloes of $L_\star$ galaxies at intermediate redshift (Tumlinson et al. 2011). Interestingly, CGM galaxies are almost exclusively systems with active star formation; this gas appears to be circulating through the halo rather than escaping in a large-scale wind.

The baryon inventory is highly uncertain at the present time. The SSD budget (i) overlooks the intracluster medium in more numerous, low mass clusters ($\lesssim 10^{14}$

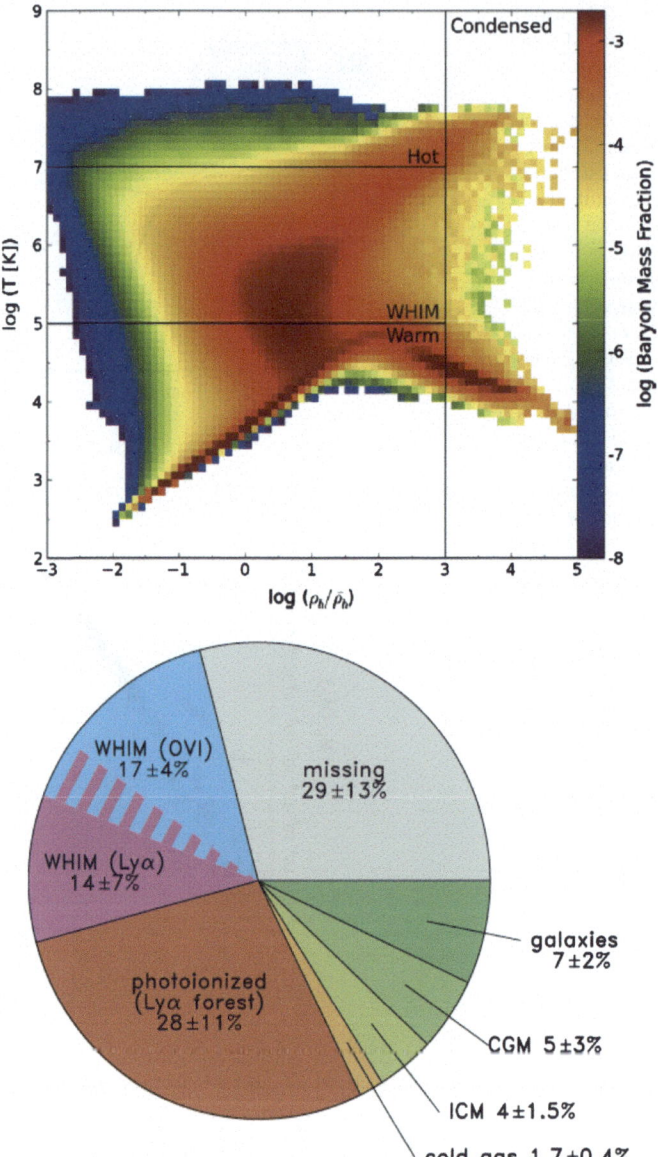

Fig. 1.8 *Upper* Physical state of the universal baryons (gas density vs. temperature) where *colour* refers to the relative baryon fraction. *Lower* Baryon inventory: warm-hot ionized medium (*WHIM*), intracluster gas (*ICM*) and circumgalactic medium (*CGM*). The overlap of the WHIM phases illustrates the overlap between these phases when detected (Shull et al. 2012). About a third of the baryons are unaccounted for. These figures are only illustrative; a lively debate continues about the relative proportions of the different phases

$M_\odot$); and (ii) the existence of a diffuse shock-heated IGM with a temperature above 10^7 K. The budget suffers from problems of double counting phases, assumes fixed chemical abundances, uses ionisation corrections from numerical simulations which may not account for all sources of heating, and so forth. Thus it remains unclear whether there are missing or invisible baryons in the universal baryon inventory (cf. Danforth and Shull 2008; Shull et al. 2012).

The dramatic evolution in the star formation rate density ($\dot{\rho}_\star$) (Hopkins and Beacom 2006) is expected to leave a signature in the baryon mass locked up in stars vs. gas with cosmic time. More of the mass is expected to be locked up in stars as the Universe ages. This is largely borne out by HST imaging of the stellar content of galaxies when compared to the cold gas content in Lyman limit systems (Wolfe et al. 2006). The corresponding evolution in the universal metal content is also observed (Ryan et al. 2009; Rafelski et al. 2012). The CIV absorbers show dramatic evolution at $z \sim 6$ suggesting a universal metallicity threshold of $Z_{IGM} \gtrsim 10^{-4} Z_\odot$ by this redshift (Ryan et al. 2009). These authors interpret this signature as fossil evidence of unseen stellar evolution at $z \sim 9$.

It is not known at what stage in cosmic time the last vestiges of pristine gas finally succumbed to stellar enrichment, but it is not inconceivable that some Pop III stars formed a billion years after the reionization epoch. This raises the tantalizing prospect that we can identify such regions in direct observations of the intermediate and high-redshift Universe. Extremely low values for the IGM metallicity have now been observed at $z \sim 3$ (Fumagalli et al. 2011). Thus we may be seeing back to the material that formed some of the most metal poor stars associated with galactic halos like our own.

1.3.4 Chemical Evolution in Galaxies

We refer the reader to the excellent lecture notes provided by F. Matteucci in this volume.

1.3.5 Milky Way and Local Group Analogues in the Real Universe

We mention one more aspect of galaxy surveys that has important implications for near field cosmology. The GAMA galaxy survey (350,000 sources at $z \lesssim 0.1$; Driver et al. 2011) is sufficiently densely sampled that we can ask questions such as 'How common are Local Group assemblies in the Universe?' This is a specific goal of the GAMA survey, i.e. to find close analogues to the Milky Way (i.e. two Magellanic companions, comparable mass) and the Local Group (two Milky Way [MW] mass galaxies with Magellanic companions). Robotham et al. (2012) find that 12 % of galaxies have a companion within a factor of two of the LMC's mass, although this drops to 3 % if we include an SMC mass companion as well. Two full analogues to

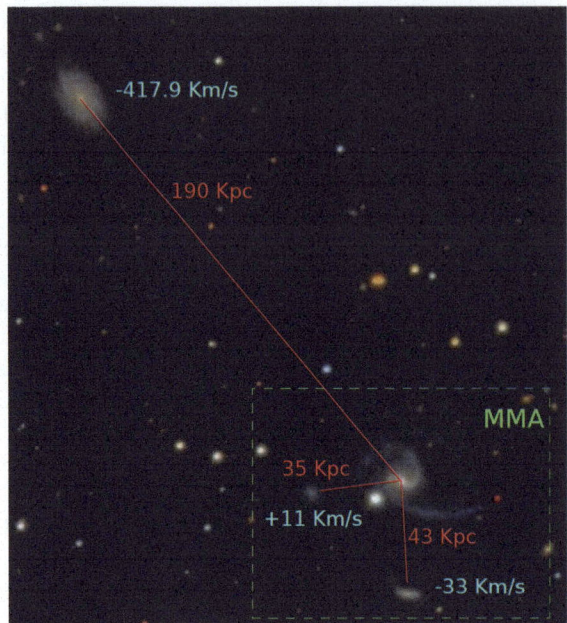

Fig. 1.9 An SDSS image of the best analogue of the Local Group to emerge from the GAMA galaxy survey (Robotham et al. 2012). The group marked MMA is remarkably similar to the Milky Way—Magellanic System where the three main galaxies are all forming stars at the present time

the Milky Way were found where a closely grouped MW-LMC-SMC type triplet are all actively star forming. This amounts to less than 1 % of galaxies in the GAMA survey. Thus while the Milky Way may be rather typical of late-type L_* galaxies, it seems to be relatively special when the satellite population is also considered even when star formation rates are ignored. The search for Local Group analogues is presently ongoing; one striking MW-M31 analogue has already been found (Fig. 1.9; Robotham et al. 2012).

1.3.6 Milky Way and Local Group Analogues in Simulated Universes

In parallel with the observations, an equally important goal is to identify realistic analogues in cosmological N-body simulations. Large-scale simulations involving up to 10^{10} particles include Millennium I (Springel et al. 2005), MareNostrum (Gottlöber et al. 2006), Millennium II (Boylan-Kolchin et al. 2009), Bolshoi (Klypin et al. 2011) and MultiDark (Prada et al. 2012). Dark matter simulations have also been carried out with exquisite detail for individual haloes (Springel et al. 2008; Stadel et al. 2009).

Some of the most important simulations to date that are specific to near field cosmology are the Constrained Local Universe Simulations (CLUES; www.cluesproject.org) that achieve realisations which are heavily constrained by local data (Hoffman and Ribak 1991; Kravtsov et al. 2002; Klypin et al. 2003; Yepes et al. 2009; Gottlöber et al. 2010). Forero-Romero et al. (2011) have recently identified what appears to be a Local Group analogue. They confirm what is known from the stellar record that the Milky Way and M31 are products of a quiet mass assembly history where the last major merger occurred more than 10 Gyr ago (Wyse 2001). Much like galaxy redshift surveys, these simulations contain a wealth of information that is yet to be studied in detail.

1.4 Gas Accretion onto Galaxies

1.4.1 Introduction

A major focus of these lectures has been to consider how gas gets into galaxies, with specific attention to the Local Group. We explore this below in terms of current fads—cooling flows, cold flows, warm flows—but we start by discussing early ideas on galactic accretion as these provide a more solid basis for understanding important physical processes.

As we have seen, it is well established that the observed baryons over the electromagnetic spectrum account for only a fraction of the universal baryon fraction f_b^o, particularly on scales of galaxies. For example, the Galaxy has numerous baryonic phases that have been studied in great detail over many years. A detailed inventory of the Galaxy reveals only a quarter of the universal fraction (Flynn et al. 2006), and the same holds true for M31 (Tamm et al. 2012). Moreover, the build-up of stars in the Galaxy requires an accretion rate of 1–3 $M_\odot \, yr^{-1}$ (Williams and McKee 1997; Binney et al. 2000), substantially larger than what can be seen from direct observation.

So how does gas get into galaxies? This is a major problem in contemporary astrophysics which remains largely unsolved. Does the gas arrive cold, warm or hot, or in the form of stars? Are the infalling baryons brought in by dark matter or as a steady rain of material? Part of the problem is that baryons have many phases (plasma, atoms, molecules, dust, metals, etc.) which call for a multiwavelength observational approach and hugely complex computer codes to track all of the ingredients. We may be a decade or more away from a physical picture that would convince most physicists of its validity.

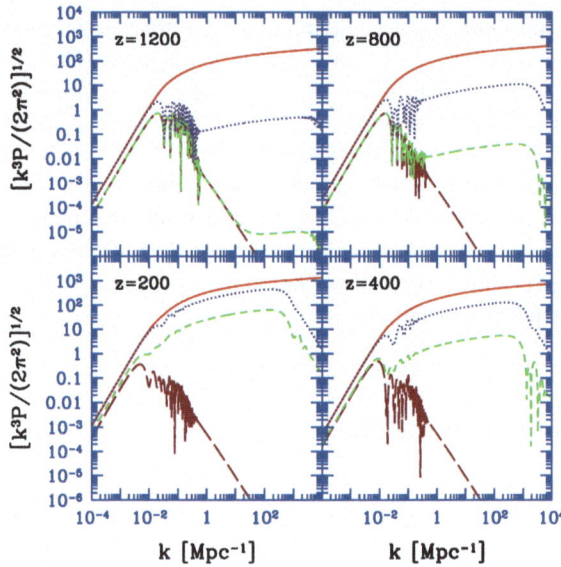

Fig. 1.10 Power spectrum plots of density and temperature fluctuations in the early universe (CMB-FAST; Barkana and Loeb 2007) as a function of comoving wavenumber $k = 2\pi/\lambda$ where λ is a physical scale in Mpc. The fluctuation spectra are shown for four redshifts: 1200, 800, 400, 200. The four curves correspond to fluctuations in ΛCDM density (*solid*), baryon density (*dotted*), baryon temperature (*short-dashed*) and photon temperature (*long-dashed*). Note how gas is well and truly accreting onto DM halos by $z = 200$

1.4.2 Earliest Epoch of Gas Accretion

Accretion prior to reionization. As we have seen, at the time of recombination, the distribution of matter was remarkably uniform, although tiny fluctuations at the level of 1 part in 100,000 can be measured from the CMB. But once the baryons and photons had become decoupled, the baryons were able to accrete onto evolving density fluctuations in the dark matter for the first time. In Fig. 1.10, the gas temperature fluctuations remain tied to the photon temperature fluctuations for a while after recombination. This is because the uncombined free electrons still experience Compton scattering with the CMB. But within the horizon scale ($k \gtrsim 0.01$ Mpc^{-1} at $z \sim 1000$), where gravity dominates, gas is clearly accreting onto dark matter halos, with a fluctuation spectrum converging on the ΛCDM spectrum by $z \sim 200$. By this time, the thermal coupling of the gas with the CMB is essentially over; the universal expansion is driving the adiabatic cooling such that the gas temperature fluctuations converge to $\sim 2/3$ of the amplitude of the gas density fluctuations.

It is interesting to ponder whether one can even observe these first accretion (fossil) events, long before the first stars formed—the so-called "Dark Ages" (Scott and Rees 1990; Madau et al. 1997). This question is of particular interest to scientists considering the impact of the Square Kilometer Array (SKA) in detecting gas phases at hese early epochs (Barkana and Loeb 2007; Furlanetto and Loeb 2004). H I 21 cm in its ground state has hyperfine splitting due to parallel spins (triplet state) of the proton and electron pair having a slightly higher energy than the anti-parallel (singlet) state. The pre-reionization era was marked by subtle variations in H ionisation fraction, gas density and spin temperature. The energy density of the redshifted 21 cm emission is expected to be about 1 % of the CMB.

Are the spatial fluctuations in the early gas distribution observable? The hyperfine levels tend to be in thermal equilibrium with the CMB. To be observable, the 21 cm level populations must correspond to a spin temperature that is different from the CMB in order to be observed in emission or absorption. Two possible ways to do this involve continuum UV from early sources of radiation (Lyα pumping), or large-scale IGM shocks or small-scale shocks (spin exchange) as gas collapses onto halos. It remains unclear which, if any, of these mechanisms is likely to lead to a successful SKA detection (Furlanetto and Loeb 2004; Shapiro et al. 2006).

In Sect. 1.2.2, we discussed how ($z \gtrsim 40$) major cooling onto a small fraction of dark matter minihalos led to the formation of the first stars. This ultimately gave rise to inhomogeneous reionization as the first stars bleached their way through the extended IGM. The suppression of low-mass galaxies during the reionization epoch (Sect. 1.2.5) is modulated by fluctuations in the timing of reionization (Babich and Loeb 2006). What are the likely effects of patchy reionization in the nearby Universe? It is conceivable that the faint end of the galaxy luminosity function varies across the hierarchy because of the varying impact of the UV field (Efstathiou 1992). But present galaxy surveys do not go deep enough to allow us to study the dwarf population far beyond the Local Group. At present, there is no clearly identified fossil signature of the reionization epoch in the Local Group galaxy population (Grebel and Gallagher 2004).

Accretion after reionization. Doroshkevich et al. (1967) first considered the impact of a UV background in quenching star formation. They realised that the photoheated gas becomes too hot to stay confined by the shallow potential well of the galaxy (see Sect. 1.2.5). The issue is particularly relevant to the Local Group because N-body simulations repeatedly overpredict the number of dwarfs that are visible today (e.g. Moore et al. 1999; Klypin et al. 1999). It is plausible that many low-mass dwarfs were suppressed for all time during the reionization epoch. Are there thousands of dark matter minihaloes moving through the Local Group devoid of baryons? There are no strong constraints at the present time.

Loeb (2010) provides three useful formulae for relating halo properties with cosmic epoch:

$$R_{\rm vir} = 1.5 \left(\frac{M}{10^8 \, {\rm M}_\odot}\right)^{1/3} \left(\frac{1+z}{10}\right)^{-1} \text{kpc} \qquad (1.18)$$

$$V_c = 17.0 \left(\frac{M}{10^8 \, {\rm M}_\odot}\right)^{1/3} \left(\frac{1+z}{10}\right)^{1/2} \text{km s}^{-1} \qquad (1.19)$$

$$T_{\text{vir}} = 1.0 \times 10^4 \left(\frac{M}{10^8 \, M_\odot}\right)^{2/3} \left(\frac{1+z}{10}\right) \, \text{K}. \qquad (1.20)$$

R_{vir} is the virial radius (see Sect. 1.2.4), i.e. the natural scale of the evolving system within which infall material is no longer collapsing (or expanding). V_c is the circular velocity at the virial radius such that $V_c = (GM/R_{\text{vir}})^{1/2}$. T_{vir} is the virial temperature at R_{vir} defined as $T_{\text{vir}} = \mu m_p V_c^2/2k$ for which μ is the mean molecular weight (0.6), m_p is the proton mass and k is Boltzmann's constant.

The critical mass m_{crit} below which galaxies are unlikely to have formed is not fully resolved. Gnedin (2000) argued that m_{crit} corresponds to the scale over which baryonic perturbations are smoothed in linear theory. But this was found to be an order of magnitude too severe by Okamoto et al. (2008) who showed that it was the gas temperature at the virial radius which determines whether a halo can accrete gas, i.e. the usual argument relating to Jean's mass where the contribution of the dark matter to the gravitational potential is included. Gas infall depends sensitively on the Jeans mass (Sect. 1.2.5). Thoul and Weinberg 1996 found a 50% reduction in the collapsed gas mass due to heating for a halo with circular velocity $V_c \sim 50$ km s^{-1} at $z \sim 2$ and complete suppression below $V_c \sim 30$ km s^{-1} (cf. Rees 1986). With the inclusion of self-shielding by the gas, Kitayama and Ikeuchi 2000 determined slightly lower velocity thresholds. Dijkstra et al. (2004) found substantially lower thresholds were possible at $z \gtrsim 10$ even while the UV background is ramping up.

In later sections, we consider in more detail just how gas manages to find its way into the evolving dark matter haloes. We start with some early ideas that continue to be relevant today because they have a solid analytic basis.

1.4.3 Early Ideas on Galaxy Accretion

The problem of accretion onto a spherical body was studied by Bondi et al. (1947; hearafter BHL). Their focus was on stellar accretion but this seminal paper is an important starting point for understanding galactic accretion. In the BHL framework, there are three characteristic accretion radii. The first, R_0, is due to the cross-sectional area of the body. A galaxy moves through the intergalactic medium, sweeping up anything in its path. The second, R_A, is the radius within which the particle velocity is less than the escape velocity and the particles are therefore accreted gravitationally onto the galaxy. This does not take into account collisions, except on the accretion axis. The final radius is the Bondi radius, R_B, which is based on a fluid dynamical treatment of the case where the galaxy is stationary relative to the gas and therefore simply accretes spherically symmetrically through gravitation. For the Galaxy, Hunt (1971) calculated the radii as $(R_0, R_A, R_B) = (12, 160, 54)$ kpc and notes that these values are within an order of magnitude and that there is therefore no single dominant process. This is in contrast with the situation of stellar accretion, where the values are (0.005, 4, 700) AU.

1 Near Field Cosmology

For galactic accretion, the dominant term is dependent on the velocity of the galaxy through the gas. Hoyle and Lyttleton (1939) showed that the rate of accretion is given by

$$A = \frac{4\pi(GM)^2\rho_\infty}{V^3} \quad (1.21)$$

given certain assumptions such as the temperature not being too high. However, this analysis was performed by assuming the density of gas to be infinite along the axis the gas was accreting onto. Bondi and Hoyle (1944) replaced this with a more physically reasonable accretion axis, which led simply to the incorporation of a multiplicative factor α, a parameter which lies between 0.5 and 1. For the case they were interested in, a star moving perpendicularly into a flat cloud, Bondi and Hoyle calculated α to be approximately 0.625. Applied to the Galaxy, using the calculated value of R_A, Eq. 1.21 gives an accretion rate of 10 $M_\odot$ yr^{-1} (Hunt 1971), a high value by modern standards.

The work discussed above ignored particle interactions except on the accretion axis. However, for accretion of the IGM onto a galaxy, the mean free path of particles is only a few parsecs, much smaller than the relevant accretion radii, which are typically in the region of tens or hundreds of kiloparsecs. Thus particle collisions can be important and therefore a fluid dynamical treatment is required. However, as Hunt (1971) noted, for the case of highly supersonic fluid flow, the kinetic energy term of the fluid equations dominates over the pressure term, meaning that the fluid dynamical treatment is unnecessary. The only place where pressure is important is on the accretion axis, where an accretion column similar to Bondi and Hoyle's is formed, so Eq. 1.21 is relevant here.

For most cases where the velocity is not highly supersonic, the full fluid dynamical treatment cannot be performed analytically. However, a case where it is possible is accretion onto a stationary body, with the flow being spherically symmetric. Bondi (1952) showed that the accretion rate for this case is given by

$$\frac{4\pi\zeta(GM)^2\rho_\infty}{c_\infty^3}, \quad (1.22)$$

where c_∞ is the sound speed in the gas at infinity and ζ is a positive number which is found by matching the pressure of the gas at the surface of the body with the surface pressure of the body. This gives an accretion rate of 1.5 $M_\odot$ yr^{-1} for the Galaxy, using the calculated R_B above, a value that is sufficient to sustain the current star formation rate in the Galaxy (Williams and McKee 1997).

The two equations above are similar, which led Bondi to suggest that the accretion rate for the intermediate case, which cannot be calculated analytically, be interpolated from the low and high velocity cases, i.e.

$$A = \frac{2\pi(GM)^2\rho_\infty}{(c_\infty^2 + V^2)^{3/2}} \quad (1.23)$$

This gives an accretion rate for the Galaxy of 2.5 $M_\odot$ yr^{-1}, similar to what is required to explain the observed rate of star formation. Hunt believed that this was the best estimate for the accretion rate of the galaxy. Dodd (1953) confirmed that this formula was at least approximately correct for the case of isothermal flow past a gravitational point source at four different supersonic velocities.

So is there observational evidence for a basic accretion process of the kind envisaged by Bondi, Hoyle and Lyttleton? What are the expected signatures? Unquestionably, the process is complicated, and the signatures will be difficult to dig out. Quite apart from the motion of the Galaxy around the Local Group barycentre (Cox and Loeb 2008), the Local Group has a comparable velocity with respect to the intergalactic medium (Tully et al. 2008, Table 3). There are no clear accretion signatures at the present time other than that inferred from the high-velocity cloud (HVC) population. The important point is that the current baryon inventory of the halo fails by a substantial factor to account for the required star formation rate in the Galaxy (Bland-Hawthorn 2009; Aumer and Binney 2009).

1.4.4 Accretion Shocks

If the Galaxy moves through ambient gas supersonically, a shock wave will appear, much like the heliopause ahead of the Sun. A shock wave is a very sudden change in the properties of the medium, usually involving a large jump in pressure, density and temperature. The condition required for supersonic accretion is that the virial temperature of the halo is greater than the temperature of the accreting gas (e.g. Binney 1977).

Models of accretion shocks show that the location of the shock depends on the relationship between cooling times and dynamical times. The warm-hot intergalactic medium is expected to be in the form of a plasma at 10^7 K, which means that cooling times are longer than dynamical times (Sutherland and Dopita 1993) and the shock is therefore expected to occur close to the virial radius (Benson 2010). A shock can either be a *bow shock*, where the shock is located in front of the body, or a *wake shock*, where the shock follows the body. A bow shock will appear if the medium can provide the sufficient pressure to maintain it, otherwise it will be a wake shock. The nature of the Galactic shock has been the subject of debate since Hunt's early work.

The Galaxy is travelling through the intragroup medium with a transonic velocity, i.e., its velocity is of the same order as the sound speed in the medium (Portnoy et al. 1993). This means that the accretion rate cannot be found analytically (Sect. 1.4.1). Hunt (1971) tested Bondi's interpolation (Eq. 1.23) using numerical integration and investigated the nature of the shock in cases where the flow is supersonic. He wrote the fluid equations as conservation laws because this allows a shock wave to appear automatically in the solutions. The shock itself does not work well with numerical methods because it is a discontinuity, but the conservation law method uses continuous quantities and the shock appears as steep rises over a few intervals, rather than as a discontinuity. Hunt found that for the subsonic case he tested, $\mathcal{M} = 0.6$, where $\mathcal{M}$ is the Mach number, the flow was largely isotropic, with a mean anisotropy of only 5%. The density is very similar to Bondi's case of spherically symmetric flow, with only 3% average deviation. His conclusion was that if the Galaxy is moving subsonically through the intergalactic medium, it would receive matter as if it were stationary.

Hunt also modelled the transonic cases $\mathcal{M} = 1.4$ and $\mathcal{M} = 2.4$ and finds that the flow exhibits a bow shock. This result is in contrast to Spiegel (1970) who argued for a tail shock because there was an insufficient pressure head to support a bow shock. Importantly, Hunt claims that a pressure head *is* formed due to the increase in density as material is accreted onto the Galaxy, which results in pressure forces acting away from the Galaxy. In other words, a constant rain of accreting material can assist in the formation of a bow shock. For $\mathcal{M} = 1.4$, Hunt finds the position of the shock at $0.87\,R_B$ in front of the body and $3.9\,R_B$ perpendicular to the direction of motion; for $\mathcal{M} = 2.4$, these numbers are 0.11 and $0.36\,R_B$. The accretion rate is found to agree with the prediction from Bondi's formula and the $\mathcal{M} = 2.4$ result is close to the expected analytic result for large Mach speed. As seen in Fig. 1.11, except for the shocked region, the flow is nearly isotropic. The deviation from the stationary case was only at the 25% level.

Shima et al. (1985) looked at the same problem as Hunt,[8] but used an integration scheme they considered to be better for capturing discontinuities. To test the accuracy of their scheme, they used it to calculate the flow of air past a rigid sphere and found that it conformed to experimental results. For the cases studied by Hunt, Shima et al. also found a bow shock but with a twofold increase in accretion rate. They suggested that the accretion rate in Bondi's interpolation formula should be increased by the same factor, approaching the formula provided by Hoyle & Lyttleton for large Mach number.

Portnoy et al. (1993) came to a different conclusion based on a model of accretion onto an *extended* gravitational source (see also Hunt 1975), rather than the point source of Hunt (1971). They found a bow shock only when they increased the mass of the Galaxy relative to its scale radius, consistent with the suggestion that they would have found a bow shock if they had used a point source. In general, they found a tail shock as suggested by (Spiegel 1970). However, including replenishment of gas from stellar halo mass loss in their model allowed a bow shock to form, with the shock supported by this gas. A bow shock appeared if the ram pressure of the outside flow and the pressure of the replenished gas were comparable.

To date, there has been little observational evidence for something akin to a 'heliopause' ahead of either the Galaxy or the Local Group. These early ideas on accretion have strong echoes in numerical simulations of cluster galaxies generating wakes as they move through the intracluster medium, although here the focus tends to be on stripping galaxies of their gas (e.g. McCarthy et al. 2008). Possibly related

---

[8] Several authors comment on the difficulty of reproducing Hunt's classic papers in 1971 and 1975. The elegant analysis stymied by typographical errors that we have identified and corrected for; it is then possible to reproduce this important work.

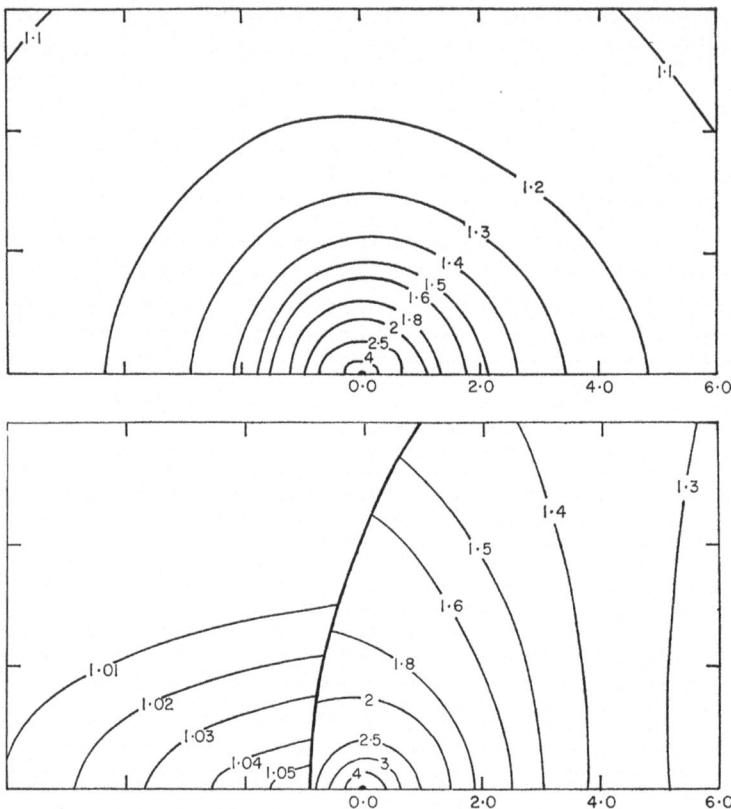

Fig. 1.11 Accretion flows onto the Galaxy centred at the origin for two different Mach numbers: *upper* $\mathcal{M} = 0.6$; *lower* $\mathcal{M} = 2.4$ (from Hunt 1971)

bow-shock structures have been observed on scales of galaxy clusters (e.g. Bagchi et al. 2006): the outer boundary of the radio continuum emission is remarkably well confined within a circle defined approximately by the virial radius of the central cluster.

1.4.5 Cooling Flows

Since the 1990s, the notion of galaxies sweeping up gas has been largely displaced by the paradigm of hot accretion. In this picture, gas and dark matter are accreted together onto central haloes, groups and clusters (Silk 1977; Rees and Ostriker 1977; White and Rees 1978). The gas becomes shock-heated to the virial temperature of the evolving dark matter system (White et al. 1993; Kaiser 1986). This picture finds support on megaparsec scales from the existence of hot coronae in groups and clusters.

Hot accretion has been a basic tenet of ΛCDM simulations for over twenty years. The belief was that the hot gas has time to cool in the central regions and accretes onto the central galaxy or group forming a 'cooling flow' (Fabian 1994). The mass cooling flux $\dot{M}$ is assumed to be steady, spherical, and subsonic, and that any gradients in the potential are small compared to the square of the sound speed. In this case the pressure is constant, and mass conservation requires that $\dot{M} = 4\pi \rho v_i r^2$ where v_i is the inflow velocity. The cooling radius r_c interior to which the cooling time t_c is smaller than the age of the dark matter halo t_h is set by the condition $t_c \sim t_h$. The flow time from r_c is $t_f \sim r_c/v \sim 4\pi \rho_c r_c^3/\dot{M}$, where ρ_c is the gas density at r_c. We assume $t_f \sim t_c$, so that the gas has time to cool before reaching $r = 0$. This sets $v \sim r_c/t_h$ at r_c. If the cooling function $\Lambda(T_e)$ does not vary rapidly with temperature T_e, the density and temperature within r_c scale nearly as $\rho \propto r_c/r$, $T_e \propto r/r_c$ (Fabian and Nulsen 1977).

On the scales of galaxies, possible evidence for hot accretion of this form has been widely discussed (Maller and Bullock 2004; Kauffmann et al. 2004) but never clearly demonstrated. In fact, beyond the Galaxy, the first reliable detections of disk galaxy coronae have only been possible in recent years for two of the most massive disk systems known (Anderson and Bregman 2011). When scaled to the mass of the Galaxy, the predicted halo accretion rate of 0.1 $M_\odot$ yr^{-1} is much too low to be a useful source of baryon accretion.

On group scales, unless the Local Group is very old, it is unlikely that a steady-state flow has been established (e.g. Tabor and Binney 1993), especially as infall of gas into the Local Group is likely to be intermittent and winding down over cosmic time (Maloney and Bland-Hawthorn 2001). Blitz et al. (1999) made the spectacular suggestion that the HVC population is evidence of such an event on megaparsec scales, but these clouds are now known to lie within 50 kpc of the Galaxy (Bland-Hawthorn et al. 1998; Putman et al. 2003).

For massive clusters, Peterson et al. (2003) carried out *XMM-Newton* spectroscopy of 14 clusters with the highest $\dot{M}$ rates estimated from using the above analysis. A wide range of ions and ionisation states were detected and the results are unambiguous, a relatively rare occurrence in extragalactic astronomy. In all cases, *the cluster gas was found to be much too hot to support cooling flows*. For some (or all?) of these systems, there is evidence that central AGNs have kept the gas hot through intermittent jet interactions with the ICM. Thus there appears to be little or no support for cooling flows on galaxy to cluster scales.

1.4.6 Cold Flows

The hot accretion picture leads to a delay in the star formation because the gas needs time to cool to trigger star formation. While being an early supporter of hot accretion, Binney (1977) also pointed out that some of the accreted gas is likely to be 'cool' (10^4K) because any dense clumps are unstable to cooling processes. High resolution simulations ultimately confirmed that much of the gas can cool and accrete through filamentary streams (Fardal et al. 2001; Birnboim and Dekel 2003; Katz et al. 2003; Keres et al. 2005; Keres and Hernquist 2009). The most recent simulations (e.g. Fig. 1.12) find that cold flows dominate accretion onto all halo masses before $z \gtrsim 3$ and exclusively galaxies less massive than $L_\star$ since that time (Keres et al. 2009; Brooks et al. 2009)—this is not a view that is universally supported (e.g. Benson and Bower 2011). The build-up of the hot halo initially helps the cold flow to slow down during accretion, but ultimately the halo can become quite dominant and shut off the accretion process.

The 'cold flow' paradigm has some interesting consequences, most notably in explaining angular momentum build-up in disks. In the early 'hot accretion' picture, the dark matter and gas in the extended halo acquire essentially all of their angular momentum through tidal torques with the surrounding mass distribution. This largely occurred at turn around ($z \sim 5$) when galaxies fell away from the Hubble flow. Within this paradigm, Fall and Efstathiou (1980) present what has become the standard paradigm for disk formation: the shocked gas settle to a disk which then smoothly accreted inwards while conserving angular momentum, originally proposed by Mestel. Thus, for example, the Galactic disk with its exponential scale length of $\sim$5 kpc has conserved angular momentum from a collapsing gaseous halo of order $\sim$300 kpc in size (see White 1984).

In the context of cold flows, Keres et al. (2005) state that the infalling streams merge onto disks "like streams of cars entering an express way" converting some fraction of their infall velocity to rotational velocity. The 'French school' (Pichon et al. 2011; Codis et al. 2012; Kimm et al. 2012) extend this idea to overturn earlier ideas on how angular momentum is advected through the virial sphere. In their simulations, they find that a typical $L\star$ galaxy has 2-3 cold flow streams although only one is particularly active at a given time. These are long-lived coherent flows that can build up a disk's spin, scaling properties and orientation. The widths of these filaments do appear to depend on the resolution of the simulation (Rosdahl and Blaizot 2012).

At the present time, no clear evidence of 'cold flows' have been observed which is somewhat surprising given that they are believed to occur, albeit at a lower level, to the present day (Keres and Hernquist 2009). An interesting trend is to associate extended Lyα emission around galaxies at high redshift with cold flow accretion (e.g. Rauch et al. 2011) although the situation is undoubtedly complicated (Rosdahl and Blaizot 2012). Other hints of cold accretion come from absorption site lines to high redshift quasars (Ribaudo et al. 2011) but the evidence remains circumstantial.

Whether cold streams survive in future hydrodynamic simulations, with higher resolution and better adaptive mesh algorithms to capture important microprocesses, remains to be seen. But it is interesting to consider what disk properties are signatures of the formation process at high redshift; this is an issue we return to below in our discussions on near-field cosmology.

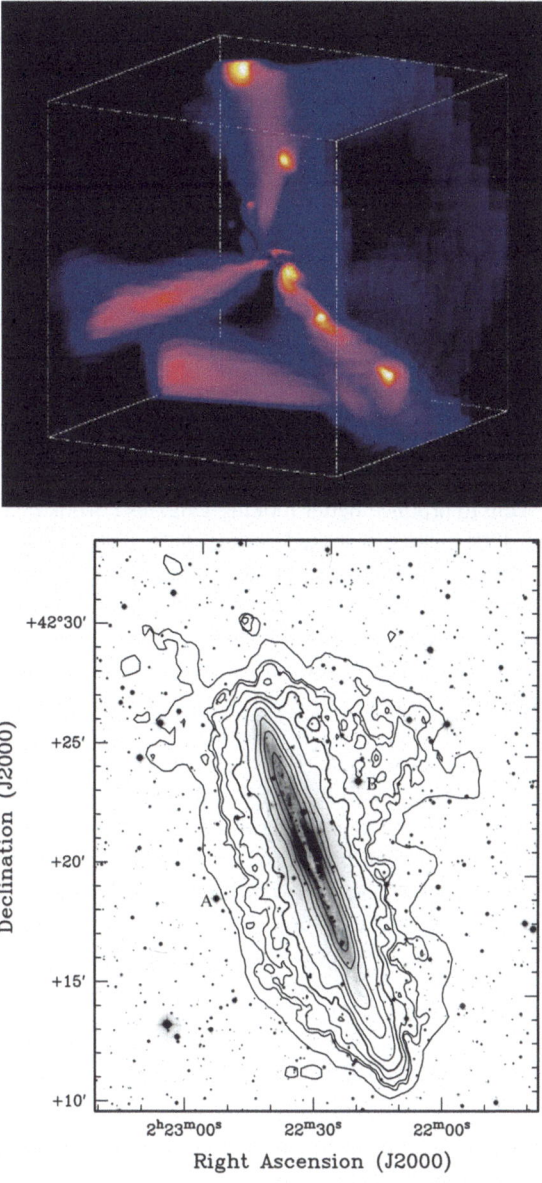

Fig. 1.12 *Upper* Cold gas flows onto a 10^{12} M$_\odot$ dark matter halo observed in a 3D hydrodynamic simulation (Dekel et al. 2009). The size of the box is about 300 kpc. Nothing like this has been observed directly although an intriguing H I filament ($\sim$25 kpc) has been observed in NGC 891 (*lower*), a galaxy that shares most of the Galaxy's attributes (Oosterloo et al. 2007)

1.4.7 Warm Flows

ΛCDM models consistently fail to produce realistic disks (Bullock et al. 2001, although see Agertz et al. 2009). Galaxy disks are thin and relatively fragile, and are easily disrupted or heated from infalling structure. Kormendy et al. (2010) showed that two thirds of his galaxy sample are pure disks with only small central bulges[9] and found no correspondence with the big-bulge thick disks arising naturally in the simulations. These are simply too violent to form the thin disks we observe today predicting very high merger rates of infalling dark matter clumps and too many clumps at all epochs (Springel et al. 2005). The random motions between simulated galaxies are much higher than observed in peaceful suburbs like our own Local Group of galaxies, and so on.

An even more extreme example of disks are the super thin disks first identified by Matthews et al. (1999). The gathering consensus is that gas must enter a disk galaxy quietly (Freeman 2008). Newly contrived models that encourage quiescent gas accretion do seem to produce better looking galaxies (Brook et al. 2011).

Marinacci et al. (2011) have proposed coronal halo accretion driven by star formation in the disk driving gas shells into the halo. This followed from the realisation that the disk-halo interaction rate in the Galaxy ($\gtrsim 10$ $M_\odot$ yr^{-1}) is much larger than the required accretion rate onto the disk (Bland-Hawthorn 2008).

Another quiescent channel is a 'warm flow' that follows the breakdown of cold gas as it disrupts during its passage through the hot halo (Bland-Hawthorn 2009; Heitsch and Putman 2009). This process is clearly observed in the disruption of the Magellanic Stream in a shock cascade (Bland-Hawthorn et al. 2007). These authors have shown that much of the Magellanic Stream is in the form of warm gas due to a shock cascade along the Stream; this warm gas is raining onto the Galaxy. The Magellanic Stream may well be the source of most high velocity clouds seen throughout the halo (Sect. 1.7.1.5). The warm gas accretion rate is at least as high as the H I accretion rate, and may even dominate. But the ionised+neutral accretion rates are still well below the levels required to sustain the present star formation rate.

1.4.8 Accretion via Major and Minor Mergers

In our discussion of gas accretion onto galaxies, little has been said about whether gas and dark matter accrete together onto galaxies. If so, is the gas carried by the dark matter, or does it enter separately? Since we do not understand the main channels of gas accretion, this is a difficult question to answer.

Major mergers. The recent review by Sancisi et al. (2008) of H I in the local Universe finds that cold gas disconnected from galaxies is always associated with interactions between galaxies—in other words, there is no evidence for isolated (primordial?) H I gas at present. Major galaxy mergers and collisions involving both

---

[9] Small central bulges are expected in pure disk evolution; see Sect. 1.6.1.

dark matter and baryons can be an important aspect of the accretion process (Hibbard and van Gorkom 1996).

The stellar record tells us that the Galaxy has not experienced a major accretion event (i.e. at least as massive as an L_* galaxy) over the past 10 Gyr (Wyse 2001). Evidence of less massive gas-rich accretion events is more easy to come by. A particularly spectacular interaction phenomenon is the Magellanic H I Stream that trails from the LMC-SMC system (10:1 mass ratio) in orbit about the Galaxy. Since its discovery in the 1970s, there have been repeated attempts to explain the Stream in terms of tidal and/or viscous forces (q.v. Mastropietro et al. 2005; Connors et al. 2006). Indeed, the Stream has become a benchmark against which to judge the credibility of N-body+gas codes in explaining gas processes in galaxies. A fully consistent model of the Stream continues to elude even the most sophisticated codes. The total gas content of the Magellanic System falls well short of the required gas accretion rate ($\lesssim 0.2\ M_\odot \mathrm{yr}^{-1}$; Putman et al. 2012).

Minor mergers. Kauffman et al. (2010) have investigated the possibility that dwarf galaxies are major sources of baryons. After compiling a large sample of isolated galaxies from the SDSS survey, they find strong correlations between the total mass of gas in satellites and the colours and specific star formation rates of central galaxies. The inferred accretion rate from the *observed* baryons is two orders of magnitude too small to account for the star formation rate in the central galaxy, which led them to postulate that dwarfs are harbouring large reserves of warm gas. An attractive aspect of the model is that warm gas is more easily concealed because its emissivity depends on the square of the gas density (Bland-Hawthorn et al. 2007), unlike H I gas where the projected column has a linear dependence on density.

Warm gas within dwarf galaxies however is easily stripped by the Galactic halo through ram pressure stripping (Nichols and Bland-Hawthorn 2009, 2011). The stripped gas becomes part of the warm-hot halo and can conceivably seed clouds for later accretion (Joung et al. 2012). An additional problem is the lack of strong evidence for massive hot haloes around dwarf galaxies. Gallagher et al. (2003) used the Wisconsin Hα Mapper (WHAM) to obtain very deep limits over a 1° beam on the presence of ionised gas in the Draco and Ursa Minor dwarf spheroidals. No warm gas was seen down to a mass limit that is of order 10 % of the stellar mass fraction in both objects.

1.4.9 Accretion of High Velocity Clouds

1.4.9.1 Decelerating Gas Clouds in a Hot Corona

We now consider the infall of individual H I clouds towards the Galactic disk. Do they survive to reach the disk or break up to become part of the hot corona? This important question bears on how much warm/hot gas is presently concealed in the Galactic corona. The question is very difficult to settle from direct observation because emission from a relatively cool Galactic corona ($T_e \approx 2 \times 10^6$ K) peaks in the extreme ultraviolet region.

The Galactic halo is home to many high velocity H I clouds (HVCs) and gas streams of unknown origin (Wakker 2001; Lockman et al. 2008), although the Magellanic Stream is a likely source for a significant fraction. The survival and stability of these clouds is a problem that has long been recognized (e.g. Benjamin and Danly 1997). It is likely that many or all halo clouds have experienced some deceleration during their transit through the hot gas in the lower halo.

To understand HVCs, we consider the drag equation for an H I cloud moving through a stationary medium,

$$\mu_c \dot{v}_c = \frac{1}{2} C_D \rho_h(z) v_c^2 - \mu_c g(z) \tag{1.24}$$

where μ_c is the surface density of the cloud. Equation 1.24 only holds as long as the cloud stays together. The drag coefficient C_D is a measure of the efficiency of momentum transfer to the cloud. For the high Reynolds numbers typical of astrophysical media, incompressible objects have $C_D \approx 0.4$ (e.g. a rough sphere) which indicates that the turbulent wake behind the plunging object efficiently transfers momentum to the braking medium. The leading face of a compressible cloud may become flattened, such that the approaching medium is brought to rest in the reference frame of the cloud; in this instance, $C_D \gtrsim 1$ may be more appropriate (we adopt $C_D = 1$ here).

Let us consider the Galactic hot corona to be an isothermal gas in hydrostatic equilibrium with the gravitational potential, $\phi(R, z)$, where R is the Galactocentric radius and z is the vertical scale height. We adopt a total potential of the form $\phi = \phi_d + \phi_h$ for the disk and halo respectively; for our calculations at the Solar Circle, we ignore the Galactic bulge. The galaxy potential is defined by

$$\phi_d(R, z) = -c_d v_{circ}^2 / (R^2 + (a_d + \sqrt{z^2 + b_d^2})^2)^{0.5} \tag{1.25}$$

$$\phi_h(R, z) = c_h v_{circ}^2 \ln((\psi - 1)/(\psi + 1)) \tag{1.26}$$

and $\psi = (1 + (a_h^2 + R^2 + z^2)/r_h^2)^{0.5}$. The scaling constants are $(a_d, b_d, c_d) = (6.5, 0.26, 8.9)$ kpc and $(a_h, r_h) = (12, 210)$ kpc with $c_h = 0.33$ (e.g. Miyamoto and Nagai 1975). The circular velocity $v_{circ} \approx 220$ km s^{-1} is now well established through wide-field stellar surveys (e.g. Smith et al. 2007).

We determine the vertical acceleration at the Solar Circle using $g = -\partial \phi(R_o, z)/\partial z$ with $R_o = 8$ kpc. The hydrostatic halo pressure follows from

$$\frac{\partial \phi}{\partial z} = -\frac{1}{\rho_h} \frac{\partial P}{\partial z} \tag{1.27}$$

We adopt a solution of the form $P_h(z) = P_o \exp((\phi(R_o, z) - \phi(R_o, 0))/\sigma_h^2)$ where σ_h is the isothermal sound speed of the hot corona. To arrive at P_o, we adopt a coronal

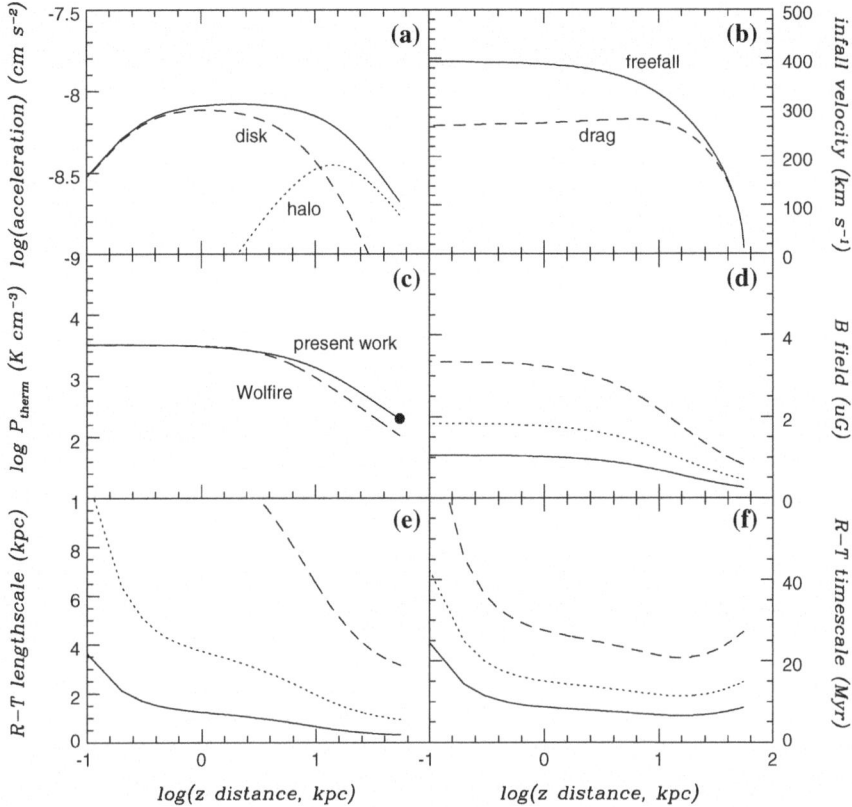

Fig. 1.13 From *top left* to *lower right*: **a** gravitational acceleration due to the disk+halo—all plots are shown as a function of vertical height above the disk at the Solar Circle; **b** infall velocity for a point mass starting from the Magellanic Stream with halo drag ($C_D = 1$; *dashed curve*) and without (*solid*); **c** coronal gas pressure for our model (*solid line*) compared to the Wolfire model (*dashed*)—the *dot* indicates the halo pressure used in our hydrodynamical model; **d** magnetic field strength for $\beta = 1$ (*dashed*), $\beta = 0.3$ (*dotted*), $\beta = 0.1$ (*solid*); **e** minimum lengthscale for RT instability; **e** timescale for RT instability

halo density of $n_{e,h} = 10^{-4}\,\text{cm}^{-3}$ at the Stream distance (55 kpc) in order to explain the Magellanic Stream Hα emission (Bland-Hawthorn et al. 2007), although this is uncertain to a factor of a few. We choose $T_h = 2 \times 10^6$ K to ensure that OVI is not seen in the diffuse corona consistent with observation (Sembach et al. 2003); this is consistent with a rigorously isothermal halo for the Galaxy. Our solution to the above equation is shown in Fig. 1.13c and it compares favorably with the pressure profile derive by others (e.g. Sternberg et al. 2002).

A solution specific to our model is shown in Fig. 1.13b where the freefall velocity is now slowed by about 35%. In practice, the cloud's projected motion can be considerably less than its 3D space velocity (e.g. Lockman et al. 2008). In all likelihood, infalling HVCs have experienced significant deceleration through ram pressure exerted by the corona. But even before the cloud reaches terminal velocity, the cloud is expected to break up (Murray and Lin 2004; Heitsch and Putman 2009).

So how do clouds resist the destructive forces of RT and shock instabilities? One prospect is the stabilizing influence of magnetic fields when a cloud passes through a magnetized medium. The halo magnetic field is poorly constrained at the present time (Sun et al. 2008). We describe the uniform magnetic field in terms of the pressure of the halo medium, or

$$\frac{B^2}{8\pi} = \beta P_h \qquad (1.28)$$

such that $B \approx 1$ μG at the distance of the Stream (55 kpc) if the field is in full equipartition with the corona. But there is evidence that the field is weaker than implied here ($\beta \approx 0.3$; Sun et al. 2008), at least within 5 kpc of the Galactic plane For the warm, denser low-latitude gas (Reynolds layer), we adopt the new parameter fits of Gaensler et al. (2008) from a re-analysis of pulsar data. The lower β value finds support from recent magnetohydrodynamic simulations of the Reynolds layer (Hill et al. 2008).

1.4.9.2 Stability Limits and Growth Timescales

We consider the surface of a high velocity cloud as a boundary between two fluids. In practice, the Galactic ionizing radiation field imparts a multiphase structure to the cloud. At all galactic latitudes within the Stream distance, HVCs with column densities of order 10^{20} cm^{-2} or higher have partially ionized skins to a column depth of roughly 10^{19} cm^{-2} for sub-solar gas due to the Galactic ionizing field. Between the warm ionized skin and the cool inner regions is a warm neutral medium of twice the skin thickness; both outer layers have a mean particle temperature of $\lesssim 10^4$ K.

The cloud is denser than the halo gas. Because of the gravitational field, RT instabilities can grow on the boundary. Furthermore, KH instabilities may also develop due to the relative motion of the cloud with respect to an external medium. Buoyant bubbles in galaxy clusters are stabilized against RT and KH instabilities by viscosity and surface tension due to magnetic fields in the boundary (De Young 2003). Here we examine whether HVC boundaries are similarly stabilized against disruption in the Galactic halo.

When there is no surface tension, no viscosity and no relative motion between the two media, the growth rate of the RT instability for a perturbation with wavenumber k is $\omega = \sqrt{gk}$, where g is the gravitational acceleration at the fluid boundary. The wavenumber is related to a perturbation length scale, $\ell = 2\pi/k$. The instability requires a few e-folding timescales to fully develop; the timescale is given by (De Young 2003)

$$t_{grow} = \omega^{-1} = \sqrt{\frac{\ell}{2\pi g}}. \qquad (1.29)$$

1 Near Field Cosmology

In the presence of a magnetic field, the transverse component (B_{tr}) provides some surface tension which can help to suppress RT instabilities below a lengthscale of (Chandrasekhar 1961)

$$\ell_{min} = \frac{B_{tr}^2}{2\rho_c g}. \tag{1.30}$$

Here ρ_c is the mass density of the denser medium, i.e. the cloud, and B_{tr} is the average value of the transverse magnetic field at the boundary.

In order to illustrate when RT instabilities become important, we assume a flat rotation curve for the Galaxy such that (Kafle et al. 2012)

$$g \approx \frac{v_{circ}^2}{R} = 1.6 \times 10^{-8} \left(\frac{v_{circ}}{220 \text{ km s}^{-1}}\right)^2 \left(\frac{R}{R_o}\right)^{-1} \text{ cm s}^{-2}. \tag{1.31}$$

This is only a rough approximation to the form expected from the coupled equations in Eq. 1.26. We stress that the actual behaviour discussed below, and shown in Fig. 1.13e, f solves for the gravitational potential correctly.

Shortly after the discovery of HVCs, it was thought that they may be self-gravitating. But this would place them at much greater distances than the Magellanic Stream (e.g. Oort 1966) which is now known not to be the case (e.g. Putman et al. 2003). Instead, we consider two cases: (i) HVCs in pressure equilibrium with the coronal gas; (ii) HVCs with parameters fixed by direct observation. In (i), because the temperature is not strongly dependent on radius, but the number density decreases rapidly with increasing radius, we expect the increased pressure to compress the clouds at lower latitudes.

We estimate the impact of RT instabilities using Eqs. 1.27 and 1.29 for a cloud temperature of $T_c = 10^4$ K in pressure equilibrium with the hot halo, the electron density is given by

$$n_{e,c} \approx n_{e,h} \frac{T_h}{T_c} \tag{1.32}$$

$$= 0.02 \left(\frac{R}{55 \text{ kpc}}\right)^{-2} \left(\frac{T_h}{2 \times 10^6 \text{K}}\right) \left(\frac{T_c}{10^4 \text{K}}\right)^{-1} \text{ cm}^{-3}.$$

We use Eqs. 1.30, 1.32 and 1.28 to estimate ℓ_{min} as a function of Galactocentric radius. The minimum length scale for instability is

$$\frac{\ell_{min}}{R} \sim \frac{8\pi\beta k_B T_c}{m_p v_{circ}^2} \tag{1.33}$$

$$= 0.004 \left(\frac{T_c}{10^4 \text{K}}\right) \left(\frac{\beta}{0.1}\right) \left(\frac{v_{circ}}{220 \text{ km s}^{-1}}\right)^{-2}$$

and its associated growth timescale using Eq. 1.29

$$t_{grow} \sim \sqrt{\frac{4\beta k_B T_c}{m_p v_{circ}^2}} \, \Omega^{-1} \tag{1.34}$$

$$= 1.1 \text{ Myr} \left(\frac{T_c}{10^4 \text{ K}}\right)^{\frac{1}{2}} \left(\frac{\beta}{0.1}\right)^{\frac{1}{2}} \left(\frac{v_{circ}}{220 \text{ km s}^{-1}}\right)^{-2} \left(\frac{R}{R_o}\right)$$

where the angular rotation rate is given by $\Omega = v_{circ}/R$.

Because we have assumed that $B^2 \propto n_{e,h}$ (equipartition) and $n_{e,c} \propto n_{e,h}$ (pressure equilibrium), neither the minimum scale length or its growth timescale depend on the halo density or temperature. They do depend on the temperature of the clouds and the ionization state. If the clouds are hotter than 10^4 K, then $n_{e,c}$ is overestimated under the assumption of pressure equilibrium. This would lead to larger minimum instability lengthscales and growth timescales. If the Galactic rotation curve drops faster than the flat profile implied by Eq. 1.27, we would have underestimated both the minimum instability scale length and its associated growth timescale at large radii.

Under the assumption of pressure equilibrium, the falling clouds become more compressed as they approach the disk which can hasten cooling. This effect may help to stabilize against break up, particularly if a cool shell develops (cf. Sternberg and Soker 2008).

In the absence of gravitational instability, the flow is stable against the KH instability if (Chandrasekhar 1961)

$$U^2 < \frac{B_{tr}^2 (\rho_c + \rho_h)}{2\pi \rho_c \rho_h} \tag{1.35}$$

where U is the relative velocity between the two fluids. When $\rho_c > \rho_h$, this requirement becomes

$$U < \sqrt{\frac{8\beta k_B T_h}{m_p}} \tag{1.36}$$

$$= 115 \left(\frac{\beta}{0.1}\right)^{\frac{1}{2}} \left(\frac{T_h}{2 \times 10^6 \text{ K}}\right)^{\frac{1}{2}} \text{ km s}^{-1}$$

and we have described the magnetic field in terms of the halo pressure using Eq. 1.28. This requirement is also independent of the halo density as we have related the magnetic field to the halo pressure, although it is dependent on the halo temperature. This requirement is nearly satisfied for HVCs if the magnetic field is near equipartition.

1.5 Near Field Cosmology

1.5.1 Introduction

The formation and evolution of galaxies is one of the great outstanding problems of astrophysics. Within the broad context of hierachical structure formation, we have only a crude picture of how galaxies like our own came into existence.[10] A detailed physical picture where individual stellar populations can be associated with (tagged to) elements of the protocloud is far beyond our current understanding.

As we have shown, important clues about the sequence of events involved in galaxy formation are already apparent from the high redshift Universe (far-field cosmology) but now we turn our attention to fossil evidence provided by the Galaxy (near-field cosmology). Detailed studies of the Galaxy lie at the core of understanding the complex processes involved in baryon dissipation. This is a necessary first step towards achieving a successful theory of galaxy formation.

What do we mean by the reconstruction of early galactic history? We seek a detailed physical understanding of the sequence of events which led to the Milky Way. Ideally, we would want to tag (i.e. associate) components of the Galaxy to elements of the protocloud—the baryon reservoir which fueled the stars in the Galaxy.

From theory, our prevailing view of structure formation relies on a hierarchical process driven by the gravitational forces of the large-scale distribution of cold, dark matter (ΛCDM). The ΛCDM paradigm provides simple models of galaxy formation within a cosmological context (Peebles 1974; White and Rees 1978; Blumenthal et al. 1984). N-body and semi-analytic simulations of the growth of structures in the early Universe have been successful at reproducing some of the properties of galaxies. Current models include gas pressure, metal production, radiative cooling and heating, and prescriptions for star formation.

The number density, properties and spatial distribution of dark matter halos are well understood within ΛCDM (Sheth and Tormen 1999; Jenkins et al. 2001). However, computer codes are far from producing realistic simulations of how baryons produce observable galaxies within a complex hierarchy of dark matter. This a necessary first step towards a viable theory or a working model of galaxy formation.

In this review, our approach is anchored to observations of the Galaxy, interpreted within the broad scope of the ΛCDM hierarchy. Many of the observables in the Galaxy relate to events which occurred long ago, at high redshift. Figure 1.14 shows the relationship between look-back time and redshift in the context of the ΛCDM model: the redshift range ($z \lesssim 6$) of discrete sources in contemporary observational cosmology matches closely the known ages of the oldest components in the Galaxy. The Galaxy (near-field cosmology) provides a link to the distant Universe (far-field cosmology).[11]

---

[10] The idea of hierarchical formation predates ΛCDM (q.v. de Vaucouleurs 1970a).

[11] These terms were first used in Bland-Hawthorn (1999).

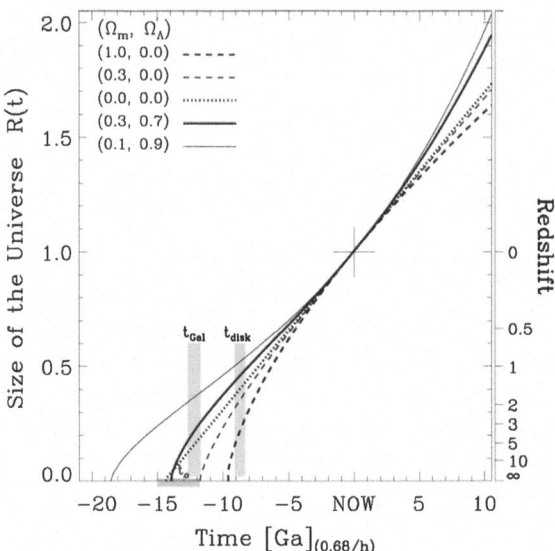

Fig. 1.14 Look-back time as a function of redshift and the size of the Universe for five different world models. The approximate ages of the Galactic halo and disk are indicated by hatched regions (we thank C. Lineweaver for modifying an earlier version of this figure for us)

Eggen et al. (1962) were the first to show that it is possible to study galactic archaeology using stellar abundances and stellar dynamics; this is probably the most influential paper on the subject of galaxy formation. ELS studied the motions of high velocity stars and discovered that, as the metal abundance decreases, the orbit energies and eccentricities of the stars increased while their orbital angular momenta decreased. They inferred that the metal-poor stars reside in a halo that was created during the rapid collapse of a relatively uniform, isolated protogalactic cloud shortly after it decoupled from the universal expansion. ELS is widely viewed as advocating a smooth monolithic collapse of the protocloud with a timescale of order 10^8 years.

The ELS picture was challenged by Searle (1977) and Searle and Zinn (1978) who noted that Galactic globular clusters have a wide range of metal abundances essentially independent of radius from the Galactic Center. They suggested that this could be explained by a halo built up over an extended period from *independent* fragments with masses of $\sim 10^8$ $M_\odot$. In contrast, in the ELS picture, the halo formed in a rapid free-fall collapse. But halo field stars, as well as globular clusters, are now believed to show an age spread of 2–3 Gyr (Marquez and Schuster 1994); for an alternative view, see Sandage and Cacciari (1990). The current paradigm, that the observations argue for a halo that has built up over a long period from infalling debris, has developed after many years of intense debate.

This debate parallelled the changes that were taking place in theoretical studies of cosmology (e.g. Peebles 1971; Press and Schecter 1974). The ideas of galaxy formation via hierarchical aggregation of smaller elements from the early Universe fit in readily with the Searle and Zinn view of the formation of the galactic halo from small fragments. The possibility of identifying debris from these small fragments was already around in Eggen's early studies of moving groups, and this is now an active field of research in theoretical and observational stellar dynamics. It offers the possibility to reconstruct at least some properties of the protogalaxy and so to improve our basic understanding of the galaxy formation process.

We can extend this approach to other components of the Galaxy. We will argue the importance of understanding the formation of the galactic disk, because this is where most of the baryons reside. Although much of the information about the properties of the protogalactic baryons has been lost in the dissipation that led to the galactic disk, a similar dynamical probing of the early properties of the disk can illuminate the formation of the disk, at least back to the epoch of last significant dissipation. It is also clear that we do not need to restrict this probing to stellar dynamical techniques. A vast amount of fossil information is locked up in the detailed stellar distribution of chemical elements in the various components of the Galaxy, and we will discuss the opportunities that this offers.

Before we embark on a detailed overview of the relevant data, we give a descriptive working picture of the sequence of events involved in galaxy formation. For continuity, the relevant references are given in the main body of the review where these issues are discussed in more detail. **In Appendix B, we provide extra information on stellar data with an emphasis on how to find important source material. We also summarise techniques and methods to assist the aspiring galactic archaeologist (near field cosmologist).**

1.5.2 A Working Model of How the Galaxy Formed

Shortly after the Big Bang, cold dark matter began to drive baryons towards local density enhancements. The first stars formed after the collapse of the first primordial molecular clouds. The Universe was ionized again in a process that is thought to have commenced ∼200 million years after the Big Bang and is observed to be complete by the time the Universe was 10 times that age. The study of this reionization is a topic for current research, but we do know that the first generations of massive stars likely played a major role. Observations of distant galaxies, seen as they were in the past because of the light travel time, show that many young galaxies were strong sources of ionizing radiation, but observations must reach still greater distances to show us the earliest generations of stars. This is a key science goal of the next generation of large telescopes, including the James Webb Space Telescope due to launch at the end of the decade.

In Fig. 1.15, we provide a cartoon of how the Galaxy formed from the initial quantum fluctuations. By a redshift of about $z \sim 10$, sufficient dark matter clumps (and associated baryons) were able to fall out of the Hubble flow to form modern day $L_\star$ galaxies. Within the context of ΛCDM, the dark halo of the Galaxy assembled first, although it is likely that its growth continues to the present time. In some

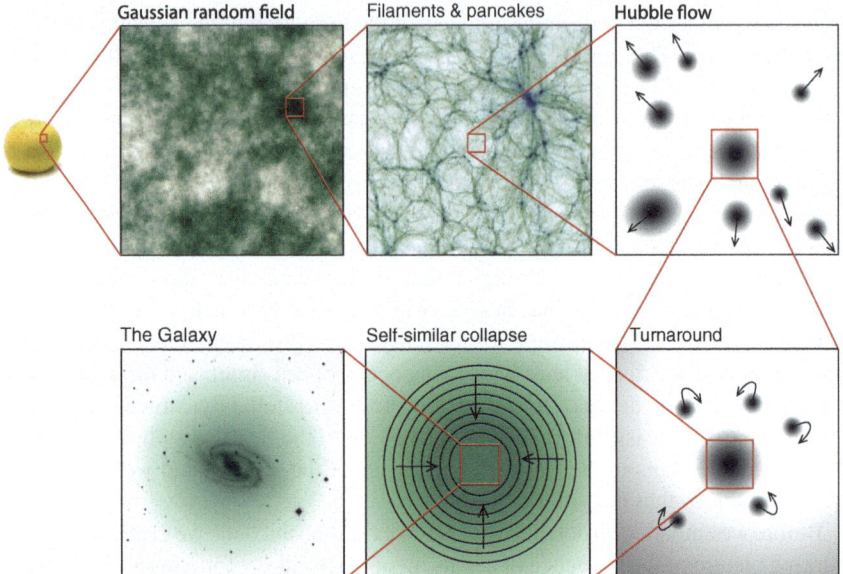

Fig. 1.15 A sequence of images depicting the major stages in the evolution of the Galaxy from the initial fluctuation spectrum—see text. This was preceded by inflation when the Universe went from sub-atomic scales to the size of a grapefruit in less than a picosecond

galaxies, the first episodes of gas accretion established the stellar bulge, the central black hole, the first halo stars and the globular clusters. In the Galaxy and similar systems, the small stellar bulge may have formed later from stars in the inner disk.

The early stages of the Galaxy's evolution were marked by violent gas dynamics and accretion events, leading to the high internal densities of the first globular clusters, and perhaps to the well-known 'black hole mass—stellar bulge dispersion' relation. The stellar bulge and massive black hole may have grown up together during this active time. We associate this era with the 'Golden Age', the phase before $z \sim 1$ when star formation activity and accretion disk activity were at their peak.

At that time, there was a strong metal gradient from the bulge to the outer halo. The metal enrichment was rapid in the core of the Galaxy such that, by $z \sim 1$, the mean metallicities were as high as [Fe/H] ~ -1 or even higher. In these terms, we can understand why the inner stellar bulge that we observe today is both old and moderately metal rich. The first halo stars ([Fe/H] ≈ -5 to -2.5) formed over a more extended volume and presumably date back to the earliest phase of the protocloud. The first globular clusters formed over a similar volume from violent gas interactions ([Fe/H] ≈ -2.5 to -1.5). We believe now that many of the halo stars and globulars are remnants of early satellite galaxies which experienced *independent* chemical evolution before being accreted by the Galaxy.

The spread in [Fe/H], and the relative distribution of the chemical elements, is a major diagnostic of the evolution of each galactic component. If the initial mass function is constant, the mean abundances of the different components give a rough indication of the number of SN II enrichments which preceded their formation, although we note that as time passes, an increasing fraction of Fe is produced by SN Ia events. For a given parcel of gas in a closed system, only a few SN II events are required to reach [Fe/H] ≈ -3, 30 to 100 events to get to [Fe/H] ≈ -1.5, and maybe a thousand events to reach solar metallicities. We wish to stress that [Fe/H] is not a clock: rather it is a measure of supernova occurrences and the depth of the different potential wells that a given parcel of gas has explored.

During the latter stages of the Golden Age, most of the baryons began to settle to a disk for the first time. Two key observations emphasize what we consider to be the mystery of the main epoch of baryon dissipation. First, there are no stars with [Fe/H] < -2.2 which rotate with the disk. Secondly, despite all the activity associated with the Golden Age, at least 80% of the baryons appear to have settled gradually to the disk over many Gyr; this fraction could be as high as 95% if the bulge formed after the disk.

About 10% of the baryons reside in a 'thick disk' which has [Fe/H] ≈ -2.2 to -0.5, compared to the younger thin disk with [Fe/H] ≈ -0.5 to $+0.3$. It is striking how the globular clusters and the thick disk have similar abundance ranges, although the detailed abundance distributions are different. There is also a similarity in age: globular clusters show an age range of 12 to 14 Gyr, and the thick disk appears to be at least 12 Gyr old. Both the thick disk and globulars apparently date back to the epoch of baryon dissipation during $z \sim 1-5$. Figure 1.16 summarises our present understanding of the complex age–metallicity distribution for the various components of the Galaxy.

It is a mystery how the thick disk and the globulars should have formed so early *and* over such a large volume from material which was already enriched to [Fe/H] ~ -2. Could powerful winds from the central starburst in the evolving core have distributed metals throughout the inner protocloud at about that time?

While half of all baryons were residing within the Galaxy by $z \sim 1$, we emphasize that 90% of the *disk baryons* have settled quiescently to the thin disk since z $\sim$ 1. The disk star formation rate during this extended period was roughly constant (Binney et al. 2000) or possibly in slow decline (Aumer and Binney 2009). The rapid decline in $\dot{\rho}_\star$ observed in deep galaxy surveys since $z \sim 1$ is not in evidence over the volume of the Local Group.

1.5.3 Timescales and Fossils

The oldest stars in our Galaxy are of an age similar to the look-back time of the most distant galaxies in the Hubble Deep Field (Fig. 1.14). For the galaxies, the cosmological redshift measured from galaxy spectra presently takes us to within a few percent of the origin of cosmic time. For the stars, their upper atmospheres provide fossil evidence of the available metals at the time of formation. The old Galactic stars and the distant galaxies provide a record of conditions at early times in cosmic history, and both harbor clues to the sequence of events which led to the

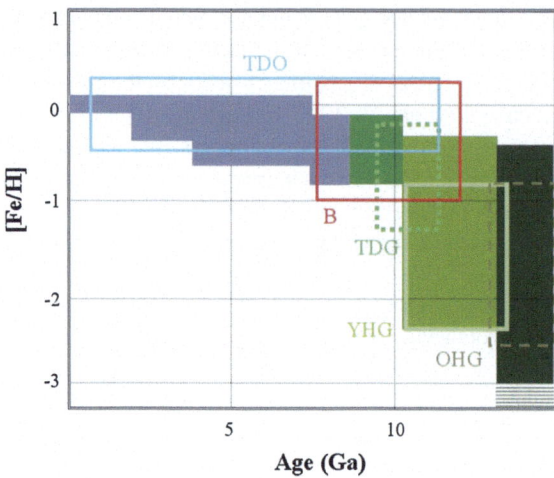

Fig. 1.16 The age-metallicity relation of the Galaxy for the different components (see text): *TDS* thin disk stars; *TDO* thin disk open clusters; *ThDS* thick disk stars; *ThDG* thick disk globulars; *B* bulge; *YHG* young halo globulars; *OHG* old halo globulars

formation of galaxies like the Milky Way. The near field also provides important evolutionary timescales for individual stars and groups of stars, as we discuss in the next section.

The key timescale provided by far-field cosmology is the look-back time with the prospect of seeing galaxies at an earlier stage in their evolution. However, this does *not* imply that these high-redshift objects are unevolved. We know that the stellar cores of galaxies at the highest redshifts ($z \sim 5$) observed to date exhibit solar metallicities, and therefore appear to have undergone many cycles of star formation (Hamann and Ferland 1999). Much of the light we detect from the early Universe probably arises from the chemically and dynamically evolved cores of galaxies.

Near-field cosmology provides a dynamical timescale, $\tau_D \sim (G\rho)^{-\frac{1}{2}}$, where ρ is the mean density of the medium. The dynamical timescale at a radial distance of 100 kpc is of order several Gyr, so the mixing times are very long. Therefore, on larger scales, we can expect to find dynamical and chemical traces of past events, even where small dynamical systems have long since merged with the Galaxy.

We note that the ΛCDM hierarchy reflects a wide range of dynamical timescales, such that different parts of the hierarchy may reveal galaxies in different stages of evolution. In this sense, the hierarchy relates the large-scale density to the morphology and evolution of its individual galaxies; this is the so-called 'morphology-density relation' (Dressler 1980; Hermit et al. 1996; Norberg et al. 2001). Over a large enough ensemble of galaxies, taken from different regions of the hierarchy, we expect different light-weighted age distributions because one part of the hierarchy is more evolved than another (see Sect. 1.3.1).

1.5.4 Stellar Age Dating

Stellar ages let us evaluate when events occurred. They are important for measuring the star formation history and for understanding how the metallicity and dynamics of different groups of stars have evolved: e.g. how has the star formation rate and the kinematics and the metallicity of the thin disk near the sun changed from 10 Gyr ago to the present time? Current evidence, based on chromospheric stellar ages, indicates that the star formation rate in the galactic thin disk near the sun has been roughly uniform over the last 10 Gyr, with episodic increases and decreases of about a factor 2 (e.g. Rocha-Pinto et al. 2000). Stellar ages are still difficult to measure and lead to much uncertainty about the evolution of the Galaxy. Measuring stellar ages is one of the most important and challenging goals for the future.

Nucleo-cosmochronology (or cosmochronometry), or the aging of the elements through radioactive decay, has a long history (Fowler and Hoyle 1960; Butcher 1987). A related technique is widely used in solar system geophysics. Independent schemes have aged the oldest meteorites at 4.53 ± 0.04 Gyr (Guenther and Demarque 1997; Manuel 2000). The small uncertainties reflect that the age dating is direct. Element pairs like Rb and Sr are chemically distinct and freeze out during solidification into different crystalline grains. The isotope ^{87}Rb decays into ^{87}Sr which can be compared to ^{86}Sr, a non-radiogenic isotope, measured from a control sample of Sr-rich grains. This provides a direct measure of the fraction of a ^{87}Rb half-life ($\tau_{1/2} = 47.5$ Gyr) since the meteorite solidified.

It appears that, until we have a precise understanding of BBNS and the early chemical evolution history of the Galaxy, geophysical precision will not be possible for stellar ages. The major problem is that, as far as we know, there is no chemical differentiation which requires that we know precisely how much of each isotope was originally present. Modern nucleo-cosmochronology compares radioactive isotope strengths to a stable r-process element (e.g. Nd, Eu, La, Pt). The thorium method (^{232}Th, $\tau_{1/2} = 14.0$ Gyr) was first applied by Butcher (1987) and refined by Pagel (1989). Other radioactive chronometers include ^{235}U ($\tau_{1/2} = 0.70$ Gyr) and ^{238}U ($\tau_{1/2} = 4.47$ Gyr) although Yokoi et al. (1983) have expressed concerns about their use (cf. Cayrel et al. 2001). Arnould and Goriely (2001) propose that the isotope pair ^{187}Re–^{187}Os ($\tau_{1/2} = 43.5$ Gyr) may be better suited for future work.

With the above caveats, we point out that several groups are now obtaining exquisite high-resolution data on stars with enhanced r-process elements (Cayrel et al. 2001; Sneden et al. 2000; Burris et al. 2000; Westin et al. 2000; Johnson and Bolte 2001; Cohen et al. 2002; Hill et al. 2002). For a subset of these stars, radioactive ages have been derived (Truran et al. 2001) normalized to the heavy element abundances observed in meteorites.

There are few other direct methods for deriving ages of *individual* stars. A promising field is asteroseismology which relies on the evolving mean molecular weight in stellar cores (Christensen-Dalsgaard 1986; Ulrich 1986; Gough 1987; Guenther 1989). The stellar oscillation frequencies depend on density distribution in stellar interior, which changes as the star ages. The asteroseismology space missions

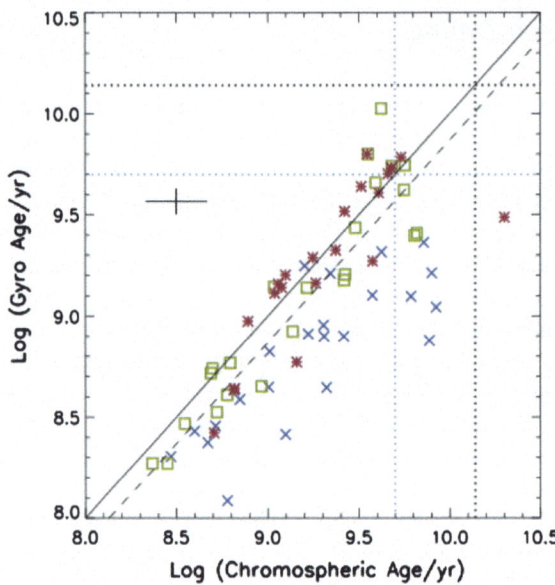

Fig. 1.17 Gyro ages versus chromospheric ages for a sample of well-studied stars (Barnes 2007). The *blue crosses* are for stars bluer than $B - V = 0.6$ and should be disregarded

(MOST, CoRoT, Kepler) will contribute greatly to deriving stellar ages with 5–10% errors. Gough (2001) has determined 4.57 ± 0.12 Gyr for the Sun which should be compared with the age of meteorites quoted above.

There are age indicators based on stellar rotation and chromospheric activity which are particularly useful for main sequence stars (see review by Barnes 2007). These indicators are calibrated on star clusters and the Sun. Stars spin down with age: their rotation periods are usually measured photometrically (Kepler, COROT) and can be used to estimate age—this technique is known as gyrochronology. Chromospheric Ca K emission has been much used as an age indicator in the past, but is believed to be less accurate for older stars. This method is based on the decrease of chromospheric activity with age. Barnes (2007) presents a figure of chromospheric ages against gyrochronology ages for a sample of well studied stars (see Fig. 1.17).

Other traditional methods rely on aging a population of stars that are representative of a particular component of the Galaxy. For example, Gilmore et al. (1989) use the envelope of the distribution in a color-abundance plane to show that all stars more metal poor than [Fe/H] = -0.8 are as old as the globular clusters. Similarly, the faint end of the white dwarf luminosity function is associated with the coolest, and therefore the oldest, stars (Oswalt et al. 1996). The white dwarf luminosity function is useful for deriving ages of a population (eg the disk or the halo) or for a globular cluster. This method uses the cooling and fading of white dwarfs as an age estimator for the population. For example, Leggett et al. (1998) used the luminosity function of white dwarfs in the nearby Galactic disk to derive an age of about 9 Gyr for the oldest stars of the disk. A comparison of the white dwarf luminosity function for the Galactic disk and two globular clusters, M4 and NGC 6397, by Hansen et al. (2007), show that the disk is clearly a few Gyr younger than the clusters.

It has long been known that disk stars span a wide range of ages from the diversity of main-sequence stars. Edvardsson et al. (1993) derived precise stellar evolution ages for nearby individual post main-sequence F stars using Strömgren photometry, and showed that the stars in the Galactic disk exhibit a large age spread with ages up to roughly 10 Gyr (Fig. 1.16). Using the 'inverse age-luminosity relation' for RR Lyrae stars, Chaboyer et al. (1996) found that the oldest globular clusters are older than 12 Gyr with 95% confidence, with a best estimate of 14.6 ± 1.7 Gyr (Chaboyer et al. 1998). But Hipparcos appears to show that the RR Lyr distances are underestimated leading to a downward revision of the cluster ages: 8.5–13.3 Gyr (Gratton et al. 1997); 11–13 Gyr (Reid 1998); 10.2–12.8 Gyr (Chaboyer 1998). For a coeval population (e.g. open and globular clusters), isochrone fitting is widely used. The ages of the galactic halo and globular clusters, when averaged over eight independent surveys, lead to 12.2 ± 0.5 Gyr (Lineweaver 1999).

Parallaxes provide the stellar luminosity L after bolometric correction, and photometry gives the effective temperature T_{eff}. When the abundance [Fe/H] is known, comparison of the location of the star in the $L - T_{\text{eff}}$ plane with theoretical isochrones can give an estimate of the age in regions where the L and T_{eff} depend on the age. For example, this method cannot be used on the unevolved main sequence, and is not ideal on the upper giant branch. The region near the main sequence is much used for isochrone ages, because these stars are common, but the isochrones in this region are complex and can give multi-valued age estimates. The subgiant branch is particularly well suited to isochrone ages. The Gaia mission will provide a huge increase in accurate parallaxes (errors of order 1% for bright stars with V < 14) and hence in precise age estimates.

If no trigonometric parallax is available, isochrones in the log-g − T_{eff} plane can be used to estimate ages, where log-g and T_{eff} can be derived from a combination of high resolution spectroscopy or low resolution spectrophotometry and broad-band photometry. Large biases can occur for isochrone ages if the errors in the observed quantities (e.g L and T_{eff}) are significant, because the underlying distributions of stellar mass and abundances are not uniform. Bayesian techniques can include these underlying distributions as priors (e.g. Pont and Eyer 2004).

For star clusters, the color-magnitude plane is often well populated, and isochrone fits to the main sequence, turnoff region and giant branch can be used to estimate the age and distance if the abundance [Fe/H] is known. The recent ACS survey of Galactic globular clusters shows that old coeval clusters are present over the whole range of cluster metallicities, but that some of the outer clusters are younger and show an age-metallicity relation (Sarajedini et al. 2007).

As Reid et al. (2007) and Barnes (2007) showed, there is still much disagreement about estimating the ages of individual stars, because measuring accurate stellar ages is difficult. It remains a vital goal for unravelling the dynamical and chemical evolution of the Galaxy. The present estimate for the age of the old thin disk population when averaged over five independent surveys is 8.7 ± 0.4 Gyr (Lineweaver 1999), although Oswalt et al argue for $9.5^{+1.1}_{-0.8}$ Gyr.

For a world model with ($\Omega_\Lambda = 0.7$, $\Omega_m = 0.3$), the Big Bang occurred 13.7 Gyr ago—in our view, there is no compelling evidence for an age crisis from a comparison of estimates in the near and far field. But the inaccuracy of age dating relative to an absolute scale does cause problems. At present, the absolute ages of the oldest stars cannot be tied down to better than about 2 Gyr, the time elapsed between $z = 6$ and $z = 2$. This is a particular handicap to identifying specific events in the early Universe from the stellar record.

1.5.5 Goals of Near Field Cosmology

We believe that the major goal of near field cosmology is to tag individual stars with elements of the protocloud. Some integrals of motion are likely to be preserved while others are scrambled by dissipation and violent relaxation. We suspect that complete tagging is impossible. However, some stars today may have some integrals of motion which relate to the protocloud at the epoch of last dissipation.

As we review, different parts of the Galaxy have experienced dissipation and phase mixing to varying degrees. The disk, in contrast to the stellar halo, is a highly dissipated structure. The bulge may be only partly dissipated. To what extent can we unravel the events that produced the Galaxy as we see it today? Could some of the residual inhomogeneities from prehistory have escaped the dissipative process at an early stage?

Far field cosmology currently takes us back to the 'epoch of last scattering' as seen in the microwave background. Cosmologists would like to think that some vestige of information has survived from earlier times (cf. Peebles et al. 2000). In the same spirit, we can hope that fossils remain from the 'epoch of last dissipation', i.e. the main epoch of baryon dissipation that occurred as the disk was being assembled.

To make a comprehensive inventory of surviving inhomogeneities would require a vast catalog of stellar properties that is presently out of reach. The Gaia space astrometry mission, set to launch at the end of the decade, will acquire detailed phase space coordinates for about one billion stars, within a sphere of diameter 20 kpc (the Gaiasphere). In Sect. 1.8, we look forward to a time when all stars within the Gaiasphere have complete chemical abundance measurements (including all heavy metals). Even with such a vast increase in information, there may exist fundamental—but unproven—limits to unravelling the observed complexity.

The huge increase in data rates from ground-based and space-based observatories has led to an explosion of information. Much of this information from the near field is often dismissed as 'weather' or unimportant detail. But in fact fundamental clues are already beginning to emerge. In what is now a famous discovery, a large photometric and kinematic survey of bulge stars revealed the presence of the disrupting Sagittarius dwarf galaxy (Ibata et al. 1994), now seen over a large region of sky and in a variety of populations (see Sect. 1.7.3.8). Perhaps the most important example arises from the chemical signatures seen in echelle spectroscopy of bulge, thick disk and halo stars. In Sect. 1.9, we envisage a time when the analysis of thousands of spectral lines for a vast number of stars will reveal crucial insights into the sequence of events early in the formation of the Galaxy.

In this review, we discuss fossil signatures in the Galaxy. A key aspect of fossil studies is a reliable time sequence. In Sect. 1.5.4, we discuss methods for age-dating individual stars and coeval groups of stars. In Sect. 1.6, we describe the main components of the Galaxy. In Sect. 1.7, we divide the fossil signatures of galaxy formation into three parts: zero order signatures that preserve information since dark matter virialized; first order signatures that preserve information since the main epoch of baryon dissipation; second order signatures that arise from major processes involved in subsequent evolution. In Sect. 1.8, we look forward to a time when it is possible to measure ages, phase space coordinates and chemical properties for a vast number of stars in the Galaxy. Even then, what are the prospects for unravelling the sequence of events that gave rise to the Milky Way? We conclude with some experimental challenges for the future.

1.6 Structure of the Galaxy

How would our Galaxy appear from the perspective of an external observer? This vexing question has a long history. Little progress has been made since de Vaucouleurs (1970b) classified the Milky Way as an SAB(rs)bc galaxy which recognizes it as weakly barred/ringed system. Several candidates have been put forward as surrogates of the Milky Way: NGC 891 (edge on) and NGC 3124, NGC 3992 or NGC 2336 (face on) are possible examples (see Fig. 1.18; Efremov 2011). There is ongoing debate about whether the Galaxy has 2 or 4 spiral arms, and whether these are asymmetric or not. Even the structure of the central bar is debated (Sect. 1.6.1).

Like most spiral galaxies, our Galaxy has several recognizable structural components that probably appeared at different stages in the galaxy formation process: a central bulge/bar, a thin disk, a thick disk and a tenuous stellar halo. These components will retain different kinds of signatures of their formation. We will describe these components in the context of other disk galaxies, and use images of other galaxies in Fig. 1.19 to illustrate the components.

1.6.1 The Bulge

First compare images of M104 and IC 5249 (Fig. 1.19c, g): these are extreme examples of galaxies with a large bulge and with no bulge. Large bulges like that of M104 are structurally and chemically rather similar to elliptical galaxies: their surface brightness distribution follows an $r^{1/4}$ law (e.g. Pritchet and van den Bergh 1994) and they show similar relations of [Fe/H] and [Mg/Fe] with absolute magnitude (e.g.

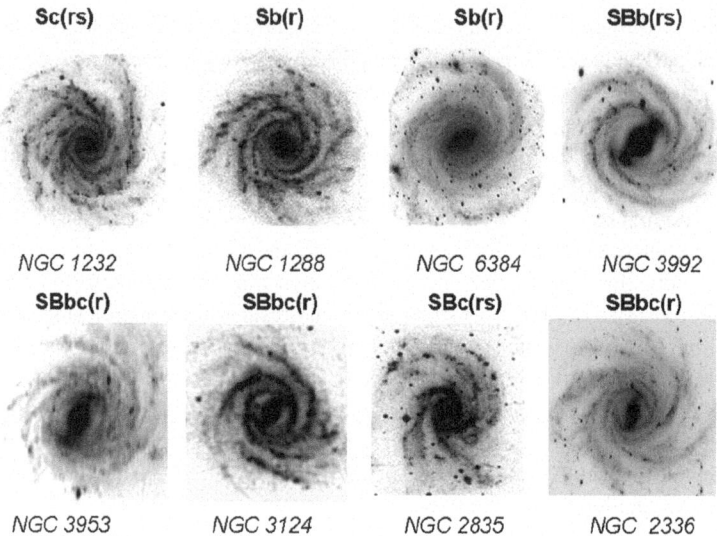

Fig. 1.18 A selection of galaxies that bear resemblance to the Milky Way (Efremov 2011); NGC 3992 is arguably the closest match to the Galaxy

Jablonka et al. 1996). These properties lead to the view that the large bulges formed rapidly. The smaller bulges are often boxy in shape, with a more exponential surface brightness distribution (e.g. Courteau et al. 1996). The current belief is that they may have arisen from the stellar disk through bending mode instabilities.

Spiral bulges are usually assumed to be old but this is poorly known, even for the Galaxy. The presence of bulge RR Lyrae stars indicates that at least some fraction of the galactic bulge is old (Rich 2001). Furthermore, the color-magnitude diagrams for galactic bulge stars show that the bulge is predominantly old. McWilliam and Rich (1994) measured [Fe/H] abundances for red giant stars in the bulge of the Milky Way. They found that, while there is a wide spread, the abundances ([Fe/H] ≈ -0.25) are closer to the older stars of the metal rich disk than to the very old metal poor stars in the halo and in globular clusters, in agreement with the abundances of planetary nebulae in the Galactic bulge (e.g. Exter et al. 2001).

The COBE image of the Milky Way (Fig. 1.19b) shows a modest somewhat boxy bulge, typical of an Sb to Sc spiral. Figure 1.19d shows a more extreme example of a boxy/peanut bulge. If such bulges do arise via instabilities of the stellar disk, then much of the information that we seek about the state of the early galaxy would have been lost in the processes of disk formation and subsequent bulge formation. Although most of the more luminous disk galaxies have bulges, many of the fainter disk galaxies do not. Bulge formation is not an essential element of the formation processes of disk galaxies.

Galactic bulges show a wide range of morphology (e.g. Kormendy and Kennicutt 2004). At one extreme are the large classical bulges like the bulge of the Sombrero

1 Near Field Cosmology

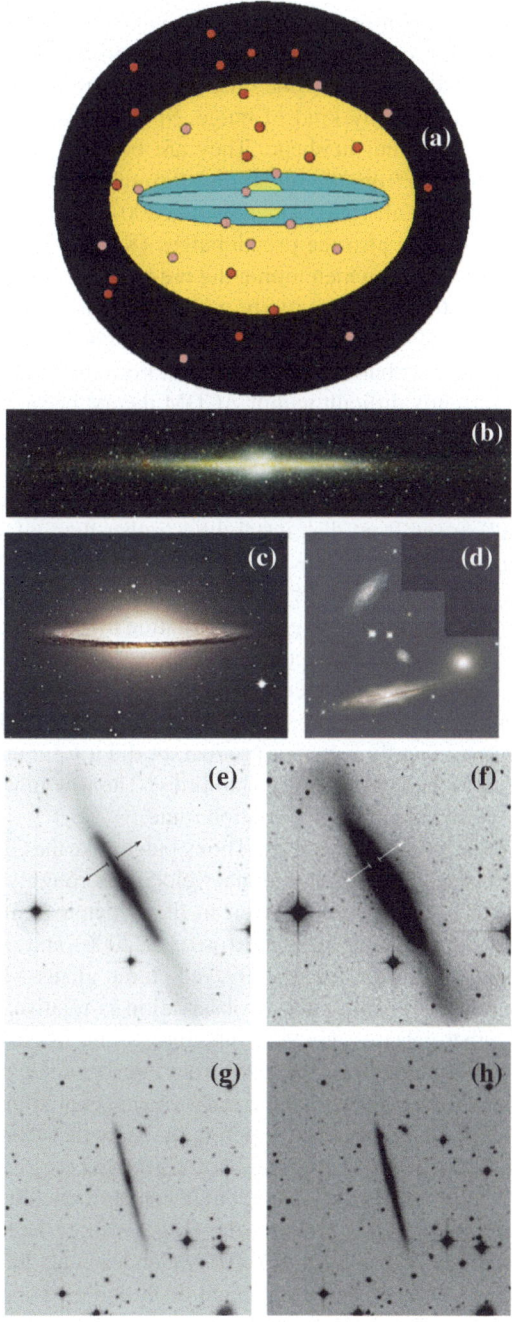

Fig. 1.19 a Sketch of Milky Way showing the stellar disk (*light blue*), thick disk (*dark blue*), stellar bulge (*yellow*), stellar halo (*mustard yellow*), dark halo (*black*) and globular cluster system (*filled circles*). The radius of the stellar disk is roughly 15 kpc. The baryon and dark halos extend to a radius of at least 100 kpc. **b** Infrared image of the Milky Way taken by the DIRBE instrument on board the Cosmic Background Explorer (COBE) Satellite (we acknowledge the NASA Goddard Space Flight Center and the COBE Science Working Group for this image). **c** M104, a normal disk galaxy with a large stellar bulge (from AAO). **d** Hubble Heritage image of the compact group Hickson 87; one galaxy has a peanut-shaped stellar bulge due to dynamical interaction with other group members. **e** Image of the S0 galaxy NGC 4762 (Digital Sky Survey) shows its thin disk and stellar bulge. **f** A deeper image of NGC 4762 (DSS) shows its more extended thick disk. The base of the arrows in **e** and **f** shows the height above the plane at which the thick disk becomes brighter than the thin disk (Tsikoudi 1980). **g** Image of the pure disk galaxy IC 5249 (DSS) shows its thin disk and no stellar bulge. **h** A deeper image of IC 5249 (DSS) shows no visible thick disk, although a very faint thick disk has been detected in deep surface photometry

galaxy, which are believed to be generated by major mergers. They share some structural, dynamical and population properties with the lower-luminosity bright elliptical galaxies. At the other extreme are the pure disk galaxies which show no evidence for any kind of bulge. Smaller bulges like the Galactic bulge are often boxy or peanut in shape. They are believed to be associated with bar structures and to arise from the stellar disk through bar-forming and bar-buckling instabilities (Combes et al. 1990; Athanassoula 2007).

There are also the pseudobulges (Kormendy 1993) which appear to be density enhancements which mimic the presence of a bulge in the radial surface brightness distribution, but lie entirely within the disk. These are currently believed to come from secular processes which transfer angular momentum and rearrange the surface density distribution within the disk. Forming large galaxies with small or no bulges is currently difficult within ΛCDM theory, because of the relatively active ongoing merger history in ΛCDM simulations.

In their recent survey of nearby galaxies, Kormendy et al. (2010) find that essentially all of the disks are 'pure'; while some have bulges, most of these bulges could have arisen from disk instabilities rather than infall. These authors state that "it is hard to understand how bulgeless galaxies could form as the quiescent tail of a distribution of merger histories. Recognition of pseudobulges makes the biggest problem with cold dark matter galaxy formation more acute: How can hierarchical clustering make so many giant, pure-disk galaxies with no evidence for merger-built bulges?"

Establishing the merger history of the Milky Way observationally is a major goal for Galactic archaeology, so we need to understand how the Galactic bulge formed. Is it even partly a merger product, or did it form entirely through internal processes such as the disk and bar instabilities? The way in which the Galactic bulge rotates is a key indicator of its formation route.

The rotation field of small boxy bulges like the Galactic bulge are characteristically cylindrical (i.e. the rotational velocity is roughly constant with height above the galactic plane). This is seen in the dynamical simulations, observations of other boxy bulges (e.g. Falcon-Barroso et al. 2006), and also in the Galactic bulge (brightest giants000Howard et al. 2009; red clump giants—Ness et al. 2012, 2013; Freeman et al. 2013). This kind of observation is relatively straightforward for the bulges of other galaxies, via their integrated light. In our own Galaxy, the morphology of the rotation field is measured from large samples of individual bulge stars, and it is important to ensure that the sample is not contaminated by stars of the Galactic disk. The available evidence supports the view that the Galactic bulge formed primarily via disk/bar instabilities and not via mergers.

The stars of the Galactic bulge appear to be old and enhanced in alpha elements; it is not yet clear whether the bulge is more alpha enhanced than the local thick disk (e.g. Fulbright et al. 2007; Melendez et al. 2008). The significance of the alpha enrichment is that the chemical evolution of the stars in the bulge must have been very rapid, taking no more than about a Gyr. The bar-forming and bar-buckling process takes a few Gyr to act after the disk has settled. In the bar-buckling scenario, the bulge structure is then probably younger than the bulge stars themselves, which would have originally been part of the inner disk. The alpha enrichment of the bulge stars would come from the rapid star formation and chemical evolution which took place in the inner disk before the instability had time to act.

ΛCDM simulations make a prediction about the distribution of the first stars which form in the environment of a large galaxy like the Milky Way. These stars form in the more prominent overdensities of the mass distribution at redshifts >12 (e.g. Moore et al. 2006), long before the large galaxy itself becomes visible. As the galaxy comes together, the first stars are concentrated to the central regions of the galaxy, and inhabit the bulge region. They are not necessarily the most metal-poor stars in the galaxy, because of the chemical evolution which takes place in these overdensities before they become part of the larger system. The first stars are interesting archaeologically, but it will be a challenge to recognize them from among the stars of the inner halo which will also be concentrated towards the Galactic bulge.

There are repeated claims that the inner short, boxy bar is embedded within a long bar, where the major axes are misaligned with one another (López-Corredoira et al. 2005). This combination of short and long bars may be a common feature of many spirals (e.g. Efremov 2011). Evidence for the long bar has been claimed from infrared star counts (Benjamin et al. 2005; López-Corredoira et al. 2007; Cabrera-Lavers et al. 2008) and stellar velocities (Gardner and Flynn 2010). The dimensions of the long bar are x:y:z ≈ 8.0:1.2:0.2 kpc and it is inclined at about 45° to our line of sight to the Galactic Centre (López-Corredoira et al. 2007). But an argument against the long bar is that there is no kinematic signature of it in the solar neighbourhood (Siebert et al. 2011). It is possible that the infrared star counts are due to small spiral arms that extend from the ends of a bar aligned with the inner short bar (e.g. NGC 3992 in Fig. 1.18; Efremov 2011). Future kinematic surveys of the inner bulge should clarify this issue.

1.6.2 The Disk

Now look at the disks of these galaxies. The exponential thin disk, with a vertical scale height of about 300 pc, is the most conspicuous component in edge-on disk galaxies like NGC 4762 and IC 5249 (Fig. 1.19e, g). The thin disk is believed to be the end product of the quiescent dissipation of most of the baryons and contains almost all of the baryonic angular momentum. For the galactic disk, which is clearly seen in the COBE image in Fig. 1.19b, we know from radioactive dating, white dwarf cooling and isochrone estimates for individual evolved stars and open clusters that the oldest disk stars have ages in the range 10–12 Gyr (see Sect. 1.5.4).

The disk is the defining stellar component of disk galaxies, and the Universe (Driver et al. 2007), and understanding its formation is in our view the most important goal of galaxy formation theory. Although much of the information about the pre-disk state of the baryons has been lost in the dissipative process, some tracers remain, and we will discuss them in the next section.

Many disk galaxies show a second fainter disk component with a larger scale height (typically about 1 kpc); this is known as the thick disk. Deep surface photometry of IC 5249 shows only a very faint thick disk enveloping the bright thin disk (Abe et al. 1999): compare Fig. 1.19g, h. In the edge-on S0 galaxy NGC 4762, we see a much brighter thick disk around its very bright thin disk (Tsikoudi 1980): the thick disk is easily seen by comparing Fig. 1.19e, f. The Milky Way has a significant thick disk (Gilmore and Reid 1983): its scale height ($\sim$1 kpc) is about three times larger than the scale height of the thin disk, its surface brightness is about 10 % of the thin disk's surface brightness, its stellar population appears to be older than about 12 Gyr, and its stars are significantly more metal poor than the stars of the thin disk. The galactic thick disk is currently believed to arise from heating of the early stellar disk by accretion events or minor mergers (see Sect. 1.7.2.3).

The thick disk may be one of the most significant components for studying signatures of galaxy formation because it presents a 'snap frozen' relic of the state of the (heated) early disk. Although some apparently pure-disk galaxies like IC 5249 do have faint thick disks, others do not (Fry et al. 1999): these pure-disk galaxies show no visible components other than the thin disk. As for the bulge, formation of a thick disk is not an essential element of the galaxy formation process. In some galaxies the dissipative formation of the disk is clearly a very quiescent process. We direct the reader to a recent review of the Galactic disk by Rix and Bovy (2012).

1.6.3 The Stellar Halo

The Galactic halo (Fig. 1.19a) is a vast ancient repository that takes us back to a time when dark matter collapsed into bound structures and the Galaxy was seeded for the first time. Helmi (2008), Tolstoy et al. (2009) and Klement (2010) provide excellent overviews of the many stellar systems and fragments that inhabit the halo. These include field halo stars (Christlieb et al. 2002; Cayrel et al. 2004; Frebel et al. 2005; Cohen et al. 2007), globular clusters (Gratton et al. 2004), dwarf spheroidals (Mateo 1998; Venn et al. 2004), ultra-faint dwarf galaxies (Simon and Geha 2007; Kirby et al. 2008), stellar streams (Ibata et al. 1995; Chou et al. 2010), stellar associations (Walsh et al. 2007) and satellites to dwarf galaxies (Coleman et al. 2004; Belokurov et al. 2009).

A formal definition of the halo is made difficult by the presence of substructure, particularly if the entire halo is ultimately shown to be made up of distinct kinematic subsystems. But we do recognise a roughly smooth component from deep photometric surveys (e.g. Bell et al. 2008). Its mass is only about 1 % of the total stellar mass (about 10^9 $M_\odot$: e.g. Morrison 1993). The surface brightness of the galactic halo, if observed in other galaxies, would be too low to detect from its diffuse light. It can be seen in other galaxies of the Local Group in which it is possible to detect the individual evolved halo stars.

The metal-poor halo of the Galaxy is very interesting for galaxy formation studies because it is so old and likely constitutes some of the first Galactic objects to form. The halo may have formed over a few billion years. Kalirai (2012) has recently determined an age 11.4 ± 0.7 Gyr from the white dwarf halo population in the solar neighbourhood. His sample is demonstrably younger than the oldest globular clusters in the halo.

The galactic halo has a power law density distribution $\rho \propto r^{-3.5}$ although this appears to depend on the stellar population (Vivas et al. 2001; Chiba and Beers 2000). Unlike the disk and bulge, the angular momentum of the halo is close to zero (e.g. Freeman 1987), and it is supported almost entirely by its velocity dispersion; some of its stars are very energetic, reaching out to at least 100 kpc from the galactic center (e.g. Carney et al. 1990).

The current view is that the galactic halo formed at least partly through the accretion of small metal-poor satellite galaxies which underwent some independent chemical evolution before being accreted by the Galaxy (Searle and Zinn 1978; Freeman 1987; Bell et al. 2007). Although we do still see such accretion events taking place now, in the apparent tidal disruption of the Sgr dwarf (Ibata et al. 1995), most of them must have occurred long ago. Accretion of satellites would dynamically heat the thin disk, so the presence of a dominant thin disk in the Galaxy means that most of this halo-building accretion probably predated the epoch of thick disk formation $\sim$12 Gyr ago. We can expect to see dynamically unmixed residues or fossils of at least some of these accretion events (e.g. Helmi and White 1999).

Of all the galactic components, the stellar halo offers the best opportunity for probing the details of its formation. There is a real possibility to identify groups of halo stars that originate from common progenitor satellites (Eggen 1977; Helmi and White 1999; Harding et al. 2001; Majewski et al. 2000). However, if the accretion picture is correct, then the halo is just the stellar debris of small objects accreted by the Galaxy early in its life. Although it may be possible to unravel this debris and associate individual halo stars with particular progenitors, this may tell us more about the early chemical evolution of dwarf galaxies than about the basic issues of galaxy formation. We would argue that the thin disk and thick disk of our Galaxy retain the most information about how the Galaxy formed. On the other hand, we note that current hierarchical ΛCDM simulations predict many more satellites than are currently observed. It would therefore be very interesting to determine the number of satellites that have already been accreted to form the galactic stellar halo.

We should keep in mind that the stellar halo accounts for only a tiny fraction of the galactic baryons and is dynamically distinct from the rest of the stellar baryons. There is mounting evidence that the stellar halo may harbour distinct abundance populations (Nissen and Schuster 2012; Schuster et al. 2012) and/or comprise discrete rotating/counter-rotating components (Carollo et al. 2007; Kafle et al. 2012, 2013; cf. Schönrich et al. 2012). We should note that the stellar halo of the Galaxy may not be typical: the halos of disk galaxies are quite diverse. The halo of M31, for example, follows the $r^{1/4}$ law (Pritchet and van den Bergh 1994) and is much more metal rich in the mean than the halo of our Galaxy (Durrell et al. 2001) although it does have stars that are very metal weak. It should probably be regarded more as the outer parts of a large bulge than as a distinct halo component.

1.6.4 The Dark Halo

The second inconspicuous component is the dark halo, which is detected only by its gravitational field. The dark halo contributes at least 90 % of the total galactic mass and its $\rho \sim r^{-2}$ density distribution extends at least to 100 kpc (e.g. Kochanek 1996; Battaglia et al. 2005). Beyond here, at some stage, the halo density must decline rapidly, as expected from theory and observed in numerical simulations. **The story of how dark matter in galaxies was established is given in Appendix A.**

Near the Sun, the density of the dark halo is about 0.008 $M_\odot$ pc^{-3} compared to the density of luminous matter of about 0.01 $M_\odot$ pc^{-3}. Its contribution to the local surface density is more significant. The total surface density of matter near the sun, within $|z| = 1.1$ kpc, is about 71 $M_\odot$ pc^{-2}, and the surface density of luminous matter alone is about 48 $M_\odot$ pc^{-2}: near the Sun, the dark matter contributes about a third of the Galactic surface density (Kuijken and Gilmore 1991).

The Galactic dark halo is believed to be spheroidal rather than disk-like (Crézé et al. 1998; Ibata et al. 2001b; see Pfenniger et al. 1994 for a contrary view). The tidal streams of the Sgr dwarf galaxy suggest that the halo is triaxial (Law et al. 2009), with axial ratios of about b/a = 0.83 and c/a = 0.67. The minor axis appears to lie approximately along the Sun-Galactic Centre line and the major axis along the direction of Galactic rotation. The contribution of the dark matter in the inner Galaxy can be estimated using models of the Galactic mass distribution, based on the COBE near-infrared light distribution and normalised by HST star counts and by the kinematics of gas in the inner Galaxy. These models consistently give a good fit to the observed optical stellar microlensing depth as measured by the large microlensing surveys like MACHO, EROS and OGLE, which implies that the gravitational field in the inner few kpc of the Galaxy is dominated by the baryonic component (e.g. Englmaier and Gerhard 2006).

It is expected that at least some satellite galaxies accreted by larger disk galaxies would be pulled down into the disk by dynamical friction before tidal disruption (e.g. Quinn and Goodman 1986; Walker et al. 1996; Abadi et al. 2003). These systems would deposit their dark matter and their stars into the disk of the large galaxy. The presence of such disk dark matter would be difficult to detect if its scale height is a few kpc, but its existence would be interesting in the context of the enhancement to direct detection of dark matter (e.g. Purcell et al. 2009). The vast samples of precise distances and transverse velocities expected from the Gaia mission offer prospects for independent estimates of total local densities and surface densities from the kinematics of disk stars in the inner disk of the galaxy.

We should emphasise here the small fraction of the Galactic mass which is in the form of stellar baryons. The mass of the dark halo, from a range of estimators, is about $1–1.5 \times 10^{12}\, M_\odot$ (Smith et al. 2007). The high value is preferred by the orbit of the Magellanic System (Nichols & Bland-Hawthorn 2011) whereas the low value is derived from halo stars (e.g. Kafle et al. 2012). With less uncertainty, the stellar mass of the Galaxy is about $6 \times 10^{10}\, M_\odot$ (see below). The stellar baryons provide only about 3 % of the total mass which agrees well with estimates from weak lensing for the stellar fraction of L_* galaxies (e.g. Mandelbaum et al. 2006). This 3% fraction is much less than the universal ratio of $f_b^o \approx 17\%$.

In the current picture of galaxy formation, the dark halo plays a very signicant role. The disk is believed to form dissipatively within the potential of the virialized spheroidal halo which itself formed through the fairly rapid aggregation of smaller bodies (e.g. White and Rees 1978). ΛCDM simulations suggest that the halo may still be strongly substructured (see Sect. 1.7.1.5). If this is correct, then the lumpy halo would continue to influence the evolution of the galactic disk, and the residual substructure of the halo is a fossil of its formation. It remains a challenge to find direct observational evidence of this dark substructure.

Within the limitations mentioned above, each of these distinct components of the Galaxy preserves signatures of its past and so gives insights into the galaxy formation process. We now discuss these signatures.

1.7 Signatures of Galaxy Formation

Our framework is that the Galaxy formed through hierarchical aggregation. We identify three major epochs in which information about the proto-hierarchy is lost:

1. the dark matter virializes—this could be a time of intense star formation but need not be, as evidenced by the existence of very thin pure disk galaxies;
2. the baryons dissipate to form the disk and bulge;
3. an ongoing epoch of formation of objects within the disk and accretion of objects from the environment of the galaxy, both leaving some long-lived relic.

We classify signatures relative to these three epochs. The role of the environment is presently difficult to categorize in this way. Environmental influences must be operating across all of our signature classes.

1.7.1 Zero Order Signatures: Information Preserved Since Dark Matter Virialized

1.7.1.1 Introduction

During the virialization phase, a lot of information about the local hierarchy is lost: this era is dominated by merging and violent relaxation. The total dark and baryon mass are probably roughly conserved, as is the angular momentum of the region of the hierarchy that went into the halo. The typical density of the environment is also roughly conserved: although the structure has evolved through merging and relaxation, a low density environment remains a low density environment (see White 1996 for an overview).

1.7.1.2 Signatures of the Environment

The local density of galaxies (and particularly the number of small satellite systems present at this epoch) affects the incidence of later interactions. For the Local Group, the satellite numbers appear to be lower than expected from ΛCDM (Moore et al. 1999; Klypin et al. 1999). However there is plenty of evidence for past and ongoing accretion of small objects by the Milky Way and M31 (Ibata et al. 1995, 2001b).

The thin disk component of disk galaxies settles dissipatively in the potential of the virialised dark halo (e.g. Fall and Efstathiou 1980). The present morphology of the thin disk depends on the numbers of small galaxies available to be accreted: a very thin disk is an indication of few accretion events (dark or luminous) after the epoch of disk dissipation and star formation (e.g. Freeman 1987; Quinn et al. 1993; Walker et al. 1996). The formation of the thick disk is believed to be associated with a discrete event that occurred very soon after the disk began to settle, at a time when about 10% of the stars of the disk had already formed. In a low density environment, without such events, thick disk formation may not occur. Since the time of thick disk formation, the disk of the Galaxy appears to have been relatively undisturbed by accretion events. This is consistent with the observation that less than 10% of the metal-poor halo comes from recent accretion of star forming satellites (Unavane et al. 1996).

The existence and structure of the metal-poor stellar halo of the Galaxy may depend on accretion of small objects. This accretion probably took place after the gaseous disk had more or less settled—the disk acts as a resonator for the orbit decay of the small objects. So again the environment of our proto-galaxy may have a strong signature in the very existence of the stellar halo, and certainly in its observed substructure. We would not expect to find a stellar halo encompassing pure disk galaxies, consistent with the limited evidence now available (Freeman et al. 1983; Schommer et al. 1992).

1.7.1.3 Signatures of Global Quantities

During the process of galaxy formation, some baryons are lost to ram pressure stripping and galactic winds. Most of the remaining baryons become the luminous components of the galaxy. The total angular momentum J of the dark halo may contribute to its shape, which in turn may affect the structure of the disks. For example, warps may be associated with misalignment of the angular momentum of the dark and baryonic components. The dark halo may have a rotating triaxial figure: the effect of a rotating triaxial dark halo on the self-gravitating disk has not yet been seriously investigated (see Bureau et al. 1999).

The binding energy E at the epoch of halo virialization affects the depth of the potential well and hence the characteristic velocities in the galaxy. It also affects the parameter $\lambda = J|E|^{\frac{1}{2}} G^{-1} M^{-\frac{5}{2}}$, where M is the total mass: λ which critical for determining the gross nature of the galactic disk as a high or low surface brightness system (e.g. Dalcanton et al. 1997).

The relation between the specific angular momentum J/M and the total mass M (Fall 1983) of disk galaxies is well reproduced by simulations (Zurek et al. 1988). Until recently, ellipticals and disk galaxies appeared to be segregated in the Fall diagram: from the slow rotation of their inner regions, estimates of the J/M ratios for ellipticals were about 1 dex below those for the spirals. More recent work (e.g. Arnaboldi et al. 1994) shows that much of the angular momentum in ellipticals appears to reside in their outer regions, so ellipticals and spirals do have similar locations in the Fall diagram. Internal redistribution of angular momentum has clearly occurred in the ellipticals (Quinn and Zurek 1988).

The remarkable Tully-Fisher law (1977) is a correlation between the HI linewidth and the optical luminosity of disk galaxies. It appears to relate the depth of the potential well and the baryonic mass (McGaugh et al. 2000). Both of these quantities are probably roughly conserved after the halo virialises, so the Tully-Fisher law should be regarded as a zero-order signature of galaxy formation. The likely connecting links between the (dark) potential well and the baryonic mass are (i) a similar baryon/dark matter ratio from galaxy to galaxy, and (ii) an observed Faber-Jackson law for the dark halos, of the form $M \propto \sigma^4$: i.e. surface density independent of mass for the dark halos.

1.7.1.4 Signatures of the Internal Distribution of Specific Angular Momentum

The internal distribution of specific angular momentum $j(h)$ of the baryons (i.e. the mass with specific angular momentum $< h$) largely determines the shape of the surface brightness distribution of the disk rotating in the potential well of the dark matter. Together with $j(h)$, the total angular momentum and mass of the baryons determine the scale length and scale surface density of the disk. Therefore the distribution of total angular momentum and mass for protodisk galaxies determines the observed distribution of the scale length and scale surface density for disk galaxies (Freeman 1970; de Jong and Lacey 2000).

Many studies have assumed that $j(h)$ is conserved through the galaxy formation process, most notably Fall and Efstathiou (1980). It is not yet clear if this assumption is correct. Conservation of the internal distribution of specific angular momentum $j(h)$ is a much stronger requirement than the conservation of the *total* specific angular momentum J/M. Many processes can cause the internal angular momentum to be redistributed, while leaving the J/M ratio unchanged. Examples include the effects of bars, spiral structure (Lynden-Bell and Kalnajs 1972) and internal viscosity (Lin and Pringle 1987).

The maximum specific angular momentum h_{max} of the baryons may be associated with the truncation of the optical disk observed at about four scale lengths (de Grijs et al. 2001; Pohlen et al. 2000). This needs more investigation. The truncation of disks could be an important signature of the angular momentum properties of the early protocloud, but it may have more to do with the critical density for star formation or the dynamical evolution of the disk. Similarly, in galaxies with very extended HI, the edge of the HI distribution may give some measure of h_{max} in the protocloud. On the other hand, it may be that the outer HI was accreted subsequent to the formation of the stellar disk (van der Kruit 2002), or that the HI edge may just represent the transition to an ionized disk (Maloney 1993; Bland-Hawthorn et al. 1997).

This last item emphasizes the importance of understanding what is going on in the outer disk. The outer disk offers some potentially important diagnostics of the properties of the protogalaxy. At present there are too many uncertainties about significance of (a) the various cutoffs in the light and HI distributions, (b) the age gradient seen by Bell and de Jong (2000) from integrated light of disks but not by Friel (1995) for open clusters in the disk of the Galaxy, and (c) the outermost disk being maybe younger but not 'zero age', which means that there is no real evidence that the disk is continuing to grow radially. It is possible that the edge of the disk has something to do with angular momentum of baryons in the protocloud or with disk formation process, so it may be a useful zero-order or first-order signature.

1.7.1.5 Signatures of the ΛCDM Hierarchy

ΛCDM predicts a high level of substructure which is in apparent conflict with observation. Within galaxies, the early N-body simulations appeared to show that substructure with characteristic velocities in the range $10 < V_c < 30$ km s^{-1} would be destroyed by merging and virialization of low mass structures (Peebles 1970; White 1976; White and Rees 1978). It turned out that the lack of substructure was an artefact of the inadequate spatial and mass resolution (Moore et al. 1996). Current simulations reveal 500 or more low mass structures within 300 kpc of an L_* galaxy's sphere of influence (Moore et al. 1999; Klypin et al. 1999). This is an order of magnitude larger than the number of low mass satellites in the Local Group. Mateo (1998) catalogues about 40 such objects and suggests that, at most, we are missing a further 15–20 satellites at low galactic latitude. Kauffmann et al. (1993) were the first to point out the 'satellite problem' and suggested that the efficiency of dynamical friction might be higher than usually quoted. However, without recourse to fine tuning, this would remove essentially all of the observed satellites in the Local Group today.

Since the emergence of the ΛCDM paradigm, an inevitable question is whether a 'basic building block' can be recognized in the near field. Moore et al emphasize the self-similar nature of ΛCDM sub-clustering and point to the evidence provided by the mass spectrum of objects in rich clusters, independent of the N-body simulations. The lure of finding a primordial building block in the near field has prompted a number of tests. If the dark mini-halos comprise discrete sources, it should be possible to detect microlensing towards a background galaxy (see Sect. 1.6.4).

The satellite problem appears to be a fundamental prediction of ΛCDM in the non-linear regime. Alternative cosmologies have been suggested involving the reduction of small-scale power in the initial mass power spectrum (Kamionkowski and

Liddle 2000), warm dark matter (Hogan and Dalcanton 2001; White and Croft 2000; Colin et al. 2000), or strongly self-interacting dark matter (Spergel and Steinhardt 2000). Several authors have pointed out that some of the direct dark-matter detection experiments are sensitive to the details of the dark matter in the solar neighbourhood. Helmi et al. (2001) (cf. Moore et al. 2001) estimate that there may be several hundred kinematically cold 'dark streams' passing through the solar neighbourhood.

If ΛCDM is correct in detail, then we have simply failed to detect or to recognize the many hundreds of missing objects throughout the Local Group. For example, the satellites may be dark simply because baryons were removed long ago through supernova-driven winds (Dekel and Silk 1986; Mac Low and Ferrara 1999). In support of this idea, x-ray halos of groups and clusters are almost always substantially enriched in metals ([Fe/H] $\geq$ -0.5; Renzini 2000; Mushotzky 1999). In fact, we note that up to 70 % of the mass fraction in metals is likely to reside in the hot intracluster and intragroup medium (Renzini 2000). Another explanation may be that the absence of baryons in hundreds of dark satellites was set in place long ago during the reionization epoch. Many authors note that the accretion of gas on to low-mass halos and subsequent star formation is heavily suppressed in the presence of a strong photoionizing background (Ikeuchi 1986; Rees 1986; Babul and Rees 1992; Bullock et al. 2000). This effect appears to have a cut-off at low galactic mass at a characteristic circular velocity close to 30 km s^{-1} (Thoul and Weinberg 1996; Quinn et al. 1996), such that the small number of visible Local Group dwarfs are those which exceed this cutoff or acquired most of their neutral hydrogen before the reionization epoch.

Blitz et al. (1999) suggested that the high-velocity H I gas cloud (HVC) population is associated with dark mini-halos on megaparsec scales within the Local Group. This model was refined by Braun and Burton (1999) to include only the compact HVCs. The HVCs have long been the subject of wide-ranging speculation. Oort (1966) realized that distances derived from the virial theorem and the H I flux would place many clouds at megaparsec distances if they are self-gravitating. If the clouds lie at about a megaparsec and are associated with dark matter clumps (Nichols and Bland-Hawthorn 2009), then they could represent the primordial building blocks. However Hα distances (Bland-Hawthorn et al. 1998) suggest that most HVCs lie within 50 kpc and are unlikely to be associated with dark matter halos (Putman et al. 2003). We note that several teams have searched for but failed to detect a faint stellar population in HVCs.

Moore et al. (1999; see also Bland-Hawthorn and Freeman 2000) suggested that ultrathin disks in spirals are a challenge to the ΛCDM picture in that disks are easily heated by orbiting masses. However, Font et al. (2001) find that in their ΛCDM simulations, very few of the ΛCDM sub-halos come close to the optical disk which may allow a very cold disk to form. At present there are real problems in reconciling the predictions of ΛCDM simulations with observations on scales of the Local Group (e.g. Kormendy et al. 2010).

1.7.2 First Order Signatures: Information Preserved Since the Main Epoch of Baryon Dissipation

1.7.2.1 The Structure of the Disk

At what stage in the evolution of the disk are its global properties defined? In part, we have already discussed this question in Sect. 1.7.1.4. The answer depends on how the internal angular momentum distribution $j(h)$ has evolved as the disk dissipated and various non-axisymmetric features like bars and formation and spiral structure came and went. Viscous processes associated with star formation, as suggested by (Lin and Pringle 1987), may also contribute to the evolution of the $j(h)$ distribution.

The global structure of disks is defined by the central surface brightnes I_o and the radial scalelength h of the disk. de Jong and Lacey (2000) evaluated the present distribution of galaxies in the (I_o, h) plane. If $j(h)$ has indeed remained roughly constant, as is often assumed for discussions of disk formation (e.g. Fall and Efstathiou 1980; Fall 1983), then the global parameters of the disk—the scale length, central surface brightness and the Tully-Fisher relation—are relics of the main epoch of baryon dissipation.

The edges of disks may hold clues to important physical processes. Until recently, it was believed that all galaxy disks undergo truncation near the Holmberg radius (26.5 mag arcsec^{-2}) and that this was telling us something important about the collapsing protocloud of gas that formed the early disk (van der Kruit 1979). This picture was challenged by our earlier finding (Bland-Hawthorn et al. 2005) of a classic exponential disk (Freeman 1970) with no break in NGC 300, out to the distances corresponding to 10 disk scale lengths in this low-luminosity late-type spiral (see also Irwin et al. 2005). The study of NGC 300 is particularly notable because the surface brightness profile was traced with resolved stars down to an effective surface brightness well below the levels achieved with surface photometry.

A recent study of the SDSS archives suggests that there are three basic classes of surface brightness profiles in galaxies (Pohlen et al. 2008). Truncations in surface brightness profiles, characterized by the smooth break between the inner shallower and the outer steeper exponential, have been known for more than two decades and detected in a large number of galaxies (van der Kruit 1979; de Grijs et al. 2001; Pohlen et al. 2002). Despite this, the origin of the break remains unclear.

The currently favoured model for the explanation of these sub-exponential truncations is the star formation threshold scenario (Kennicutt 1989; Schaye 2004; Roskar et al. 2008a, b; but see van der Kruit 2007). The threshold gas densities predicted in these models (3–10 $M_\odot$ pc^{-2}; Schaye 2004) broadly agree with the observed surface brightnesses at the radius of the break which correspond to stellar mass densities of 10–20 $M_\odot$ pc^{-2} (Bakos et al. 2008). The second class of surface brightness profiles, so-called upbending or super-exponential breaks, may be due to minor mergers, supported both observationally (Ibata et al. 2005) and by N-body/smoothed particle hydrodynamics simulations (Younger et al. 2007). In the next section, we show how disk edges are being exploited to test new ideas on disk evolution.

1.7.2.2 Can Disks Preserve Fossil Information?

Here, we consider radial and vertical fossil gradients in the disk, in particular of abundance and age. Our expectation is that much of the information will be diluted through the dynamical evolution and radial mixing of the disk.

For spirals, different mechanisms may be at work to establish gradients (Molla et al. 1996): (a) a radial variation of the yield due either to the stellar metal production or to the initial mass function, (b) a radial variation of the timescale for star formation, (c) a radial variation for the timescale of infall of gas from outside the disk. Once gradients are established, these can be amplified or washed out by radial mixing (Edmunds 1990; Götz and Köppen 1992).

Most stars are born in large clusters numbering hundreds or even thousands of stars. Some clusters stay together for billions of years, whereas others become unbound shortly after the initial starburst, depending on the star formation efficiency. When a cloud disperses, each star suffers a random kick superimposed on the cloud's mean motion. Thereafter, stars are scattered by transient spiral arm perturbations and star-cloud encounters.

These perturbations allow the star to migrate in integral space. During interaction with a single spiral event of pattern speed Ω_p, a star's energy and angular momentum change while it conserves its Jacobi integral: in the (E, J) plane, stars move along lines of constant $I_J = E - \Omega_p J$. The star undertakes a random walk in the (E, J) plane, perturbed by a series of spiral arm events (Sellwood 1999; Dehnen 2000). N-body models of disk evolution indicate that radial mixing is strong (Sellwood and Binney 2002; Lynden-Bell and Kalnajs 1972). This is believed to be driven by *transient* spiral waves that heat the in-plane motions, although the process is not yet well understood. Long-term spiral arms produce no net effect. Remarkably, a *single* spiral wave near co-rotation can perturb the angular momentum of a star by $\sim$20% without significant heating: the star is simply moved from one circular orbit to another, inwards or outwards, by $\sim$2 kpc (Sellwood and Binney 2002). Substantial variations in the angular momentum of a star are possible over its lifetime.

In addition to radial heating, stars experience vertical disk heating: their vertical velocity dispersion increases as they age. This is believed to occur through a combination of in-plane spiral-arm heating and scattering off giant molecular clouds (e.g. Spitzer and Schwarzschild 1953; Carlberg and Sellwood 1985). The in-plane heating is most effective at the inner and outer Lindblad resonances and vanishes at corotation. In the vertical direction, an 'age-velocity dispersion' relation is observed for stars younger than about 3 Gyr, but older disk stars show a velocity dispersion that is independent of age. Thus, the vertical structure does depend on the mean age of the population for $\tau < 3$ Gyr (Edvardsson et al. 1993, confirmed from Hipparcos data by Gomez et al. 1997).

As the amplitude of the random motions increases, the star becomes less vulnerable to heating by transient spiral waves, and the heating process is expected to saturate. This probably happens after about 3 Gyr (Binney and Lacey 1988; Jenkins and Binney 1990), consistent with observation. This is important for our purpose here. It means that dynamical information is preserved about the state of the thin disk at an early epoch, or roughly $\tau_L - 3 \approx 7$ Gyr ago, for which τ_L is the look-back time when the disk first began to form.

The survival of old open clusters like NGC 6791, Berkeley 21 and Berkeley 17 (Friel 1995; van den Bergh 2000) is of interest here. The oldest open clusters exceed 10 Gyr in age and constitute important fossils (Phelps and Janes 1996). Both old and young open clusters are part of the thin disk. If the heating perturbations occur over a lengthscale that significantly exceeds the size of an open cluster, it seems likely that the cluster will survive. A large spiral-arm heating event will heat many stars along their I_J trajectories. The trace of the heating event is likely to survive for a very long time but be visible only in integral space. We note that vertical abundance gradients have not been seen among the open clusters (Friel and Janes 1993).

About 4% of disk stars are super metal-rich (SMR) relative to the Hyades (Castro et al. 1997). SMR stars of intermediate age appear to have formed a few kpc inside of the Solar circle from enriched gas. The oldest SMR stars appear to come from the Galactic Center: their peculiar kinematics and outward migration may be associated with the central bar (Carraro et al. 1998; Grenon 1999).

In summary, our expectation is that fossil gradients within the disk are likely to be weak. This is borne out by observations of both the stars and the gas (Chiappini et al. 2001). The vertical structure of the disk preserves another fossil—the 'thick disk'— which we discuss in the next section. Like the open clusters, this component also does not show a vertical abundance gradient (Gilmore et al. 1995). In later sections, we argue that this may be the most important fossil to have survived the early stages of galaxy formation.

1.7.2.3 Disk Heating by Accretion: The Thick Disk

Heating from discrete accretion events also imposes vertical structure on the disk (Quinn and Goodman 1986; Walker et al. 1996). Such events can radically alter the structure of the inner disk and the bulge (see Fig. 1.19d for an example), and are currently believed to have generated the thick disk of the Galaxy.

The galactic thick disk was first recognized by Gilmore and Reid (1983). It includes stars with a wide range of metallicity, from $-2.2 \leq$ [Fe/H] ≤ -0.5 (Chiba and Beers 2000): most of the thick disk stars are in the more metal-rich end of this range. The velocity ellipsoid of the thick disk is observed to be $(\sigma_R, \sigma_\phi, \sigma_z) = (46 \pm 4, 50 \pm 4, 35 \pm 3)$ km s^{-1} near the sun, with an asymmetric drift of about 30 km s^{-1}. For comparison, the nearby halo has a velocity ellipsoid $(\sigma_R, \sigma_\phi, \sigma_z) = (141 \pm 11, 106 \pm 9, 94 \pm 8)$ km s^{-1} and its asymmetric drift is about 200 km s^{-1}.

The mean age of the thick disk is not known. From photometric age-dating of individual stars, the thick disk appears to be as old as the globular clusters. Indeed, the globular cluster 47 Tuc (age 12.5 ± 1.5 Gyr: Liu and Chaboyer 2000) is often associated with the thick disk.

After Quinn and Goodman 1986, Walker et al. (1996) showed in detail that a low mass satellite could substantially heat the disk as it sinks rapidly within the potential well of a galaxy with a 'live' halo. The conversion of satellite orbital energy to disk thermal energy is achieved through resonant scattering. Simulations of satellite accretion are important for understanding the survival of the thin disk and the origin of the thick disk. This is particularly relevant within the context of ΛCDM. The satellites which do the damage are those that are dense enough to survive tidal disruption by the Galaxy. We note that even dwarf spheroidals which appear fluffy are in fact rather dense objects dominated by their dark matter.

It is fortuitous that the Galaxy has a thick disk, since this is not a generic phenomenon. The disk structure may be vertically stepped as a consequence of past discrete accretion events. The Edvardsson et al. (1993) data appears to show an abrupt increase in the vertical component of the stellar velocity dispersion at an age of 10 Gyr: see also Strömgren (1987). Freeman (1991) argued that the age–(velocity dispersion) relation shows three regimes: stars younger than 3 Gyr with $\sigma_z \sim 10$ km s^{-1}, stars between 3 and 10 Gyr with $\sigma_z \sim 20$ km s^{-1}, and stars older than 10 Gyr with $\sigma_z \sim 40$ km s^{-1}. The first regime probably arises from the disk heating process due to transient spiral arms which we described in the previous section. The last regime is the thick disk, presumably excited by an ancient discrete event.

Can we still identify the disrupting event that lead to the thick disk? There is increasing evidence now that the globular cluster ω Cen is the stripped core of a dwarf elliptical (see Sect. 1.7.3.6). It is possible that the associated accretion event or an event like it was the event that triggered the thick disk to form.

In summary, it seems likely that the thick disk may provide a snap-frozen view of conditions in the disk shortly after the main epoch of dissipation. Any low level chemical or age gradients would be of great interest in the context of dissipation models. In this regard, Hartkopf and Yoss (1982) argued for the presence of a vertical abundance gradient in the thick disk, although Gilmore et al. (1995) find no such effect. Because stars of the thick disk spend relatively little time near the galactic plane, where the spiral arm heating and scattering by giant molecular clouds is most vigorous, radial mixing within the thick disk is unlikely to remove all vestiges of a gradient. If our earlier suggestions are right (Sect. 1.7.1.4), we might expect to see a different truncation radius for the thick disk compared to the thin disk.

1.7.2.4 Is There an Age–Metallicity Relation?

Some fossil information has likely been preserved since the main epoch of baryon dissipation. The inner stellar bulge is a striking example. It is characterized by old, metal-rich stars which seems to be at odds with the classical picture where metals accumulate with time (Tinsley 1980). However, the dynamical timescales in the inner bulge are very short compared to the outer disk, and would have allowed for rapid enrichment at early times. This is consistent with the frequent occurrence of metal-rich cores of galaxies observed at high redshift (Hamann and Ferland 1999). The dynamical complexity of the Galactic bulge may not allow us to determine the sequence of events that gave rise to it. We anticipate that this will come about from far-field cosmology (Robertson et al. 2010).

The existence of an age-metallicity relation (AMR; Fig. 1.16) in stars is a very important issue, about which there has long been disagreement. Twarog (1980) and Meusinger et al. (1991) provide evidence for the presence of an AMR, while Carlberg et al. (1985) find that the metallicity of nearby F stars is approximately constant for stars older than about 4 Gyr. More recently it has become clear that an AMR is apparent only in the solar neighbourhood and is strictly true only for stars younger than 2 Gyr and hotter than log $T_{\rm eff} = 3.8$ (Feltzing et al. 2001). Edvardsson et al. (1993) demonstrate that there is no such relation for field stars in the old disk. Similarly, Friel (1995) shows that there is no AMR for open clusters (see Sect. 1.7.3.5): she goes on to note that

Apparently, over the entire age of the disk, at any position in the disk, the oldest clusters form with compositions as enriched as those of much younger objects.

In fact, it has been recognized for a long time (e.g. Arp 1962; Eggen and Sandage 1969; Hirshfeld et al. 1978) that old, metal-rich stars permeate the galaxy, throughout the disk, the bulge and the halo. We regard the presence of old metal-rich stars as a first-order signature. An age–metallicity relation which applies to all stars would have been an important second-order signature, but we see no evidence for such a relation, except among the young stars.

1.7.2.5 Effects of Environment and Internal Evolution

Environmental influences are operating on all scales of the hierarchy and across all stages of our signature classification, so our attempts to classify signatures are partly artificial. Within ΛCDM, environmental effects persist throughout the life of the galaxy.

The parameters that govern the evolution of galaxies are among the key unknowns of modern astrophysics. Are the dominant influences internal (e.g. depth of potential) or external (e.g. environment) to galaxies? We consider here the effects of environment and internal evolution on the validity of the first-order signatures of galaxy formation (i.e. the properties that may have been conserved since the main epoch of baryon dissipation).

The well-known 'G dwarf problem' indicates that external influences are important. A simple closed box model of chemical evolution predicts far too many metal-poor stars in the solar neighbourhood (Tinsley 1980). This problem is easily remedied by allowing gas to flow into the region (Lacey and Fall 1983, 1985, Clayton 1987, 1988; Wyse & Silk 1989; Matteucci and Francois 1989; Worthey et al. 1996). In the context of ΛCDM, this is believed to arise from the continued accretion of gas-rich dwarfs (e.g. Cole et al. 1994; Kauffmann and Charlot 1998).

Environment is clearly a key factor. Early type galaxies are highly clustered compared to late type galaxies (Hubble and Humason 1931; Dressler 1980). Trager et al. (2000) find that for a sample of early-type galaxies in low-density environments, there is a large spread in the Hβ index (i.e. age), but little variation in metallicity. For galaxies in the Fornax cluster, Kuntschner (2000) finds the opposite effect: a large spread in metallicity is present with little variation in age. This probably reflects strong differences in environment between the field and the cluster.

Another likely environmental effect is the fraction of S0 galaxies in clusters, which shows a rising trend with redshift since $z \approx 0.4$ (Jones et al. 2000). Furthermore, S0 galaxies in the Ursa Major cluster show age gradients that are inverted compared to field spirals, in the sense that the cores are young and metal-rich (Tully et al. 1996; Kuntschner and Davies 1998). Both of these effects involve more recent phenomena and would be properly classified as second-order signatures.

Internal influences are also at work. A manifestation is the color-magnitude relation (CMR) in early type (Sandage and Visvanathan 1978) and late type (Peletier and de Grijs 1998) galaxies. The CMR does not arise from dust effects (Bell and de Jong 2000) and must reflect systematic variations in age and/or metallicity with luminosity. In the case of ellipticals, the CMR is believed to reflect a mass-metallicity dependence (Faber 1973; Bower et al. 1998). The relation is naturally explained by supernova-driven wind models in which more massive galaxies retain supernova ejecta and thus become more metal rich and redder (Larson 1974; Arimoto and Yoshii 1987). The CMR is presumably established during the main phase of baryon dissipation and is a genuine first-order signature.

Concannon et al. (2000) analyzed a sample of 100 early type galaxies over a large range in mass. They find that lower mass galaxies exhibit a larger range in age than higher mass galaxies. This appears to show that smaller galaxies have had a more varied star formation history, which is at odds with the naive ΛCDM picture of low-mass galaxies being older than high mass galaxies (Baugh et al. 1996; Kauffmann 1996). The work of Concannon et al. (2000) shows the presence of a real cosmic scatter in the star formation history. It is tempting to suggest that this cosmic scatter relates to different stages of evolution within the hierarchy. In this sense, we would regard the Concannon et al result as a first-order manifestation of galaxy formation (see Sect. 1.5.3).

Spiral galaxies commonly show color gradients that presumably reflect gradients in age and metallicity (Peletier and de Grijs 1998). Faint spiral galaxies have younger ages and lower metallicities relative to bright spirals. In a study of 120 low-inclination spirals, Bell and de Jong (2000) found that the local surface density within galaxies is the most important parameter in shaping their star formation and chemical history. However, they find that metal-rich galaxies occur over the full range of surface density. This fact has a remarkable resonance with the distribution of the metal rich open clusters that are found at any position in the Galactic disk (see Sect. 1.7.2.4). Bell and de Jong argue that the total mass is a secondary factor that modulates the star formation history. Once again, these authors demonstrate the existence of cosmic scatter that may well arise from variations in environment.

1.7.3 Second Order Signatures: Major Processes Involved in Subsequent Evolution

1.7.3.1 Introduction

Here we consider relics of processes that have taken place in the Galaxy since most of the baryonic mass settled to the disk. There are several manifestations of these processes, probably the most significant of which is the star formation history of the disk, for which the open clusters are particularly important probes. There is a wealth of detail relating to anomalous populations throughout the Galaxy, discussed at length by Majewski (1993). Examples include an excess of stars on extreme retrograde orbits (Norris and Ryan 1989; Carney et al. 1996), metal-poor halo stars of intermediate age (Preston et al. 1994) and metal-rich halo A stars (Rodgers et al. 1981).

In an earlier section, we discussed observational signatures of the ΛCDM hierarchy in the Galactic context. In fact, detailed observations in velocity space are proving to be particularly useful in identifying structures that have long since dispersed in configuration space. In external galaxies, related structures are showing up as low surface brightness features. We do not know what role globular clusters play in the galaxy formation picture, but we include them here because at least one of them appears now to be the nucleus of a disrupted dwarf galaxy.

1.7.3.2 Star Formation History

The star formation history (SFH) of the Galaxy has been very difficult to unravel. This is intimately tied up with how gas accretion takes place over billions of years—see Sect. 1.4. So how do galaxies accrete their gas? Is the infalling gas confined by dark matter? Does the gas arrive cold, warm or hot? Does the gas rain out of the halo onto the disk or is it forced out by the strong disk-halo interaction? These issues have never been resolved, either through observation or through numerical simulation, but we do see evidence today for all of the above processes. For example, the Magellanic System is direct evidence for the accretion of stars, multiphase gas and dark matter onto the Galaxy within the next few Gyr. But as we have seen (Sect. 1.3.5), the Milky Way-Magellanic relationship appears to be relatively unusual.

Derived star formation histories range from a roughly uniform star formation rate over the history of the disk (Binney et al. 2000) to a SFH that was highly peaked at early times (e.g. Twarog 1980; Rocha-Pinto et al. 2000; Just 2001; Aumer and Binney 2009). Galaxies of the Local Group show a great diversity in SFH (Grebel 2001), although the average history over the Local Group appears consistent with the mean cosmic history (Hopkins et al. 2001). The present emphasis is on star formation studies that make use of the integrated properties of external galaxies, but it should be noted that this is necessarily weighted towards the most luminous populations. Key results for external galaxies are reviewed in Sect. 1.7.2.5. It was concluded that environmental effects are very significant in determining the SFH for individual galaxies.

The conventional approach to the study of chemical evolution in galaxy disks is to consider the solar neighbourhood a closed box, and to assume that it is representative of all disks. Simple mathematical formulations have developed over the past 40 years (van den Bergh 1962; Schmidt 1963; Pagel and Patchett 1975; Talbot and Arnett 1971; Tinsley 1980; Twarog 1980; Pitts and Tayler 1989). Most observations are interpreted within this framework. The SFH is quantified in terms of stellar age, stellar (+gas) metallicity and, to a lesser extent, the existing gas fraction.

The use of broadband photometry coupled with stellar population synthesis is a well-established technique for probing the SFH of galaxy populations from integrated light. The power of the method is its simplicity, although it cannot uniquely disentangle the age-metallicity degeneracy (Bica et al. 1990; Charlot and Silk 1994).

Another widely used technique is the Lick index system (Burstein et al. 1984) further refined in Worthey et al. (1994) and Trager (1998). In this system, the $H\beta$ index is the primary age-sensitive spectral indicator, whereas the Mg and Fe indices are the primary metallicity indicators. The Lick indices have well known limitations: they correspond to low spectroscopic resolution (8–9Å), require difficult corrections for internal galaxy motions, and are not calibrated onto a photometric scale. Furthermore, two of the most prominent Lick indices—Mg_2 $\lambda 5176$ and Fe $\lambda 5270$—are now known to be susceptible to contamination from other elements, in particular Ca and C (Tripicco and Bell 1995).

How best to measure galaxy ages is a subject with a long history. The most reliable methods to date involve the low order transitions ($n < 4$) of the Balmer series. Ages derived from the $H\gamma$ equivalent width have been used by Jones and Worthey (1995). Rose (1994) and Caldwell and Rose (1998) have pioneered the use of even higher order Balmer lines to break the age-metallicity degeneracy (Worthey 1994). These higher order lines are less affected by Balmer line emission from the interstellar medium. They develop a line ratio index Hn/Fe which is a sum over $H\gamma$, $H\delta$ and H8 lines with respect to local Fe lines (e.g. Concannon et al. 2000). Ultimately, full spectrum fitting matched to spectral synthesis models holds the most promise (Vazdekis 1999). The new models, which have a fourfold increase in spectroscopic resolution compared to the Lick system, show that the isochrone or isochemical grid lines overlaid on a plot of two Lick indices are more orthogonal than the Worthey models.

1.7.3.3 Low Surface Brightness Structures in Galaxies

Dynamical interaction between galaxies lead to a range of structures including stellar shells (Malin and Carter 1980; Quinn 1984), fans (Weil et al. 1997), and tidal streamers (Gregg and West 1998; Calcaneo-Roldan et al. 2000; Zheng et al. 1999; Zucker et al. 2004). Some excellent examples are shown in Fig. 1.20. We see evidence of multiple nuclei, counter-rotating cores, and gas in polar orbits. At low light levels, the outermost stellar contours of spiral disks appear frequently to exhibit departures

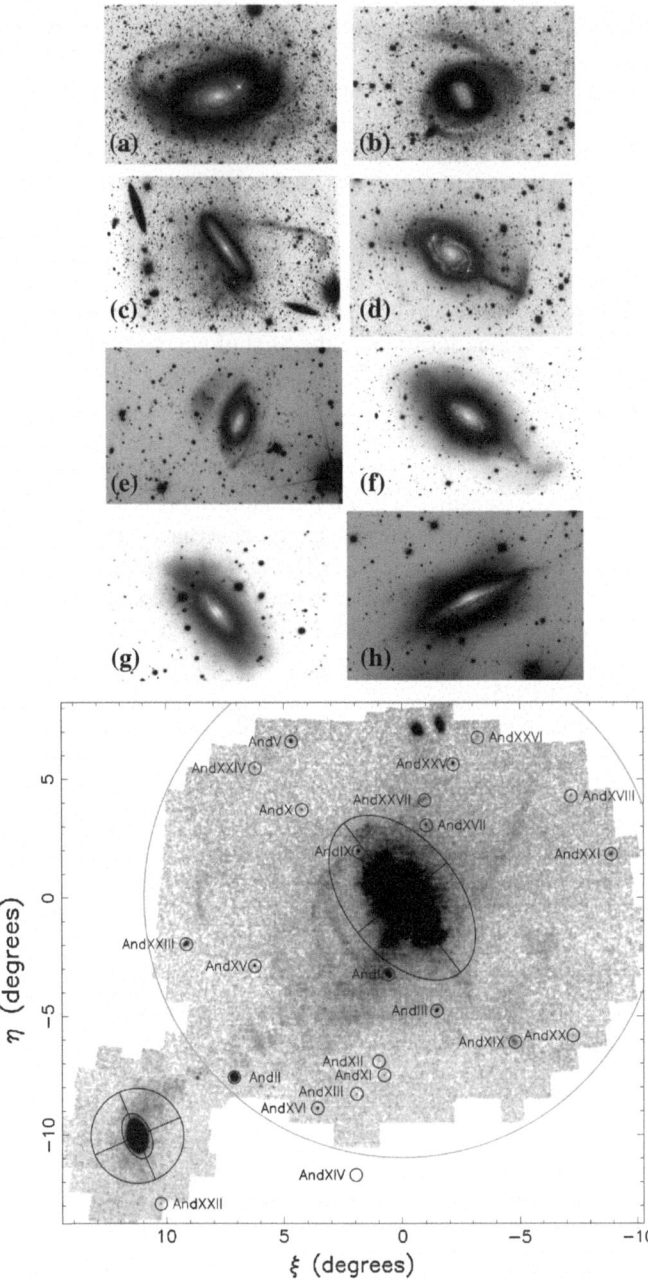

Fig. 1.20 *Upper* Examples of normal spirals with faint stellar streamers in the outer halo taken from a recent survey by Martinez-Delgado et al. 2010. *Lower* M31 and M33 system (PANDAs survey; Richardson et al. 2011); many associated dwarfs are also shown

from axisymmetry (Rix and Zaritsky 1995). The same is true for spiral arms in all Hubble types (Schönmakers et al. 1997; Sanchez-Gil et al. 2011).

The stellar streamers are particularly interesting as these may provide important constraints on galaxy models, particularly as kinematic measurements become possible through the detection of planetary nebulae. More than a dozen stellar streams are already known and this is probably indicative of a much larger population at very low surface brightness. Johnston et al. (2002) show that stellar streamers can survive for several Gyr and are only visible above the present optical detection limit ($\sim$30 V mag arcsec^{-2}) for roughly 4×10^8 yr. A few galaxy groups (e.g. the Leo group) do show largescale H I filaments that can remain visible for many Gyr.

Deep CCD imaging has revealed a stellar loop around NGC 5907 (Shang et al. 1998) and a stellar feature extending from NGC 5548 (Tyson et al. 1998). The technique of photographic amplification has revealed stellar streamers in about ten sources (Malin and Hadley 1997; Calcaneo-Roldan et al. 2000; Weil et al. 1997). For these particular observations, the limiting surface brightness is $\mu_V \approx 28.5$ mag arcsec^{-2}. For all of these systems, we estimate that the total stream luminosities are in the range $3 - 20 \times 10^7 L_\odot$.

Wide-field CCD cameras have revealed stellar streamers through multiband photometry of millions of individual sources. A pointillist image can then be reconstructed in narrow color intervals so as to enhance features with respect to the field. This has lead to the discovery of stellar streams in M31 (Ibata et al. 2001a; McConnachie et al. 2009) and tidal tails extending from the globular cluster Pal 5 (Odenkirchen et al. 2001). This technique has the potential to push much deeper than the direct imaging method described above. The low surface brightness Universe is notoriously difficult to observe. Modern telescope and instrument designs are simply not optimized for this part of parameter space. Many claims of diffuse light detections in the neighbourhood of galaxies have been shown to arise from internal scattered light.

Looking farther afield, we see evidence for discrete accretion events in the making. The Galaxy is encircled by satellite galaxies that appear confined to one or two great streams across the sky (Lynden-Bell and Lynden-Bell 1995). The most renowned of these are the Magellanic Clouds and the associated HI Magellanic stream. All of these are expected to merge with the Galaxy in the distant future, largely due to the dynamical friction from the extended halo.

In Sect. 1.7.3.8, we discuss 'moving groups' identified within the Galaxy from proper motion and spectroscopic surveys. Their projected surface brightness is $\mu_r = 30 - 34$ mag arcsec^{-2}, below the limit of modern imaging techniques.

1.7.3.4 Star Clusters as a Probe of the Environment

A key characteristic of star formation is the creation of star clusters. The initial cluster mass function (ICMF), if it can be determined reliably, provides important information on the physical conditions under which the clusters formed (Elmegreen 2010; Larsen 2009; Bastian et al. 2012). In principle, this function describes the creation of all clusters, from globulars to open clusters, to loose associations. The ICMF gives the number of clusters formed per unit mass, N_c, where present-day cluster formation is often described with a power-law with index γ: $dN_c/dm_c \propto m_c^\gamma$. A flat ICMF ($\gamma \gtrsim -1.5$), which favours massive clusters, appears to be indicative of high-pressure environments (e.g. starbursts, mergers), whereas a steep ICMF ($\gamma \lesssim -2$) is typically associated with quiescent environments (e.g. solar neighbourhood).

In recent years, much of the discussion on cluster formation has focussed on how many stars are born in bound versus unbound systems. In galaxies with low star formation rates, the fraction born in bound clusters appears to be close to 10 % but this increases dramatically in starburst systems (Bastian et al. 2012). But we stress that the fraction born in bound versus unbound clusters is secondary to the question of what fraction of stars are born in *instantaneous* clusters. As long as the collapsing cloud is homogenised before the first stars are formed, and therefore carries a unique chemical imprint, the long-term survival of the cloud is of lesser importance. It is possible that a much higher fraction of stars in all environments form in *instantaneous* clusters, most of which have long since diffused into the general background.

As we show in Sect. 1.8, the ICMF is of fundamental importance to chemical tagging, and unravelling the original star formation history. A few large clusters are easier to 'tag' than many smaller clusters, assuming that these are all formed in homogenised collapsing clouds. Thus, the success of chemical tagging depends critically on how a particular component of the Galaxy formed. But first we discuss some specific types of bound stellar populations found throughout the Local Group— open clusters, globular clusters, dwarf galaxies.

1.7.3.5 Open Clusters

In the context of near-field cosmology, we believe that the thick disk and the old open clusters of the thin disk are among the most important diagnostics. The open clusters are the subject of an outstanding and comprehensive review by Friel (1995). Here, we summarize the properties that are most important for our purpose.

Both old and young clusters are part of the thin disk; see Fig. 1.21 for some excellent examples of these beautiful objects. Their key attribute is that they provide a direct time line for investigating change, which we explore in the next section. The oldest open clusters exceed 10 Gyr in age and constitute important fossils (Phelps and Janes 1996). The survival of these fossil clusters is an interesting issue in its own right. Friel (1995) finds no old open clusters within a galactocentric radius of 7 kpc; these are likely to have disrupted or migrated out of the central regions (van den Bergh and McClure 1980). It has long been recognized that open clusters walk a knife edge between survival and disruption (King 1958).

Like field stars in the disk, Janes and Phelps (1994) find that the old cluster population (relative to Hyades) is defined by 375 pc scale height exponential distribution, whereas young clusters have a 55 pc scale height. Again, like the field stars, vertical abundance gradients have not been seen in open clusters (Friel and Janes 1993), although radial gradients are well established (Friel 1995; van den Bergh 2000). For

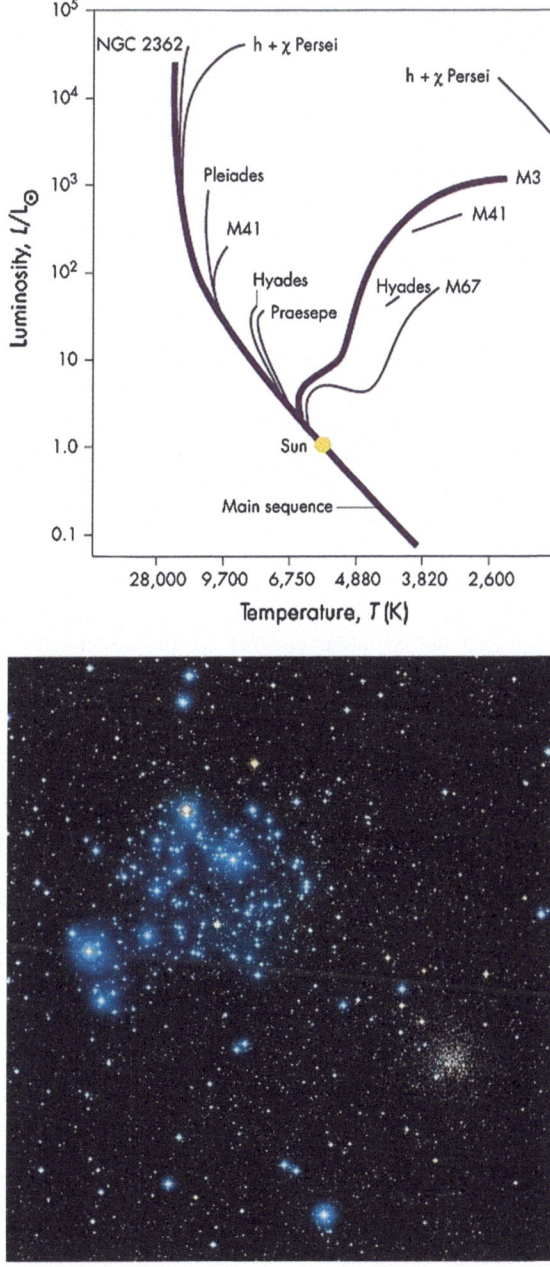

Fig. 1.21 *Upper* Hertzsprung-Russell diagram for stellar clusters over a wide range in age. *Lower* CFHT imaging of two open clusters M35 (NE) and NGC 2158 (SW) in Gemini. The young cluster M35 (NGC 2168) has a mass of roughly 2500 $M_\odot$ and is 850 pc from the Sun. The massive cluster NGC 2158 is four times further away and about 1 Gyr old

old open clusters, Twarog et al. (1997) claim evidence for a stepped radial metallicity distribution where [Fe/H] ≈ 0 within 10 kpc, falling to [Fe/H] ≈ -0.3 in the outer disk. However, this effect is not seen in young objects, e.g. HII regions and B stars (Henry 1998).

Both the old and young open clusters show essentially the same radial trend in metallicity. After reviewing the available observations, Friel (1995) finds no evidence for an age-metallicity relation for open clusters. In agreement with Eggen and Sandage (1969), she notes that over the entire age of the disk, at any position in the disk, the oldest clusters form with compositions as enriched as those of much younger objects.

These remarkable observations appear to indicate that shortly after the main epoch of baryon dissipation, the thin disk was established at least as far out as 15 kpc. The oldest open clusters approach the age of the thick disk. Since, in Sect. 1.7.2.3, we noted that the thick disk is likely to be a 'snap frozen' picture of the thin disk shortly after disk formation, we would expect the truncation of the thick disk to reflect the extent of the thin disk at the epoch of the event that puffed up the thick disk.

1.7.3.6 Globular Clusters

We have long suspected that globular clusters are the fossil remnants of violent processes in the protogalactic era (Peebles and Dicke 1968; Conroy and Spergel 2011). But there is a growing suspicion that globulars are telling us more about globulars than galactic origins (Harris 2002). The Milky Way has about 150 globular clusters with 20% lying within a few kiloparsecs of the Galactic Center. They constitute a negligible fraction of the light and mass (2%) of the stellar halo today. Their significance rests in their age. The oldest globular clusters in the outer halo have an age of 13 ± 2.5 Gyr (90% confidence).

The ages of the oldest globular clusters in the inner and outer halo, the Large Magellanic Cloud and the nearby Fornax and Sgr dwarf spheroidal galaxies show a remarkable uniformity. To a precision of ± 1 Gyr, the onset of globular cluster formation was well synchronized over a volume centered on our Galaxy with a radius >100 kpc (Da Costa 1999).

Globular cluster stars are older than the oldest disk stars, e.g. white dwarfs and the oldest red giants. These clusters are also more metal poor than the underlying halo light in all galaxies and at all radii (Harris 1991), but again there are exceptions to the rule. Since Morgan's (1959) and Kinman's (1959) classic work, we have known that there are two distinct populations of globular clusters in the Galaxy. The properties that we associate with these two populations today were derived by Zinn (1985) who showed that they have very different structure, kinematics and metallicities. The halo population is metal-poor ([Fe/H] < -0.8) and slowly rotating with a roughly spherical distribution; the disk population is metal-rich ([Fe/H] > -0.8) and in rapid rotation.

A major development has been the discovery of young globular clusters in disturbed or interacting galaxies, e.g., NGC 1275 (Holtzman et al. 1992), NGC 7252

(Whitmore et al. 1993) and the Antennae Whitmore and Schweizer (1995). Schweizer (1987) first suspected that globular clusters were formed in mergers. Later, Ashman and Zepf (1992) predicted that the HST would reveal young globular clusters through their compact sizes, high luminosities and blue colors. The very high internal densities of globular clusters today must partly reflect the conditions when they were formed. Harris and Pudritz (1994) present a model for globular clusters produced in fragmenting giant molecular clouds, which are of the right mass and density range to resemble accretion fragments in the Searle-Zinn model.

Globular clusters have been elegantly referred to as 'canaries in a coal mine' (Arras and Wasserman 1999). They are subject to a range of disruptive effects, including two-body relaxation and erosion by the tidal field of their host galaxy, and the tidal shocking that they experience as their orbits take them through the galactic disk and substructure in the dark halo. In addition to self-destruction through stellar mass loss, tidal shocking may have been very important in the early Universe (Gnedin et al. 1999). If globular clusters originally formed in great numbers, the disrupted clusters may now contribute to the stellar halo (Norris and Ryan 1989; Oort 1965). Halo field stars and globular clusters in the Milky Way have similar mean metallicities (Carney 1993); however the metallicity distribution of the halo field stars extends to much lower metallicity ([Fe/H] $\simeq -5$) than that of the globular clusters ([Fe/H] $\simeq -2.2$). We note again the remarkable similarity in the metallicity range of the globular clusters and the thick disk ($-2.2 \lesssim$ [Fe/H] $\lesssim -0.5$).

In the nucleated dwarf elliptical galaxies (Binggeli et al. 1985), the nucleus typically provides about 1 % of the total luminosity; globular clusters could be considered as the stripped nuclei of these satellite objects without exceeding the visible halo mass (Zinnecker and Cannon 1986; Freeman 1993). It is an intriguing prospect that the existing globular clusters could be the stripped relics of an ancient swarm of protogalactic stellar fragments, i.e. the original building blocks of the Universe.

In the Searle-Zinn picture, globular clusters are intimately linked to *gas-rich*, protogalactic infalling fragments. Multiple stellar populations have recently been detected in ω Cen, the most massive cluster in the Galaxy (Lee 1999). How did ω Cen retain its gas for a later burst? It now appears that it was associated with a gas-rich dwarf, either as an *in situ* cluster or as the stellar nucleus. The present-day cluster density is sufficiently high that it would have survived tidal disruption by the Galaxy, unlike the more diffuse envelope of this dwarf galaxy. The very bound retrograde orbit supports the view that ω Cen entered the Galaxy as part of a more massive system whose orbit decayed through dynamical friction.

If globular clusters are so ancient, why are the abundances of the most metal-poor population as high as they are? Because it does not take much star formation to increase the metal abundance up to [Fe/H] $= -1.5$ (Frayer and Brown 1997), the cluster abundances may reflect low levels of star formation even before the first (dark+baryon) systems came together.

Contrary to a common assumption, old age is not uniquely associated with low metallicity (cf. Sect. 1.5.3). In hindsight, we have known this for many years. Brown and van den Bout (1991) discovered a strong CO signal from a galaxy at $z \approx 2.3$. Molecule formation requires the presence of dust which in turn requires substantial enhancements in metallicity. Hamann and Ferland (1999) demonstrate that stellar populations at the highest redshifts currently observed ($z \gtrsim 5$) appear to have solar or super-solar metallicity; CO gas has also been detected at these times (e.g. Yun et al. 2000).

We believe that there is no mystery about high abundances at high redshift. The dynamical times in the cores of these systems are short, so there has been time for multiple generations of star formation and chemical enrichment. In this sense, the cores of high redshift galaxies need not be relevant to the chemical properties of the globular clusters, although both kinds of objects were probably formed at about the same time.

The first generation of globular clusters may have been produced in merger-driven starbursts when the primordial fragments came together for the first time. If at least some fragments retained some of their identity while the halo was formed, a small number of enrichment events per fragment would ensure a Poissonian scatter in properties between globular clusters, and multiple populations within individual clusters (Searle and Zinn 1978).

1.7.3.7 Dwarf Galaxies

The formation and long-term evolution of dwarf galaxies are topics of great interest today. Substantial populations are still being identified around the Galaxy and M31 (e.g. Zucker et al. 2006, 2007). A defining characteristic is that these are baryons confined by a dark matter halo, although this definition may not cover 'tidal dwarfs' if these are ultimately shown to be free of dark matter (Barnes and Hernquist 1992). Very comprehensive reviews of the dwarf population can be found in Tolstoy et al. (2009) and McConnachie (2012).

Dwarf galaxies are possibly the best probes of the first stars within the framework of near-field cosmology (Tolstoy et al. 2009; Bland-Hawthorn et al. 2010b; Karlsson et al. 2012, 2013). A substantial fraction of the dwarf galaxy population of the Local Group is expected to have already merged with the halos of M31 and the Galaxy (see Fig. 1.22). This is supported by the evidence that the stellar Galactic halo has several distinct chemical similarities that it shares with the existing population of dwarf galaxies (Kirby et al. 2008; Tolstoy et al. 2009 and references therein; Starkenburg et al. 2010; Frebel et al. 2010a, b; Norris et al. 2010a, b). Despite these similarities, the systems are not chemically identical and exactly how the merging process progressed, how the dwarf galaxies that survived are connected to their disrupted counterparts, and how these still-surviving galaxies relate to each other in a cosmological context, is not understood. The tidal streams in Sgr and Carina (Ibata et al. 1994; Battaglia et al. 2012) provide clear evidence that dwarfs are being disrupted on infall. Furthermore, all but the most massive dwarfs within the virial radii of M31 and the Galaxy show strong evidence of gas depletion (Grcevich and Putman 2009).

The study of old stellar populations in these galaxies provides us with a link to the high-redshift Universe. All Local Group dwarf galaxies, including the recently discovered ultra-faint dwarfs, are known to contain ancient, metal-poor stars (Grebel

1 Near Field Cosmology

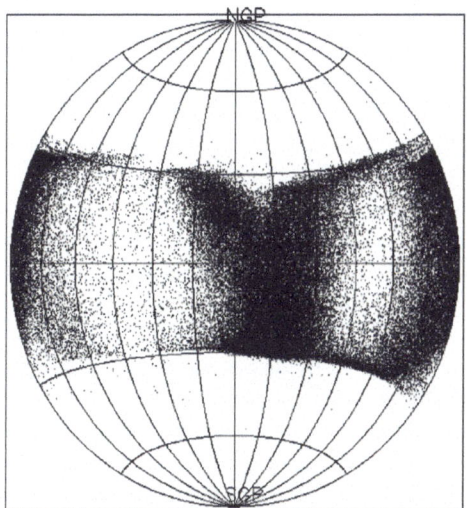

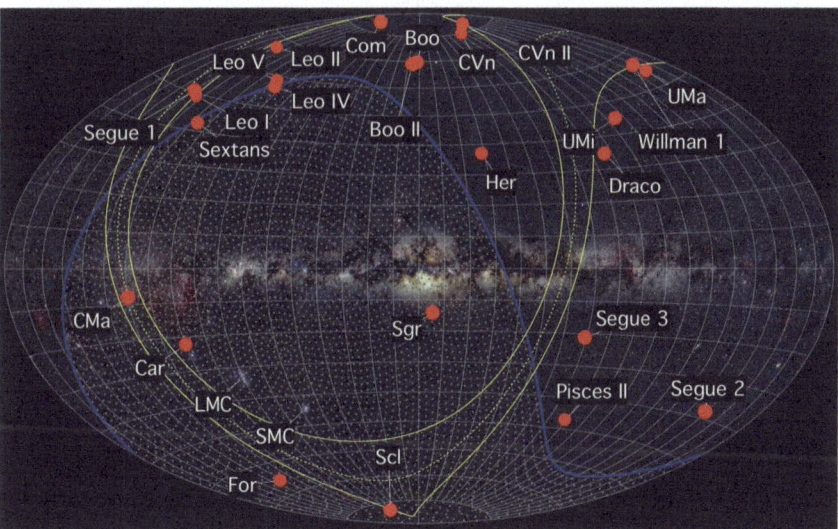

Fig. 1.22 *Upper* Aitoff projection of a satellite in orbit about the Milky Way as it would appear after 8 Gyr. While stars from the disrupted satellite appear to be dispersed over a very wide region of sky, it will be possible to deduce the parameters of the original event using special techniques (see text) (we acknowledge A. Helmi and S. White for this image). *Lower* Aitoff projection of most of the known dwarfs in orbit about the Galaxy. For decades, astronomers have argued that these are broadly confined to a plane rather than spherically distributed, an effect that is also seen in the PANDAs survey for the dwarfs in M31 (we acknowledge A. Frebel for this image)

and Gallagher 2004; Tolstoy et al. 2009). The star formation efficiency in dwarfs is very low which means that there have been relatively few star-formation events over cosmic time. The effects of stochasticity therefore plays a relatively big role (Audouze and Silk 1995; Karlsson 2005; Karlsson and Gustafsson 2005). This aspect must be addressed when seeking to understand the evolution of dwarf galaxies, particularly at early times.

Dwarf galaxies show clear evidence that star clusters have formed within them. At the present time, the most metal-poor star clusters are known have an iron abundance just below [Fe/H] $= -2$. One of the globular clusters (Cluster 1) in the Fornax dSph currently holds the record with a metallicity [Fe/H] $= -2.5$ (Letarte et al. 2006). But not all clusters have survived as gravitationally bound objects to the present epoch. Much like the halo, dwarf galaxies have stars with metallicities well below [Fe/H] $= -3$ (Kirby et al. 2008; Starkenburg et al. 2010). The relatively simple environments of the low mass dwarfs raises the prospect of identifying disrupted star clusters at much lower metallicity than has been possible before. For example, Karlsson et al. (2012) have identified through chemical tagging a disrupted star cluster in Sextans with [Fe/H] $= -2.7$. If this result is confirmed by a larger survey sample, this will constitute the most metal poor star cluster ever observed. Cluster reconstruction of this kind provides us with a unique tool to probe the formation and early evolution of the present dwarf galaxy population and to determine the sequence of events that led to the formation of the stellar halo of the Milky Way (Bland-Hawthorn et al. 2010b).

1.7.3.8 Structures in Phase Space

One class of systems that exhibit coherence in velocity space are the open clusters associated with the disk. Here the common space motion of the stars with respect to the Sun is perceived as a convergence of the proper motions to a single point (strictly speaking, minimum volume) on the sky (Boss 1908; see de Zeeuw et al. 1999 for a recent application). More than a dozen such systems have been identified this way. However, these are all young open clusters largely associated with the Gould belt. With sufficiently precise kinematics, it may be possible to identify open clusters that have recently dispersed, particularly if the group is confined to a specific radial zone by resonances in the outer disk. For example, Feltzing and Holmberg (2000) show that the metal-rich ([Fe/H] ≈ 0.2) moving group HR 1614, thought to be 2 Gyr old, can be identified in the Hipparcos data set.

Recently, attention has turned to a diverse set of 'moving groups' that are thought to be associated with the stellar halo and in some instances are clearly fossils associated with accretion events in the distant past. The evidence for these groups dates back to the discovery of the halo itself. Shortly before the publication of the landmark ELS paper, Eggen and Sandage (1959) discovered that the nearby high-velocity star, Groombridge 1830, belongs to a moving group now passing through the Galactic disk.

In a long series of papers, Eggen went on to 'identify' a number of moving groups, some of which appear to encompass the solar neighbourhood, and others that may be associated with the halo. The relevant references are given by Taylor (2000). Various authors have noted that many of the groups are difficult to confirm (Griffin 1998; Taylor 2000). More systematic surveys over the past few decades have identified a number of moving populations associated with the halo (Freeman 1987; Majewski 1993) although the reality of some of these groups is still debated. The reality of these groups is of paramount importance in the context of halo formation. Majewski et al. (1996) suspect that much or all of the halo could exhibit phase-space clumping with data of sufficient quality.

In recent years, the existence of kinematic sub-structure in the galactic halo has become clear. Helmi et al. (1999) identified 88 metal-poor stars within 1 kiloparsec of the Sun from the Hipparcos astrometric catalogue. After deducing accurate 3-D space motions, they found a highly significant group of 8 stars that appear clumped in phase space and confined to a highly inclined orbit. In fact, more than a dozen stellar streams have now been identified in the Galactic halo, the most recent being the Aquarius stream identified in the RAVE stellar survey (Williams et al. 2011); for an updated list, see Helmi (2008).

The most dramatic evidence is surely the highly disrupted Sagittarius dwarf galaxy identified by Ibata et al. (1994, 1995). These authors used multi-object spectroscopy to uncover an elongated stellar stream moving through the plane on the far side of the Galaxy. The Sgr dwarf is a low mass dwarf spheroidal galaxy about 25 kpc from the Sun that is presently being disrupted by the Galactic tidal field. The long axis of the prolate body (axis ratios $\sim$3:1:1) is about 10 kpc, oriented perpendicular to the Galactic plane along $\ell = 6°$ and centered at $b = -15°$. Sgr contains a mix of stellar populations, an extended dark halo (mass $\geq 10^9 M_\odot$) and at least four globular clusters (Ibata et al. 1997). The Sgr stream has since been recovered by several photometric surveys (Vivas et al. 2001; Newberg et al 2002; Ibata et al. 2001c).

N-body simulations have shown that stellar streams are formed when low mass systems are accreted by a large galaxy (e.g. Harding et al. 2001). Streamers remain dynamically cold and identifiable as a kinematic substructure long after they have ceased to be recognizable in star counts against the vast stellar background of the galaxy (Tremaine 1993; Ibata and Lewis 1998; Johnston 1998; Helmi and White 1999).

Within the Galaxy, moving groups can be identified with even limited phase-space information (de Bruijne 1999; de Zeeuw et al. 1999). This also holds for satellites orbiting within the spherical halo, since the debris remains in the plane of motion for at least a few orbits (Lynden-Bell and Lynden-Bell 1995; Johnston et al. 1996). But a satellite experiencing the disk potential no longer conserves its angular momentum and its orbit plane undergoes strong precession (Helmi and White 1999). In Fig. 1.22, we show the sky projection of a satellite 8 Gyr after disruption. These more complex structures are usually highly localized, and therefore easy to recognize in the space of conserved quantities like energy and angular momentum for individual stars.

The evolution in phase space of a disrupting satellite is well behaved as its stars become phase mixed. Its phase space flow obeys Liouville's theorem, i.e. the flow is incompressible. Highly intuitive accounts are given elsewhere (Carlberg 1986; Tremaine 1999; Hernquist and Quinn 1988). It should be possible to recognize partially phase-mixed structures that cover the observed space, although special techniques are needed to find them (e.g. Antoja et al. 2012).

The astrometric mission Gaia will derive 6-dimensional phase space coordinates and spectrophotometric properties for millions of stars within a 20 kiloparsec sphere—the Gaiasphere. The ambitious Gaia mission will obtain distances for up to 90 million stars with better than 5 % accuracy, and measure proper motions with an accuracy approaching *micro* arcsec per year. If hierarchical ΛCDM is correct, there should be thousands of coherent streamers that make up the outer halo, and hundreds of partially phase-mixed structures within the inner halo. A satellite experiencing the disk potential no longer conserves its angular momentum and its orbit plane undergoes strong precession (see Fig. 1.23c, d). In Fig. 1.23a, b Helmi et al. (1999) demonstrate the relative ease with which Gaia will identify substructure within the stellar halo.

1.8 Reconstructing the Past Through Chemical Tagging

1.8.1 Unravelling a Dissipative Process

Most galaxies from about $z \sim 2$ through to the present day appear to be dominated by disks. During the most active period ($z \sim 1$–3), galaxies appear very clumpy and display turbulent kinematics. The early star-forming clumps which are particularly prominent in deep HST images produced stars that must reside within galaxies today (Elmegreen and Elmegreen 2006). Did they break up and form an old puffed-up thick disk (Kroupa 2002) or surviving fossil streams (FBH), did they spiral into the galaxys centre to form a bulge (Ceverino et al. 2010), or did they migrate across the galaxy due to the formation of transient bars or spiral waves (Minchev et al. 2012)? Answers to these questions continue to elude us: the most advanced simulations are unable to resolve individual star clusters and observations at many cosmic epochs are difficult to interpret. But the HERMES survey opens up a new window to the fate of these ancient star clusters by providing detailed chemical information for vast numbers of individual stars. So how did the Galactic disk form and can the sequence of events ever be unravelled from the vast stellar inventory? This will require that some of the residual inhomogeneities from prehistory escaped the dissipative process at an early stage. Fossil hunting to date has concentrated mostly on the stellar halo, but a key source of information will be the thick disk. This is believed to be a 'snap frozen' relic which formed during or shortly after the last major epoch of dissipation, or it may have formed from infalling systems early in the life of the Galaxy.

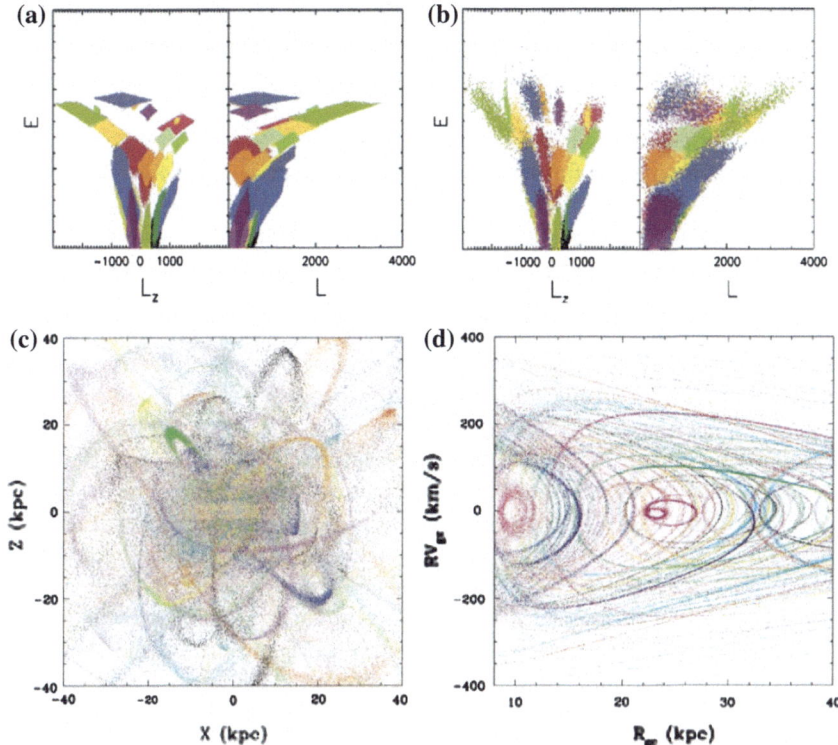

Fig. 1.23 *Upper* **a** Initial distribution of particles for 33 systems falling into the Galactic halo is integral of motion space. **b** The final distribution of particles in (**a**) after 12 Gyr after convolution with the errors expected for Gaia (we acknowledge A. Helmi for these images). A simulation of the baryon halo built up through accretion of 100 satellite galaxies. **c** The different *colors* show the disrupted remnants of individual satellites. **d** This is the same simulation shown in a different coordinate frame, i.e. the orbit radius (*horizontal*) plotted against the observed radial velocity (*vertical*) of the star (*lower*) (we acknowledge P. Harding and H. Morrison for these images)

In an earlier review (Freeman and Bland-Hawthorn 2002), we raised the prospect of 'chemical tagging' as a means to explore the early history of the halo and the thick disk by looking for discrete substructures in chemical C-space. We proposed for the first time that this will require high signal to noise, echelle spectroscopy of up to a million stars. This idea is now central to major new and proposed stellar surveys, in particular, the HERMES project (Freeman and Bland-Hawthorn 2008) and the 4MOST project (de Jong et al. 2012).

Our program has a short-term and a long-term goal. The short-term goal is to quantify the size and structure of the multi-dimensional chemical abundance space (C-space) for all major components of the Galaxy. We seek to establish how many axes in C-space are decoupled and have large intrinsic dispersions. A critical test of chemical tagging in the short term is that stellar streams in the halo, identified from detailed phase space information, are highly localized in $\mathcal{C}$-space, or are confined to chemical tracks. These trajectories presuppose that stars form in a closed box through progressive enrichments of the gas, leading to stars dispersed along a narrow track in a complex chemical space.

The long-term goal is to identify unique chemical signatures in the thick disk, originating from different formation sites, for star clusters which have long since dispersed. This will require precise chemical abundances for heavy elements such that a star can be localized to a discrete point in $\mathcal{C}$-space. If the star clusters originally formed outside the Galaxy in a bound infalling system, the stellar abundances may fall along a chemical track, rather than a discrete point in $\mathcal{C}$-space.

The ultimate goal of cosmology, both near and far, must be to explain how the Universe has arrived at its present state. It is plausible—although difficult to accept—that nature provides fundamental limits of knowledge, in particular, epochs where the sequence of events are scrambled. Our 'intuition' is that any phase dominated by relaxation or dissipation probably removes more information than it retains. But could some of the residual inhomogeneities from prehistory have escaped the dissipative process at an early stage? We may not know the answer to this question with absolute certainty for many years. In the absence of certainty, we consider what might be the likely traces of a bygone era prior to the main epoch of baryon dissipation.

In order to follow the sequence of events involved in dissipation, we propose that the critical components which need to be re-assembled are the individual star clusters which formed at each stage. Since most stars are born in dense clusters, the formation and evolution of galaxies today must involve millions of discrete cluster events throughout their history. We would like to establish the evolving mass function of star clusters, their chemical composition, formation and survival rate as a function of cosmic time. Galaxy-wide enrichment from the fall-out of nuclear winds or mergers would be evident in the fossil record of reconstructed star clusters, assuming these provide an unbiased sampling of cosmic time regardless of the star formation history.

1.8.2 How Many Star Clusters?

If we consider our volume of interest to be the Solar torus, defined as the annulus 2 kpc thick in radius centered on the Solar circle at $R_o \approx 8$ kpc, then the cluster birth rate within this region is

$$\frac{d^2\mathcal{N}}{dt\, d\ln(M_*/M_\odot)} = 3 \times 10^3 \dot{M}_0 \left(\frac{M_*}{10^4\ M_\odot}\right)^{-1} \text{Gyr}^{-1}, \qquad (1.37)$$

where $\dot{M}_0 = \dot{M}/(M_\odot\ \text{yr}^{-1})$ and $\dot{M}$ is the total star formation rate in the Galaxy and the upper and lower cluster mass limits are $(M_{*,\min}, M_{*,\max}) = (50\ M_\odot, 2 \times 10^5\ M_\odot)$. Thus, for example, in the last Gyr we expect $\sim 10^4$ clusters whose masses are within a factor of three of 10^4 $M_\odot$ to have formed within the Solar torus if the star formation rate over this time has been close to its present value of ~ 3 $M_\odot$ yr^{-1} (McKee and Williams 1997). This result is quite insensitive to the assumed limits on the cluster mass function or the extent of the star-forming disk. Changing $M_{*,\mathrm{min}}$ or $M_{*,\mathrm{max}}$ by a factor of 10 alters the estimated cluster birth rate by only $\sim 20\%$.

All but a small fraction of clusters formed are unbound, and the stars born within them drift off to become part of the field star population. By comparing catalogs of open and embedded clusters, Lada and Lada (2003) estimate that less than 10% of clusters survive more than 10 Myr, and that only 4–7% survive for 100 Myr.[12] Moreover, analysis of extragalactic cluster populations (albeit with masses significantly larger than the typical Milky Way cluster) suggests that the survival fraction is nearly independent of mass until two-body relaxation becomes the dominant destruction process at times $\gg 100$ Myr after cluster formation (Fall et al. 2005, 2009; Fall 2006; Whitmore et al. 2007). During this first 100 Myr, roughly 80–90% of clusters are disrupted independent of mass, while two-body relaxation at longer times preferentially destroys low mass clusters. Since we do not strive for better than 10% accuracy in our prediction of the disrupted cluster birth rate, we will not introduce the added complication of a mass- or time-dependent disruption rate. Instead we simply assume that a fixed 95% of all clusters that are born are disrupted, independent of mass and time. This will slightly overestimate the number of disrupted clusters that formed less than 100 Myr, but for a relatively quiescent star formation history such as the Galaxy's, these clusters constitute a negligible fraction of the total population. Thus, if we target stars older than 100 Myr, the birth rate of now-disrupted clusters is given by Eq. 1.37 to within 10%.

1.8.3 Cluster Chemistry

Since most stars are not born in clusters that remain bound, and thus do not end up in open clusters more than 10 Myr later, the clusters observed by De Silva and collaborators presumably represent a biased sample. They are either the innermost remnants of clusters that lost most of their members and produced only a small bound core (e.g. Kroupa 2001), or they are clusters that remained bound because they had star formation efficiencies significantly higher than the typical 10–30% (Lada and Lada 2003). We would therefore like to define some rough theoretical expectations about the chemical homogeneity of the disrupted cluster population.

Mixing of heavy elements in protoclusters is governed by several time scales. The dominant mode of chemical mixing in molecular clouds is turbulent diffusion (Murray and Lin 1990). The time scale associated with this process is $t_{\mathrm{diff}} \sim H^2/K$ (Xie et al. 1995), where $H = (H_i^{-1} - H_{\mathrm{H}_2}^{-1})^{-1}$ is a composite between the scale heights of the H_2 molecules, $H_{\mathrm{H}_2} = n_{\mathrm{H}_2}/|\nabla n_{\mathrm{H}_2}|$, and of the species i whose diffusion

---

[12] We point out, however, that the distribution must have a long tail because a few open clusters have ages $\gg 1$ Gyr (Friel 1995).

time we are computing, $H_i = n_i/|\nabla n_i|$. Here n_{H_2} and n_i are the number densities of the two species. The quantity K is the diffusion coefficient, which is of order σL, where σ is the turbulent velocity dispersion and L is the correlation length of the turbulence. Observations indicate that the turbulent motions in molecular clouds are correlated on the scale of the entire cloud (Heyer and Brunt 2004), so L is roughly the size of a cloud. If we consider a species that is distributed with an inhomogeneity comparable to or greater than that of the molecular hydrogen, $H_i \lesssim H_{H_2}$, and a composition gradient on the scale of the entire cloud, $H_i \sim L$, then $t_{\text{diff}} \sim L/\sigma = t_{\text{cr}}$, where t_{cr} is the cloud crossing time. Thus, the time required for turbulence to smooth out a large-scale chemical gradient in a protocluster gas cloud is of order the cloud crossing time; smaller scale gradients disappear more quickly, with the time required to smooth out the gradient varying as the square of its characteristic size scale.

Star formation in a molecular cloud also appears to begin in no more than a crossing time after its formation (Tamburro et al. 2008), so there is no clear separation in time scales between chemical homogenization and the onset of star formation. This suggests that molecular clouds cannot be too far from homogeneous when they are assembled, but that the homogenization need not be total, since small residual chemical gradients will be wiped out as star formation proceeds. Thereafter, changes in element abundances, and inhomogeneities in the resulting stars, may be produced by supernovae within the cloud. In principle, even a single supernova is sufficient to change the chemical signature measurably. For example, De Silva et al. (2007b) find that scatter in Fe abundance in the HR 1614 moving group is roughly 0.01 dex. The Fe content of the Sun is $\approx 10^{-3}$ M$_\odot$ so the measured De Silva et al. (2007b) scatter corresponds to a scatter in mass fraction of $\sim 10^{-5}$. A supernova from a 15 M$_\odot$ star produces $\sim 10^{-1}$ M$_\odot$ of Fe (Woosley and Weaver 1995), so if that supernova occurs inside a star-forming cloud that continues forming stars thereafter, the change in iron abundance will be measurable even if the supernova ejecta are mixed with 10^4 M$_\odot$ of pre-supernova gas. The time required for a very massive star to evolve from formation to explosion defines the supernova time scale, $t_{\text{SN}} \approx 3$ Myr. Thus, while the "initial conditions" for cluster formation are well-mixed, we only expect star clusters to be homogeneous if they are assembled on time scales shorter than t_{SN}.

1.8.4 Chemical Homogeneity

To determine under what conditions this requirement is satisfied, we must compare t_{SN} to the cluster formation timescale t_{form}. There is considerable debate about whether star cluster formation occurs in a single crossing time (Elmegreen et al. 2000; Hartmann et al. 2001; Elmegreen 2007; Pflamm-Altenburg and Kroupa 2007) or 3–4 crossing times (Tan et al. 2006; Huff and Stahler 2006; Krumholz and Tan 2007; Jeffries 2007; Nakamura and Li 2007; Matzner 2007). Since the latter is more restrictive, we adopt it. This leaves the problem of estimating the crossing time. To do so, we use a convenient form that expresses it in terms of a cloud's mass M and column density Σ (Tan et al. 2006)

1 Near Field Cosmology

$$t_{cr} = \frac{0.95}{\sqrt{\alpha_{vir} G}} \left(\frac{M}{\Sigma^3}\right)^{1/4}, \quad (1.38)$$

where α_{vir} is the cloud's virial ratio (Bertoldi and McKee 1992), essentially its ratio of kinetic to gravitational potential energy up to geometric factors. Observed star-forming clouds all have $\alpha_{vir} \approx 1$–2 (McKee and Ostriker 2007) so we take $\alpha_{vir} = 1.5$ as typical, but Σ is somewhat more difficult to estimate. Partly-embedded clusters detected in nearby star-forming regions using 2MASS and *Spitzer*/IRAC have surface densities of a few $0.01\,\mathrm{g\,cm^{-2}}$ (Adams et al. 2006; Allen et al. 2007), but because these samples are selected in the near-infrared they are strongly biased against more deeply-embedded and thus presumably younger clusters. Indeed, these samples are all dominated by stellar rather than gas mass, and thus have likely undergone significant expansion. Observations of gas-dominated cluster-forming regions show column densities of ~ 0.1–$3\,\mathrm{g\,cm^{-2}}$ (Plume et al. 1997; Mueller et al. 2002; Shirley et al. 2003; Faúndez et al. 2004; Fontani et al. 2005); the data are summarized in Fig. 1 of Fall et al. (2010). We therefore adopt $\Sigma = 0.3\,\mathrm{g\,cm^{-2}}$ as a fiducial value. This is toward the low end of the observed range, giving a high (and thus conservative) estimate of the crossing time. Together with an assumed formation time $t_{form} = 4 t_{cr}$ gives

$$t_{form} \approx 3.0 \left(\frac{\epsilon}{0.2}\right)^{-1/4} \left(\frac{M_*}{10^4\,M_\odot}\right)^{1/4} \mathrm{Myr}, \quad (1.39)$$

where $\epsilon = M_*/M$ is the star formation efficiency, i.e. the fraction of the initial cloud mass that is converted into stars. Observations suggest this is typically 10–30 % (Lada and Lada 2003).

Since t_{form} is the time required to form all the stars, we conclude that for clusters $10^4\,M_\odot$ or smaller, almost all of the stars will form before the first supernova occurs within the cluster, so the gas will be chemically homogeneous. This is probably somewhat conservative, since only the most massive stars go supernova in 3 Myr, and such a massive star is unlikely to be found in a cluster containing only $10^4\,M_\odot$ of stars. Moreover, t_{SN} is the time delay between formation of a massive star and the resulting supernova, and a massive star is not necessarily the first to form in a given cluster. If we assume that massive star formation occurs on average halfway through the cluster formation process, then we only require $t_{form} < 2 t_{SN}$ to ensure there is no self-pollution, which corresponds to a cluster mass $M_* = 1.6 \times 10^5\,M_\odot$. We therefore conclude[13] that clusters smaller than $10^4\,M_\odot$ should essentially all be chemically homogeneous, while a significant fraction of clusters up to $\sim 10^5\,M_\odot$ will also be homogeneous. This is to be expected if the surface density of their more massive cluster-forming regions are higher than for the lower-mass clusters.

---

[13] It is interesting to note what happens in the case of globular clusters. These form at a density of $\Sigma \sim 3\,\mathrm{g\,cm^{-2}}$ which leads to a sixfold drop in the dynamical time and a uniformity mass limit of $\gtrsim 10^7\,M_\odot$. Apart from a few light elements, globular clusters are chemically uniform, with few exceptions (e.g. ωCen; Gratton et al. 2004).

1.8.5 Unique Chemical Signatures

The analysis in Sect. 3.2 allows us to derive an order-of-magnitude estimate of the number of unique chemical signatures that are needed to identify stars with their parent clusters. Within the Solar toroid, the total number of tags required comes from integrating Eq. 1.37 over the mass limits, viz.

$$\frac{d\mathcal{N}}{dt} = \int_{\ln M_{*,min}}^{\ln M_{*,max}} \frac{d^2\mathcal{N}}{dM\,dt}\, d(\ln M_*) = 6 \times 10^5 \text{ Gyr}^{-1} \quad (1.40)$$

Thus, assuming that we could select stars with an age similar to the Sun ($t_o \approx$ 4.57±0.5 Gyr), we would need of order 6×10^5 unique chemical signatures. Bland-Hawthorn and Freeman (2004) suggested going after dozens of independent elements that could be measured accurately enough to distinguish a "weak" line strength from a "strong" one. We can express this as $N_c = N_L^m$ where N_L is the number of distinct line strengths and m is the number of independent elements or element groups. After detailed consideration of what is possible in the context of a fibre-fed high-resolution spectrograph, the HERMES project has settled on 8 independent groups with 5 measurably distinct abundance levels, i.e. $N_c \sim 5^8 \approx 4 \times 10^5$ unique chemical signatures. This is the target specification for the million-star Galactic Archaeology survey at the Anglo-Australian Telescope (AAT) due to commence in 2012 (Freeman 2008). The independent groups (e.g. alpha elements) include one or more elements that show some variation with respect to [Fe/H].

The volume of the potential chemical space ($\mathcal{C}$-space; Bland-Hawthorn and Freeman 2004) accessible to the HERMES survey is barely adequate for our proposed study. But further constraints will come from stellar ages at least for some stars. An important population is the subgiants for which ages can be derived from differential spectroscopy. In particular, a differential precision in surface gravity of $\Delta \log g \lesssim 0.1$ can be used to determine ages to ~ 1 Gyr from isochone fitting in the surface gravity versus effective temperature plane. However the analysis does emphasize the potentially vast amount of information that awaits us in systematic high-resolution surveys of millions of stars.

1.8.6 Primary Requirements of Chemical Tagging

But how are we to reconstruct star clusters which have long since dispersed? It will be necessary to tag individual stars to their parent cloud through unique chemical signatures shared by these stars, assuming these exist. We now discuss our basic strategy for 'chemical tagging' (Bland-Hawthorn and Freeman 2004; Bland-Hawthorn et al. 2010a, b). High resolution spectroscopy at high signal-to-noise ratio of many stellar types reveals an extraordinarily complex pattern of spectral lines. The spectral lines carry key information on element abundances that make up the stellar atmosphere.

Many of these elements cannot arise through normal stellar evolution, and therefore must reflect conditions in the progenitor cloud at the time of its formation.

Our long-term goal is to chemically tag stars into coeval groups, i.e. to identify individual members of star clusters which have long since dispersed. For unique chemical signatures to exist, there are several key requirements:

1. Most stars must be born in dense star clusters.
2. Most dense star clusters must be chemically uniform in key elements.
3. Key chemical elements must reflect the cloud composition of the progenitor cloud.
4. Key chemical elements must not be rigidly coupled (i.e. vary in lock step), and there must be sufficient abundance dispersion in key elements to allow for unique groups (reflecting unique sites of formation) to be readily identified.
5. There must be accessible spectral window(s) which contain the necessary information on key elements for chemical tagging.

We discuss each of these requirements, or conditions, in turn. Conditions 1, 2, 3 and 5 appear to be supported by observation. Condition 4 is the most uncertain largely because stellar abundance surveys to date target either too few stars or too few chemical elements.

Condition 1. Most stars are born within rich clusters of many hundreds to many thousands of stars (Clarke et al. 2000; Carpenter 2000). This essential fact is supported by many studies from optical, infrared, millimeter and radio surveys.

Condition 2. The widely held view that open clusters are chemically uniform can be traced to early work on Hyades by Conti et al. 1965. But, until recently, very little was known about detailed heavy element abundance work on open clusters with rigorous membership established by reliable astrometry or accurate radial velocities (<0.5 km/s), so as to minimize 'pollution' from stars not associated with the cluster.

There have been many recent studies of open clusters which have established that open clusters are indeed highly uniform. In a detailed analysis of 55 F–K dwarfs in Hyades, De Silva et al. (2007a) find that the abundance variations in Si, Ti, Na, Mg, Ca and Zn with respect to Fe, and in [Fe/H], are within the measurement errors (<0.04 dex). To date, all open clusters have been shown to be extremely chemically uniform within the errors.

Are we to expect chemical uniformity among stars in an open cluster from a theoretical perspective? This was addressed for the first time by Bland-Hawthorn et al. (2010a) who derive a 'homogeneity equation' for a star cluster of a given mass which depends only on the surface density of the cluster. Their basic condition is that the gas cloud collapses before the first intercloud supernova explodes, thereby polluting the cloud. This sets a minimum collapse time of about 3 Myr. These authors find that most open clusters up to a mass of about 10^5 $M_\odot$ are likely to be homogeneous, as will be most *globular* clusters up to a mass of about 10^7 $M_\odot$, assuming the high mass extremes even exist. (Globular clusters are highly chemically uniform in [Fe/H] but show marked anticorrelations between O, Mg and Al, Na.)

McKee and Tan (2002) propose that high-mass stars form in the cores of strongly self-gravitating and turbulent gas clouds. Two possible routes to chemical uniformity is that all stars form at the same instance from a chemically uniform cloud, or that the low mass stars form outside of the core shortly after the supernovae have uniformly enriched the cloud. The precise sequence of events which give rise to open clusters is a topic of great interest and heated debate in contemporary astrophysics (e.g. Stahler et al. 2000).

Condition 3. It is generally believed that r-process elements (e.g. Sm, Eu, Gd, Tb, Dy, Ho) cannot be formed during quiescent stellar evolution. While some doubts remain, the most likely site for the r-process appears to be Type II supernova (SN II), as originally suggested by Burbidge et al. (1957). Therefore, r-process elements measured from stellar atmospheres reflect conditions in the progenitor cloud. The same is believed to be true for most of the α elements since these are produced in the hydrostatic burning phase of the pre-supernova star.

In contrast, the s-process elements (e.g. Sr, Zr, Ba, Ce, La, Pb) are thought to arise from the He-burning phase of intermediate to low mass (AGB) stars ($M < 10\,M_\odot$), although at the lowest metallicities, trace amounts are likely to arise from high mass stars (Burris et al. 2000; Rauscher et al. 2002).

Condition 4. During the past four decades, evidence has gradually accumulated for a large dispersion in metal abundances [X/Fe] (particularly n-capture elements) in low to intermediate metallicity stars relative to solar abundances. Elements like Sr, Ba and Eu show 1–2 dex dispersion, although [α/Fe] dispersions are typically an order of magnitude smaller. Ting et al. (2012) find evidence that certain element families are decoupled and provide useful axes in C-space (see also Mitschang et al. 2013).

These observations have been used to constrain detailed supernova models, which in turn show how different yields arise as a function of progenitor mass, progenitor metallicity, mass cut (what gets ejected compared to what falls back towards the compact central object), and detonation details. These models help to explain the smaller dispersions in α elements: the α yields are not dependent on the mass cut or details of the fallback/explosion mechanism, which leads to a smaller dispersion at low metallicity.

The large scatter of [n-capture/Fe] in metal poor stars (Wallerstein et al. 1997; Sneden et al. 2000) means that n-capture element abundances in ultra-metal poor stars are products of one or very few prior nucleosynthesis events that occurred in the very early, poorly mixed galactic halo, a theme that has been developed by many authors (Gilroy et al. 1988; Audouze and Silk 1995; Bland-Hawthorn et al. 2010b).

The oversimplified schematic in Fig. 1.24 is illustrative of the expected abundance variation in successive generations of low mass stars in a closed box from gas enriched by successive generations of supernovae. The upper envelope is determined by enrichment from low mass supernovae, the lower envelope from high mass supernovae. From left to right, the black squares indicate the number of enrichment events, i.e. 1, 10, 100, 1000 events. Just ten enrichments from high mass supernovae are sufficient to enrich a cloud to [Fe/H] $= -2$. Note the rapid convergence in this simple model above [Fe/H] $= -1$.

In the simulation, we consider only core collapse supernovae; subsequent enrichment by Type Ia supernovae causes the converging stream to dip down above

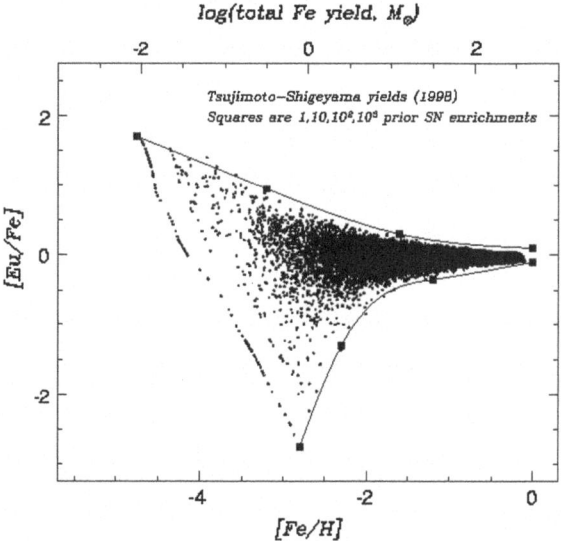

Fig. 1.24 An illustration of how successive chemical enrichments by massive supernovae can lead to low mass stars with a wide scatter in chemical abundances which converge to a universal value. The model assumes a constant star formation rate in a closed box. The stars are formed according to a Salpeter mass function and the yields are taken from models by Tsujimoto and Shigeyama (1998). Only 1 % of the stars produced in the simulation are shown after the initial burst; we show all the stars at t = 0. The contribution of Type Ia supernovae is not included for clarity

[Fe/H] = −1. It is not clear at what [Fe/H] the r-process elements become swamped by the ubiquitous Fe-group and s-process elements. Travaglio et al. (1999) suggest that the s-process does not become significant until [Fe/H] ∼ −1 because of the need for pre-existing seed nuclei, although Pagel and Tautvaisiene (1977) argue for some s-process production at [Fe/H] ∼ −2.5. In any event, Fig. 1.26 makes clear that, at sufficiently high resolution, both the Sun and Arcturus reveal a rich network of heavy elements, many of which arise from the r-process.

An important point to appreciate is that the substantially smaller scatter in [α/Fe] compared to heavy element scatter (Carretta et al. 2002) presumably argues that the scatter is unlikely to arise from chemical differentiation, i.e. where metal yields from a given supernova are distributed in different amounts, compared to Fe, to the surrounding ISM. In other words, the mixing is sufficiently uniform such that any given cell receives the same proportion of Eu/Fe from the supernova. This puts limits on the degree to which asymmetric supernovae must differentially enrich the surrounding gas (e.g. Maeda et al. 2002).

Condition 5. Any practical experiment involving chemical signatures in millions of stars will require a wide-field, multi-fibre spectrograph operating at echelle resolutions. Since rayleigh scattering renders optical fibres rather lossy below 4000 Å, it is important to establish that there exist contiguous spectral windows which contain abundance information on a large number of chemical elements. Since it is necessary

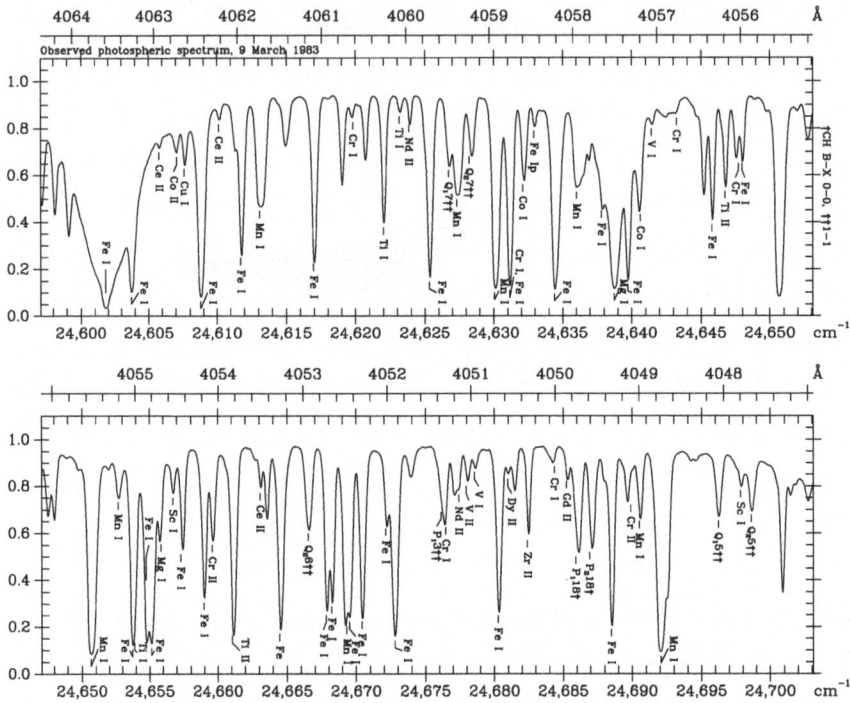

Fig. 1.25 17 Å spectral window (4047–4064 Å) of the Sun obtained with the Fourier Transform Spectrograph at the National Solar Observatory revealing detailed chemistry on Fe, Cr, Ti, V, Co, Mg, Mn, Nd, Cu, Ce, Sc, Gd, Zr, Dy. The *lower values* are wavenumbers (cm^{-1}). The data are taken from http://www.nso.edu/

to maximize the number of object fibres at the detector, this information will need to exist within a single narrow band ($\sim \lambda/35$) echelle order. The optimal parameters for a 'galaxy genesis' machine are determined below.

Do there exist narrow band spectral windows, with chemical information on dozens of elements, which could be utilized by a fibre-fed echelle spectrograph? Until recently, this has not been possible to answer objectively. Figure 1.25 shows such a window that was revealed by this study. An optimal window needs information on more than one ionization state for as many elements as possible in order to accurately determine the stellar surface gravity. Various ionic transitions in combination must also provide accurate information on the stars luminosity and surface temperature. Some regions of the spectrum are dominated by telluric features; in cool stars, molecular bands complicate the spectrum.

We concentrate our analysis on digitized atlases of the Sun ([Fe/H] = 0.0) and for the luminous giant Arcturus ([Fe/H] = -0.56). Our solar line list derives from data supplied by P. Hall (NOAO) transcribed from the Moore et al. (1996) solar spectrum (2935–8770 Å). The atlas includes information on 25,000 absorption lines covering 63 elements, for which 28 are detected in two ionization states. The Arcturus line list comes from the digitized atlas (3570–7405 Å) of Wallace et al. (1998). The line list includes 7700 atomic transition for 42 elements, for which 18 are detected in two ionization states. Both lists are dominated by Fe I (45%) and light α elements (30%).

Figure 1.26 shows the number of unique ions detectable as a function of wavelength for the Sun and for Arcturus. At each wavelength, the accessible window is $\Delta\lambda$ where $\lambda/\Delta\lambda = 35$, such that at $\lambda = 3500$ Å, the unique ion count is made over a 100 Å. The linear dependence of the window on wavelength is expected for spectral coverage in a single echelle order limited by the detector size. Note the rising count towards with decreasing wavelength down to 4000 Å, both in terms of the number of unique ions, and elements with two ionization states.

In the lower panels of Fig. 1.26, we demonstrate that this rise is largely due to a rising fraction of neutron capture elements; the rise in iron peak elements is not as dramatic. It is noteworthy that we obtain similar results for a solar metallicity main sequence star and a metal weak giant. In Fig. 1.27, we show the effect of resolving power on the fraction of lines which can be resolved as a function of wavelength. The optimal window at 4100 Å requires a spectral resolving power of $R = 40,000$ in order to resolve 80% of the lines. As the resolving power decreases, the loss of information on heavy elements is particularly dramatic.

An analysis of this kind led to the development of the first galactic archaeology machine—HERMES. This instrument (www.aao.gov.au/HERMES; Freeman and Bland-Hawthorn 2008) uses the existing 2dF positioner at the Anglo-Australian Telescope is used to obtain high-resolution spectroscopy on sufficiently bright stars ($g < 14$) over the two-degree field. The 400 fibres will allow up to a million stars to be observed over five years starting in 2013. The HERMES instrument exploits four distinct spectral windows: (blue) 4718–4903 Å; (green) 5649–5873 Å; (red1) 6481–6739 Å; (red2) 7590–7890 Å. Up to 30 distinct chemical elements have been identified across these windows, where the actual number depends on spectral type and luminosity class. We estimate that the HERMES data set will contain *at least* 8 distinct axes in $\mathcal{C}$-space.

1.8.7 Candidates for Chemical Tagging

Chemical tagging is not possible for all stars. In hot stars and young star clusters, our ability to measure abundances is reduced by the stellar rotation and lack of transitions for many ions in the optical. In very cool stars, very little light emerges in the optical or near-IR due to complex molecular and dust opacities. The ideal candidates are evolved FGK stars that are intrinsically bright, FGK subgiants and dwarfs. These populations probably account for 10% of all stars.

Giants can be observed at $R = 40,000$ over the full Gaiasphere, i.e. the 20 kpc diameter sphere centred on the Sun which will be surveyed in great detail by the GAIA satellite. Dwarf stars will only be observable within 1–2 kpc. While this is only a

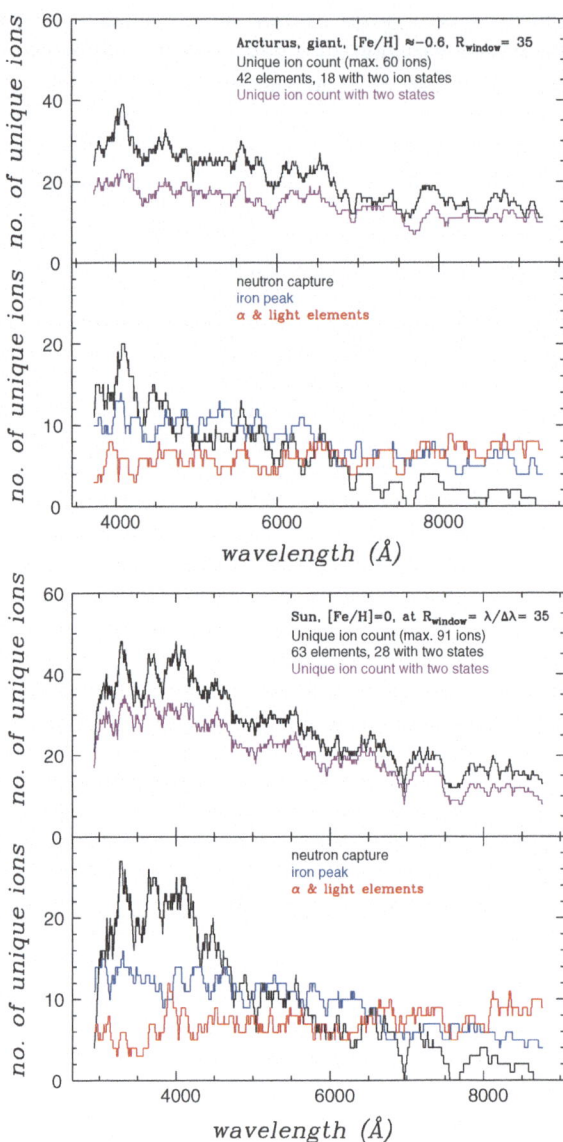

Fig. 1.26 A demonstration of where most of the chemical information lies in a highly resolved spectrum of Arcturus (*upper*) and the Sun (*lower*). The *black lines* in the top plots give the total ion count in a spectral window $\Delta\lambda = \lambda/35$. The *magenta lines* show what fraction of this count is made up of elements with two ionization states. The *bottom panels* of the plots separate these into element types. Note the rapid rise of heavy elements to the blue in both stars; note also that the solar spectral information extends to bluer wavelengths

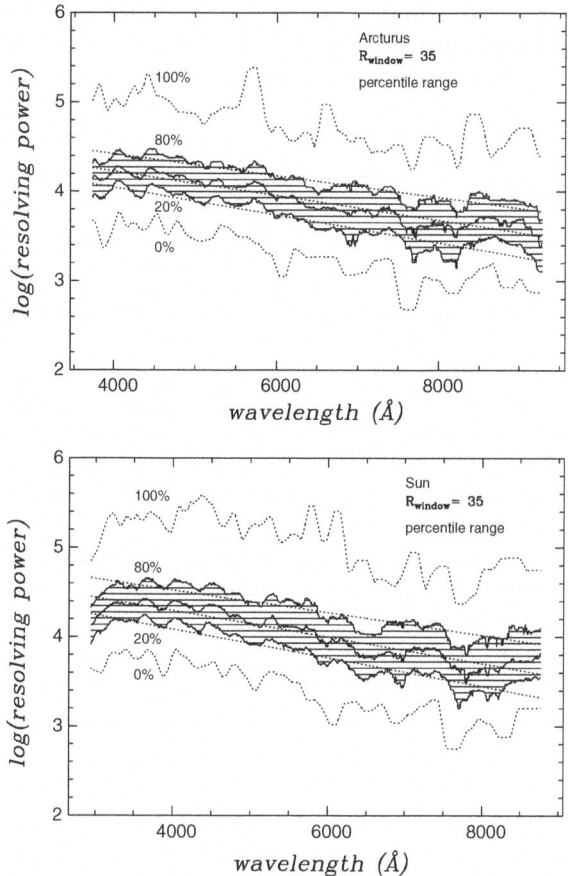

Fig. 1.27 The effect of resolving power on the fraction of spectral lines which can be resolved as a function of wavelength. In order to resolve 80% of lines at 4000 Å, we need R = 40,000; this falls to R = 10,000 at 8000 Å. Once again, both stars reveal the same general trends; note also that the solar spectral information extends to *bluer* wavelengths

small fraction of the available volume, the local volume may include a representative sample of all old disk stars, regardless of their point of origin.

For many stars, certain spectral regions are dominated by thermally broadened profiles, particularly for light elements. In general, heavier atoms show narrow profiles appropriate to their mass, although odd numbered atomic isotopes are susceptible to hyperfine splitting (due to the non-zero nuclear magnetic moment) which produces a broadened asymmetric line profile.

Giants have deep, low density atmospheres that produce strong low-ionization absorption lines compared to higher gravity atmospheres. Even in the presence of significant line blending, with sufficient signal, it should be possible to derive abundance information by comparing the fine structure information with accurate stellar synthesis models. Detailed abundances of large numbers of F and G subgiants would be particularly useful, if it becomes possible to make such studies, because direct relative ages can be derived for these stars from their observed luminosities.

1.8.8 Short-Term Goal: Size and Structure in a Multi-Dimensional C-Space

An intriguing prospect is that reconstructed star clusters can be placed into an evolutionary sequence, i.e. a family tree, based on their chemical signatures. Let us suppose that a star cluster has accurate chemical abundances determined for a large number n of elements (including isotopes). This gives it a unique location in an n-dimensional space compared to m other star clusters within that space. We write the chemical abundance space as $\mathcal{C}(\text{Fe/H}, X_1/\text{Fe}, X_2/\text{Fe}, \ldots)$ where $X_1, X_2 \ldots$ are the independent chemical elements that define the space (i.e. elements whose abundances are not rigidly coupled to other elements).

Our simple picture assumes that a cloud forms with a unique chemical signature, or that shortly after the cloud collapses, one or two massive SN II enrich the cloud with unique yields which add to the existing chemical signature. The low-mass population forms with this unique chemical signature. If the star-formation efficiency is high ($\geq 30\,\%$), the star group stays bound although the remaining gas is blown away. If the star-formation efficiency is low, the star cluster disperses along with the gas. In a closed box model, the dispersed gas reforms a cloud at a later stage.

In the closed box model, each successive generation of supernovae produce stellar populations with progressive enrichments. These will lie along a trajectory in $\mathcal{C}$-space (see Fig. 8, Karlsson and Gustafsson 2001). The overall distribution of the trajectories will be affected by fundamental processes like the star formation efficiency, the star formation timescale, the mixing efficiency, the mixing timescale, and the satellite galaxy infall rate.

A critical test of chemical tagging is that stellar streams in the halo, identified from detailed phase space information, are highly localized in $\mathcal{C}$-space, or are confined to chemical tracks—this is a key short term goal. There may already be evidence for accreted halo stars from their distinct chemical signatures (e.g. suppressed Mg) since those with estimated orbital parameters are found to have large apogalacticon distances (Carretta et al. 2002). This may suggest that these stars originated in lower mass stellar systems with very different chemical histories from that of the Galaxy. Several moving groups have now been identified with unique chemical signatures: these include U Ma (Castro et al. 1999), HR 1614 (De Silva et al. 2007b), Sgr (Chou et al. 2010), Wolf 630 (Bubar and King 2010) the Aquarius Stream (Wylie de Boer et al. 2012) and the Argus association (De Silva et al. 2013).

As we approach solar levels of metallicity in [Fe/H], the vast number of trajectories will tend to converge. By [Fe/H] ~ -2.5, AGB stars will have substantially raised the s-process element abundances; by [Fe/H] ~ -1, Type Ia supernovae will have raised the Fe-group abundances. Star clusters that appear to originate at the same location in this C-space may simply reflect a common formation site, i.e. the resolution limit we can expect to achieve in configuration space.

Even with a well established family tree based on chemical trajectories in the chemical C-space, this information may not give a clear indication of the original location within the protocloud or Galactic component. This will come in the future from realistic baryon dissipation models. Forward evolution of any proposed model must be able to produce the observed chemical tree. The C-space will provide a vast amount of information on chemical evolution history. It should be possible to detect the evolution of the cluster mass function with cosmic time (Kroupa 2002), the epoch of a starburst phase and/or associated mass ejection of metals to the halo (Renzini 2002), and/or satellite infall (Noguchi 1998). The chemical tracks could conceivably be punctuated by discontinuities due to dramatic events like galactic infall or large-scale winds (Chiappini et al. 1999).

Early estimates of the size and dimensionality of C-space have been made by Ting et al. (2012); Mitschang et al. (2013). Roughly speaking, if we are able to measure 5 distinct abundance levels across 8 dimensions (see Sect. 1.8.6), then in principle we can measure 400,000 unique chemical signatures, assuming the space is uniformly filled.

As we go back in time to the formation of the disk, we approach the chemical state laid down by Pop. III stars (Karlsson et al. 2012). These stars presumably had a top-heavy IMF and therefore no remnants are expected to have survived to the present day. To date, no distinctive signature of Pop. III has been clearly identified except that a large fraction of the most metal poor stars have now been shown to have a strong excess in [C/Fe] (Norris et al. 2007). The rarity of stars below [Fe/H] ~ -5 suggests that the protocloud was initially enriched by the first generation of stars (Argast et al. 2000) or maybe that stars moving through the ISM today have a minimum threshold metallicity due to Bondi-Hoyle accretion. If one could unravel the abundances of heavy elements at the time of disk formation, this would greatly improve the precision of nucleo-cosmochronology . Important information is beginning to emerge from echelle observations of damped Lyα systems at high redshift (Pettini et al. 2003; Cooke et al. 2011).

1.8.9 Long-Term Goal: Reconstructing Ancient Star Groups from Unique Chemical Signatures

The abundance dispersion in α and heavy elements provides a route forward for tagging groups of stars to common sites of formation. With sufficiently detailed spectral line information, it is feasible that the 'chemical tagging' will allow temporal sequencing of a large fraction of stars in a manner analogous to building a family tree through DNA sequencing. We provide a worked example below.

Consider the (extraordinary) possibility that we could put many coeval star groups back together over the entire age of the Galaxy. This would provide an accurate age for the star groups either through the color-magnitude diagram, or through association with those stars within each group that have [n-capture/Fe] $\gg$ 0, and can therefore be radioactively dated. This would provide key information on the chemical evolution history for each of the main components of the Galaxy.

There is no known age-metallicity relation that operates over a useful dynamic range in age and/or metallicity. (This effect is only seen in a small subset of hot metal-rich stars). Such a relation would require the metals to be well mixed over large volumes of the ISM. For the foreseeable future, it seems that only a small fraction of stars can be dated directly (Freeman and Bland-Hawthorn 2002).

Ideally, we would like to tag a large sample of representative stars with a precise time and a precise site of formation. Can we identify the formation site? The kinematic signatures will identify which component of the Galaxy the reconstructed star group belongs, but not specifically where in the Galactic component (e.g. radius) the star group came into existence. For stars in the thin disk and bulge, the stellar kinematics will have been much affected by the bar and spiral waves; it will no longer be possible to estimate their birthplace from their kinematics. Our expectation is that the derived family tree will severely restrict the possible scenarios involved in the dissipation process. In this respect, a sufficiently detailed model may be able to locate each star group within the simulated time sequence.

Our ability to detect structure in $\mathcal{C}$-space depends on how precisely we can measure abundance differences between stars. It may be possible to construct a large database of differential abundances from echelle spectra, with a precision of 0.05 dex or better; differential abundances are preferred here to reduce the effects of systematic error. Differential precisions of 0.01 dex have been achieved on solar twin studies working at very high resolution and signal-to-noise (Ramirez et al. 2010).

We have simulated in detail how many formation sites can be detected throughout the thin and thick disk in the Galaxy (Fig. 1.28). We have used the *Galaxia* model to simulate all components of the Galaxy (Sharma et al. 2011; Appendix B) and have applied a realistic sampling appropriate to the HERMES survey. The results are shown for two different ICMF slopes, i.e. $\gamma = -2.5$ consistent with quiescent star formation, and $\gamma = -1.5$ consistent with high pressure activity.

The total number of signatures for the thick disk is an order of magnitude less than the thin disk, consistent with its lower mass. The impact of migration, as expected, is that many more signatures are required for strong migration, and the sizes of the reconstructed clusters are characteristically smaller. The limits of the cluster mass function play an important role: high mass limits (lower figures) greatly reduces the number of required signatures.

Kroupa (2002) has argued that the thick disk formed from super star clusters as observed in the Hubble Ultra Deep Field (HUDF). In this case, a flat ICMF with high mass limits indicates that as few as 300–1000 signatures will be needed, although this range is an order of magnitude larger in the presence of strong migration.

It is an intriguing thought that one day we may be able to identify hundreds or thousands of stars throughout the Gaiasphere that were born within the same cloud

Fig. 1.28 The expected number of chemical signatures needed to reconstruct stellar clusters in the thin (*upper*) and thick (*lower*) disk. In each plot, the *black dashed line* refers to a cluster mass function with $\gamma = -2.5$, and the *red dotted line* to $\gamma = -1.5$. The x-axis is the number of reconstructed star clusters, and the y-axis is the cumulative total number of signatures required. In each panel of 4 sub-plots, the *left hand figures* assume that all stars are uniformly scattered throughout the disk through stellar migration; the *right figures* assume no migration. The *top figures* assume cluster mass limits of $m_c \in (10^2, 10^5)$ M$_\odot$; the *bottom figures* assume $m_c \in (10^5, 10^7)$ M$_\odot$

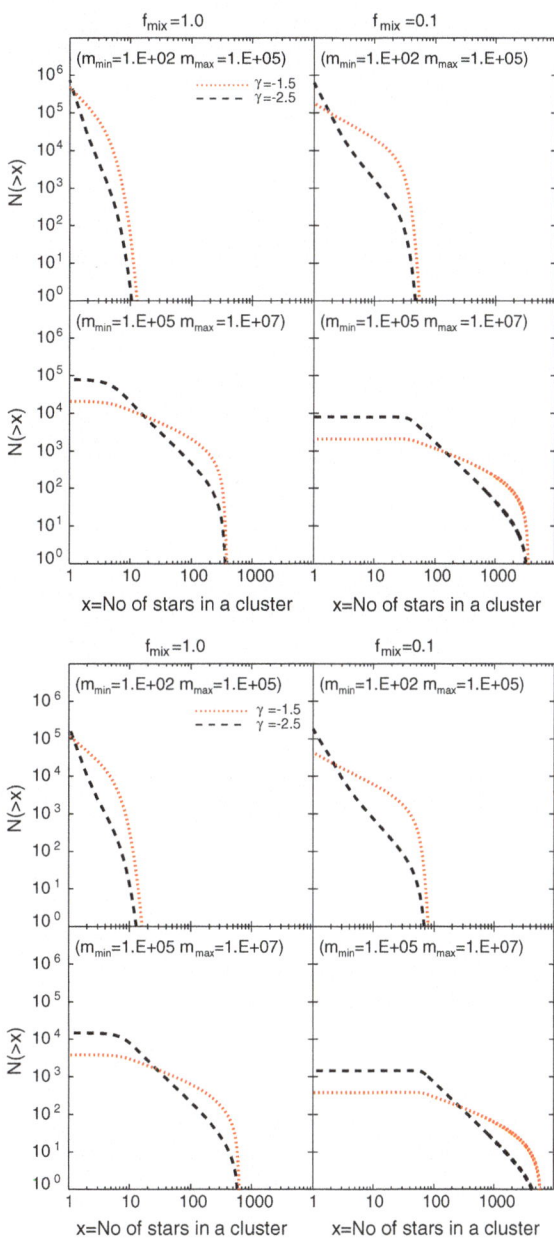

as the Sun. We note from Fig. 1.28 that some of the reconstructed star clusters could have hundreds or even thousands of members. In this instance, assuming the reconstructed clusters have a well defined age from the Hertzsprung-Russell diagram, their spatial distribution throughout the Galaxy is a direct measure of stellar migration with cosmic time (Bland-Hawthorn et al. 2010a).

1.9 Epilogue: Challenges for the Future

1.9.1 The Limitations of Near Field Cosmology: Are We Really Putting ΛCDM to the Test?

In recent reviews, Silk and Mamon (2012) and Kroupa et al. (2012) highlight the many challenges to ΛCDM due to the mismatch of observations and the most advanced numerical simulations. An issue that arises frequently is whether these discrepancies are a consequence of the lack of microphysics in numerical simulations or whether a new theory of gravity is justified (e.g. MOND). ΛCDM is widely cited as being extremely successful on the largest scales supported by "precision cosmology" measurements, particularly those that have come from CMB experiments. On scales smaller than galaxy groups, it becomes increasingly difficult to relate observations to the results of numerical simulations.

ΛCDM is a very useful hierarchical framework that acts over many orders of magnitude in physical scale (Davis et al. 1985). In this respect, it has a lot in common with statistical theories of turbulence. But the turbulence description breaks down on the smallest scales due to microphysics. This does not lead us to argue against the usefulness of the statistical models. Rather, we simply recognize that the description breaks down on the smallest scales due to other mechanisms. Our views of ΛCDM fall along the same lines. It seems unlikely that any of the existing claims that purport to challenge ΛCDM will hold up in the decades to come as we treat the complex physical processes more carefully.

Inevitably, when we find ourselves swimming in a sea of complex data, one begins to worry about fundamental limits of knowledge. But in many applied fields, important clues to fundamental physics often emerge from complex data, particularly when complex physical processes are at work (cf. beam-line experiments in particle physics). In fields where we are far from understanding key physical processes, as in the study of galaxy formation and evolution, this effort must be worth it.

Astronomers now recognize this: every large survey has unveiled new lines of enquiry or revealed something important about our environment. We talk in terms of data mining, virtual observatories, and so forth. Furthermore, the numerical simulators push down to ever decreasing scales, and work to include new algorithms that capture an important process. At what stage do we declare that galaxy formation is basically understood? Such a declaration becomes possible when one is able to reproduce the salient features of galaxies today, in a host of different environments,

1 Near Field Cosmology

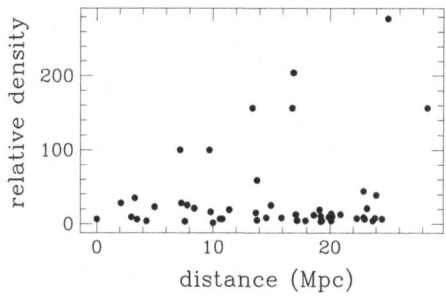

Fig. 1.29 The dependence of linear group density with linear distance from a sample of 50 galaxy groups taken from Tully (1987). Most groups are much like the Local Group in their average density (Bland-Hawthorn and Freeman 2006)

with a theory that is firmly rooted (presumably) in ΛCDM. This same axiomatic theory should be able to reproduce observations of galaxies at different epochs out to high redshift, until we reach an epoch where objects no longer look like modern day galaxies. The HUDF indicates that this appears to happen at about $z \sim 2$.

Any complete theory of galaxy formation must seek to explain the properties of the Local Group and tell us if they are relatively unusual or typical of the Universe as a whole. Recent compilations from Tully's Catalog of Galaxies (1987), and the GAMA survey (Sect. 3), suggest that "Local Group" collections of galaxies are relatively common throughout the local Universe (see Fig. 1.29). Therefore, it would be a surprise to discover decades from now that our large-scale surveys were seriously misleading us in our quest to understand galaxy formation and evolution.

A word of caution however is that to properly survey the effects of environment (or universal overdensity), we need to study resolved stellar populations out to at least 20 Mpc (Bland-Hawthorn and Freeman 2006). In Fig. 1.29, over these distances, the universal density contrast increases by more than two orders of magnitude compared to the Local Group. A physical scale of 20 Mpc allows us to as to cover the full range of galaxy environments, from voids to massive groups and clusters. This is what we refer to as the Local Universe or Local Volume, now recognized by the International Astronomical Union (Division H). This sphere encloses all galaxies that have accurate distances determined from the tip-of-the-red giant (TRGB) and surface brightness fluctuation methods. We consider the Local Universe to be a useful and physically motivated working definition. Over this volume, all galaxies to be at roughly the same cosmic epoch. We have a very good picture of the 3D distribution of galaxies provided in part by Tully Catalog of Galaxies and detailed 3D flow models for all galaxies are now possible. In Fig. 1.30, this volume falls within the domain of the Constrained Local Universe Simulations (CLUES; http://www.clues-project.org). All galaxies with masses equivalent to the LMC or larger can be imaged in most wavelength bands (e.g. x-rays, infrared, radio). The most detailed and complete observations of galaxies will always come from this volume, and will therefore continue to dominate studies of physical processes in galaxies. In time, we fully expect near-field cosmological studies to extend to the Local Volume.

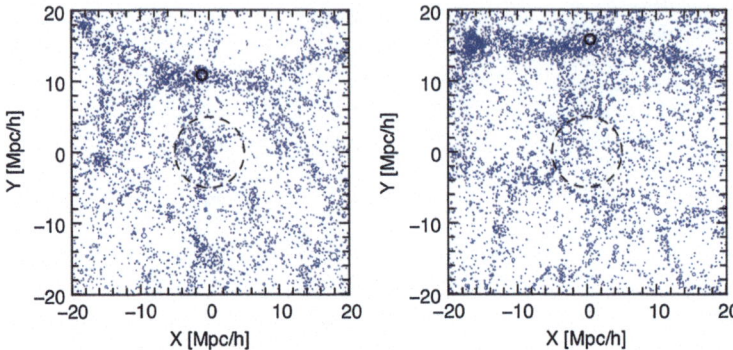

Fig. 1.30 Two realizations from the Constrained Local Universe Simulations (CLUES) collaboration (Forero-Romero et al. 2011). The *small thick circle* is identified with the Virgo Cluster. The 5 h^{-1} Mpc sphere associated with the Local Group is shown as a *dashed circle*. Only haloes with masses greater than halo mass of the LMC are shown. The slice has a depth of 25 h^{-1} Mpc centred at the Local Group

1.9.2 Future Surveys

Throughout this review, we have identified fossil signatures at all cosmic epochs, with a focus on those relating to galaxy formation and evolution accessible within the Galaxy. We have raised many questions and challenges relating to Local Group archaeology throughout the text. Fossil signatures allow us to probe back to early epochs. We believe that the near-field Universe has the same level of importance as the far-field Universe for a comprehensive understanding of galaxy formation and evolution.

We have argued that understanding galaxy formation is primarily about understanding baryon dissipation within the ΛCDM hierarchy; to a large extent, this means understanding the formation of disks. The question we seek to address is whether this can ever be unravelled in the near or far field. Dynamical information was certainly lost at several stages of this process, but we should look for preserved signatures of the different phases of galaxy formation.

Far-field cosmology can show how the light-weighted, integrated properties of disks change with cosmic time. While light-weighted properties provide some constraint on simulations of the future, they obscure some of the key processes during dissipation. The great advantage of near field studies is the ability to derive ages and detailed abundances for individual stars within galaxies of the Local Group.

It is hard to overstate the impact of ESA Gaia on the future of near-field cosmology, assuming all goes well later this year (2013). The anticipated orders-of-magnitude gain in astrometric accuracy and survey size is illustrated in Fig. 1.31. We have addressed the issue of information content within the Gaiasphere. The detailed information that is possible on ages, kinematics, and chemical properties for a billion stars—which we see as the limit of observational knowledge over the next two

1 Near Field Cosmology

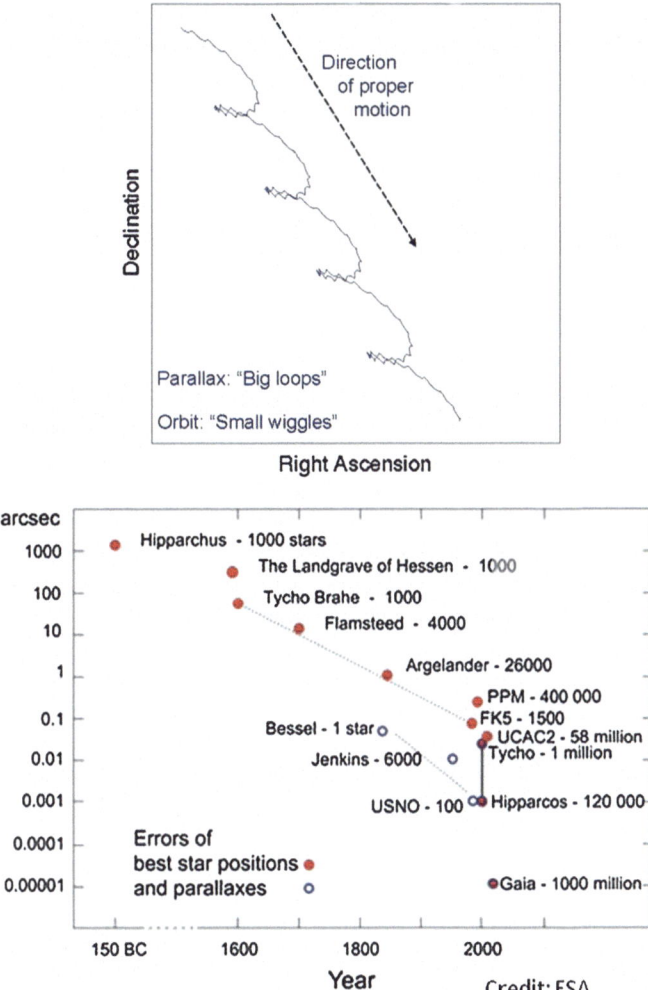

Fig. 1.31 *Upper* The components of proper motion, parallax and barycentric wobble in a star's projected motion across the sky (we acknowledge the NASA Gravity Probe B team for this figure). *Lower* The improvement in measurement accuracy of stellar positions (*red dots*) and parallaxes (*blue dots*) over the centuries. The size of each survey is also indicated. The huge expected gain in astrometric accuracy from ESA Gaia is evident. The satellite will determine proper motions for 1–2 billion stars, a few percent of all stars in the Galaxy (we acknowledge the ESA Gaia team for this figure)

decades—may reveal vast complexity throughout the disk. We may not be able to perceive the sequence of events directly. However, we are optimistic that future dissipational models may provide unique connections with the observed complexity.

We believe that detailed high resolution abundance studies of large samples of galactic stars will be crucial for the future of fossil astronomy. Gaia will provide accurate distances, ages and space motions for a vast number of stars, separate with great precision the various Galactic components, and identify most of the substructure in the outer bulge and halo. High resolution spectrographs like VLT UVES, Subaru HDS and Keck HIRES have already begun to reveal the rich seam of information in stellar abundances. Instruments like AAT HERMES, SDSS APOGEE (and VISTA 4MOST) are now poised to undertake much larger stellar surveys with uniform chemical abundances.

It is an extraordinary fact that we can probe back to the first billion years from observations of the *local* universe. We can say with absolute certainty that stars existed at this time. These were responsible for the first chemical elements (Ryan et al. 2006) and for reionizing the fog of hydrogen that permeated the early Universe (Fan et al. 2002). Precisely when the first star clusters formed is unknown. It seems likely, however, that gas was able to fragment at very high density even at primordial abundance levels (Clark et al. 2008).

It may be possible to directly probe these environments in an era of the Atacama Large Millimetre Array and the James Webb Space Telescope. But we believe that some of the most important insights, particularly with regard to progenitor yields, will undoubtedly come from near-field cosmology. We have already made a case for the importance of dwarf galaxies, particularly the ultra faint dwarfs, in preserving ancient chemical signatures (Karlsson et al. 2013). To this end, it will be necessary to equip the next generation of ELTs with wide-field multi-object spectrographs that operate at high spectroscopic resolution ($R \gtrsim 20{,}000$). The instrument will be expensive and technically challenging, but we believe this must be tackled if we are to ever unravel the formation of the Galaxy.

Acknowledgments "Discursive" is an interesting adjective that has two opposing meanings: either 'proceeding logically' or 'jumping from topic to topic.' Only the reader can decide which of these two interpretations best describes our review. The document has been compiled from lectures that were originally delivered in 2007 at Mürren in Switzerland as part of the Saas Fee series. But the lectures were further developed for seminar series delivered at the University of Wisconsin (2011), the University of Bologna (2011) and the University of Sydney (2012). Powerpoint slides for these lecture series are available upon request. Updated text from our 2002 *Annual Reviews of Astronomy and Astrophysics* article is also included.

We are indebted to Mark Krumholz, David Webster, Volker Bromm and Torgny Karlsson for important contributions to our published work, some aspects of which are repeated here. We have learnt a great deal from conversations with Alice Quillen, Sanjib Sharma, James Binney, Jerry Sellwood, Jerry Ostriker, Simon White, Scott Tremaine and Paul Nulsen. JBH is supported by a Federation Fellowship from the Australian Research Council, and further acknowledges a Merton College Fellowship, a Leverhulme Fellowship (Oxford), and a Brittingham Scholarship (Wisconsin). Finally, we owe a huge debt of gratitude to Ben Moore and Francesca Matteucci for their patience and forbearance over the years.

Appendix A: The Discovery of Dark Matter in Galaxies

In the past year, there have been excellent reviews that revisit how we got to where we are today. Who discovered that the Universe is expanding (Peacock 2013)? Who came up with Big Bang cosmology (Belenkiy 2012)? and so forth. Speakers introduce topics with a compressed narrative that tends to write out important contributions through the process of oversimplification. This is certainly the case for who discovered dark matter in galaxies.

One of the cornerstones of modern astrophysics is the cold dark matter paradigm. We know today that essentially all galaxies are encompassed by dark matter haloes that are an order of magnitude larger in diameter than what we observe with our telescopes. While the nature of "dark matter" remains a mystery, it is widely believed to be made up of one or more unidentified sub-atomic particles.

A question of more than historical interest is to ask how the existence of dark matter in galaxies was firmly established. Dark matter drives the large-scale dynamics of matter in the Universe and is the scaffolding on which individual galaxies are built. The role of dark matter in individual galaxies was finally recognized in the 1970s, as we discuss. We find that the history of this discovery is chequered with early hints, followed by definitive demonstrations, and ultimately a more complete understanding of its implications. In this respect, it is not easy to attribute the discovery of dark matter in galaxies to specific astronomers. The roll call includes Zwicky, Smith, Kahn and Woltjer for the discovery of dark matter in groups and clusters, followed by Freeman, Ostriker, Rubin and Bosma for the discovery of dark matter in galaxies. (These astronomers mostly published their results in multi-author papers, but credit is given to those that carried their arguments forward in later papers.)

The presence of some kind of "unseen matter" was first inferred by Zwicky (1933) in the Coma cluster, and by Smith (1936) in the Virgo cluster. Two decades on, Kahn and Woltjer (1959) showed that the total mass of the Local Group greatly exceeds the sum of the stellar masses of M31 and the Milky Way. In these early papers, it was not possible to establish how the unseen matter was distributed.

There were alternative interpretations that galaxy clusters were "birth sites" caught in the act of flying apart (Ambartsumian 1958; Neyman et al. 1961). But the problem with these ideas is that the core of the Coma cluster, for example, is 250 kpc in diameter, its 3D velocity dispersion is $1500 \,\mathrm{km\,s^{-1}}$ and hence its dispersal time is $\sim 10^8$ yrs. The stellar populations, however, are obviously much older. (At about this time, tidal arms in interacting galaxies were being argued as an ejection phenomenon, an interpretation that was dismissed for the same reasons.) On the basis of more reliable velocity data, Peebles (1970) and Rood et al. (1972) both derived M/L values for Coma which exceeded those for any known stellar population.

Tremaine (1999) notes that deriving galaxy masses was a thriving industry in the 1960s, but these studies vastly underestimated the true masses due to their limited radial extent and because the connection to unseen matter in clusters had not been made. Ostriker et al. (1974) and Einasto et al. (1974) were some of the first to state clearly (cf. Freeman 1970; Roberts and Rots 1973) that unseen matter in clusters is probably distributed in extensive halos around individual galaxies. This made any interpretation of "dark matter" rather colder than, say, an invisible medium permeating clusters.

Ostriker in particular had fully understood the early signs from galaxian HI rotation curves that were more flat than Keplerian, and was therefore guided by this result, at least in part. It took several decades to establish the reality of super-Keplerian rotation. The effect was already clear in Babcock's (1939) early observations confirmed decades later with more extensive observations by Roberts (1967). In hindsight, we can say that this was a necessary but not a sufficient condition because the rotation curves did not extend far enough in radius. Flat or slowly declining rotation over the inner 6 or so scale lengths is easily accounted for in terms of disk components (e.g. van Albada et al. 1985). Kalnajs (1983) questioned evidence for DM for some of the optical rotation curves presented by Rubin at the Besancon conference, where he found the stellar components alone could explain the raised rotation curves.

Throughout the 1970s, few were thinking of unseen haloes in terms of elementary particles (cf. Tremaine and Gunn 1979). Essentially all papers discussed unseen haloes in terms of baryons, especially low mass stars extending down to brown dwarfs (e.g. Roberts 1975; White and Rees 1978). For example, a string of influential papers on galaxy formation considered how gas accretes and cools within a gravitating potential well (Rees and Ostriker 1977; Binney 1977; Silk 1977). Gunn (1977) and White and Rees (1978) argued that the inferred dark haloes formed naturally from the infall of collisionless dark matter onto bound structures in an expanding universe. Interestingly, at the same time, Searle and Zinn (1978) realized that the Galactic system of globular clusters must have accreted onto the halo over billions of years, rather than in the rapid collapse envisioned by Eggen et al. (1962). As far as one can determine, these papers implicitly assumed that the gravitating haloes were baryonic in nature, to the extent that galaxy haloes were discussed by them.

Flat rotation curves. Flat rotation curves in individual galaxies, with their small velocity scales, provide some of the most compelling evidence we have for the existence of (cold) dark matter haloes. In her history of the discovery of dark matter, Trimble (1987) notes that "Freeman (1970) was among the first to notice that such non-Keplerian rotation curves were a widespread phenomenon and to deduce that there might be considerable gravitating mass outside the observed region." This observation is in connection to his comment in his Appendix A that:

> For NGC 300 and M33, the 21 cm data give turnover points near the photometric outer edges of these systems. These data have relatively low spatial resolution; if they are correct, then there must be in these galaxies additional matter which is undetected, either optically or at 21 cm. Its mass must be at least as large as the mass of the detected galaxy, and its distribution must be quite different from the exponential distribution which holds for the optical galaxy.

In the centennial issue of the *Astrophysical Journal*, Tremaine (1999) comments on the importance of the Ostriker et al. (1974) paper but notes that "the first clear statement that rotation curves demand dark matter" is Freeman's (1970) statement above. This point is made again in van Albada et al.'s (1985) classic study of NGC 3198.

An influential paper in the same year as Freeman's classic work was Rubin and Ford's (1970) optical study of M31. They show the galaxy has a non-Keplerian (although declining) rotation curve to a radius of 20 kpc. No definitive conclusions were drawn from their careful study about unseen matter because their disk models were able to produce some of the slow decline in rotation over the inner disk. They state that "it does not appear possible, from the presently available data, to infer anything about the mass beyond 24 kpc in M31."

In a published Research Note, on the basis of three extended rotation curves (M81, M101, and Rubin's data for M31), Roberts and Rots (1973) noted that extended rotation curves indicate "a significant amount of matter at these large distances and imply that spiral galaxies are larger than found from photometric measurements." These authors discussed the possible role of mass to light variations in galaxies and between galaxies. In a published Comment, Roberts (1976) drew attention to the striking phenomenon of flat rotation curves (see also Roberts 1975).

In the mid 1970s, there was some discussion that the few extended HI rotation curves that existed (e.g. M81, M31, M33) are possibly affected by dynamical interactions with neighbours (e.g. conference discussions by Sancisi, van der Kruit, van Albada, Allen). Another point of discussion was the contribution of poorly calibrated side lobes (Salpeter) and beam smearing arising from the Arecibo observations.

But the critical study that put the role of galaxian dark matter beyond doubt was Bosma's (1978) PhD thesis which included 21 cm rotation curves for a large sample of galaxies, including about 20 disk galaxies to unprecedented radial scales. This thesis was widely circulated at the time and remains the 2nd most cited PhD in the history of astronomy. Bosma concludes "The mass models indicate that in the outer parts of a spiral the mass-to-light ratio is higher than in the inner parts. Perhaps a substantial fraction of the mass is not distributed in a disk at all."

From 1978 to 1985, Rubin and collaborators wrote influential papers presenting a large sample of optical rotation curves, many of which exhibit flat rotation on radial scales comparable to Bosma's HI data and consistent with dark matter (cf. Kalnajs 1983). Rubin considered both disk and spherical models and recognized the possible implication of massive halos around spirals to large radius, an insight that they credit to the Einasto and Ostriker papers. In the same year as Bosma's thesis, Rubin et al. (1978) state:

> These results take on added importance in conjunction with the suggestion of Einasto et al. (1974) and Ostriker et al. (1974) that galaxies contain massive halos extending to large radius. Such models imply that the galaxy mass increases significantly with increasing radius which in turn requires that rotational velocities remain high for large radius. The observations presented here are thus a necessary but not sufficient condition for massive halos. The developments of the 1970s, in large part because of the phenomenon of flat rotation curves, paved the way for the cold dark matter paradigm in the following decade.

Today, the nature of dark matter remains unclear although one or more sub-atomic particles appear to be the most likely candidates. Examples include the hypothetical axion, neutralino, or a weakly interacting massive particle (WIMP), perhaps 100 times the mass of the proton, predicted by extensions of the Standard Model. WIMPs may be their own antiparticles and therefore capable of mutual annihilation, an effect that can be searched for with gamma ray satellites. The most recent ΛCDM simulations involving as many as 1010 particles reveal that dark haloes are likely to be highly structured and clumpy. Recent experiments with the Fermi gamma ray satellite to search for decaying dark matter (assuming a decay channel involving photons) has drawn a blank. A recent review by Abbasi et al. (2012) puts stringent limits on all dark matter candidates to date.

In summary, Ostriker was arguably the most persistent champion of dark haloes in the early 1970s. While no single result was compelling, taken together, the case for dark haloes appeared to be strong, as discussed in his 1974 paper. Most but not all of the arguments have held up with the passage of time (Tremaine 1999). For example, Turner's (1976a, b) arguments based on binary galaxies were undermined by White (1981, 1983): the derived masses are plagued by uncertain orbits, contamination, selection effects and so on. In the same vein, Ostriker and Peebles (1973)[14] wrote an influential paper on the stabilizing influence of dark haloes against bar modes. They fully understood that a disk could be stabilized by its internal velocity dispersion (e.g. Hohl 1970) but the Q parameter required is much larger than observed. Bars are not a solution either: the stable barred galaxies that they found were again much hotter than observed bar galaxies. (What could conceivably undermine their result is a rotation curve with a sharp bend near the centre, i.e. with a very small uniformly rotating core. This was demonstrated by Toomre and Sellwood in the 1980s.) Today we know that bars occur in a high fraction of disk galaxies (Eskridge et al. 2000) so it seems unlikely that dark haloes are successful in stabilizing disks against bar modes.

Appendix B: Stellar Data: Sources and Techniques

B.1 Data Needed for Galactic Archaeology

Our goal is to use relic or fossil information to evaluate the state of the Galaxy and understand how it got to be in this state. We want to answer questions such as

- What were the major formation events?
- How important was the role of mergers?
- What was the infall history and the star formation history?
- What were the major dynamical and chemical evolution processes?

To answer these questions, we need stellar data to compare with theoretical predictions and to guide the theory. The basic stellar data that we need are the magnitudes and colours of the stars, their distances, motions and chemical properties, and their ages. We will look first at which each kind of data can do for us, and then at the techniques for acquiring the data.

---

[14] This paper is routinely cited (incorrectly) as arguing for the existence of dark haloes but they only studied the dynamics of the Galaxy inside of the Solar Circle.

1 Near Field Cosmology

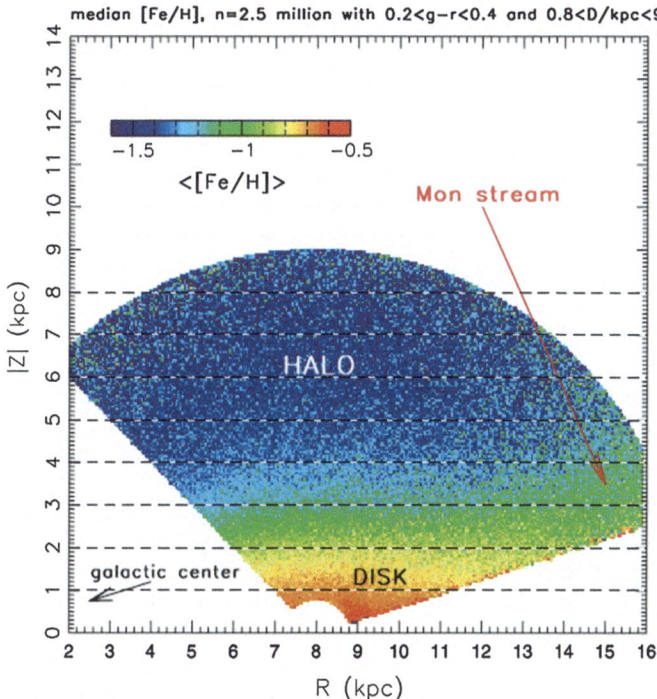

Fig. 1.32 Distribution of metallicity in the Galaxy from SDSS photometry (from Ivezic et al. 2008)

B.1.1 Stellar Photometry: Magnitudes and Colours of Stars

Photometric catalogues are essential input data for stellar observational programs. They give magnitudes and colours for up to billions of stars over the whole sky or large fractions of the sky. The catalogues typically have stellar coordinates and photometry in two or more optical or near-IR bands, at different levels of accuracy.

They can be used to make preliminary estimates of stellar parameters like temperature and chemical abundance and distance for vast numbers of stars. From photometric catalogues alone, it is possible to derive useful information about the structure of the Galaxy. For example, the 5-band *ugriz* photometry from the Sloan Digital Sky Survey has been used to derive the distribution of stars and their overall chemical abundances in the Galaxy. Ivezic et al. (2008) estimated photometric distances and [Fe/H] abundance for 2.5 million FG stars from the SDSS to derive the distribution of abundance and of positions (R, z) out to distances of about 8 kpc from the Sun. Their map (Fig. 1.32) nicely shows the planar stratification of [Fe/H] in the Galaxy, decreasing with height above the Galactic plane. The Monoceros concentration of relatively metal-poor stars at a Galactocentric distance of about 16 kpc can also been seen in their data.

B.1.2 Stellar Distances: Where Do Stars Lie

We need stellar distances to

- measure transverse velocities of stars from their proper motions
- map substructure in the halo and disk
- calibrate the luminosities of different kinds of stars
- measure the structure and dynamics of the Galactic components.

B.1.3 Stellar Motions

The basic data are the stellar radial velocities and proper motions. With an estimate of distance, it is possible to calculate the 3D space motions. The motions of stars are used to

- measure how energetic the orbits of particular kinds of stars are, and how far they are from circular motion
- measure the sense of their angular momentum: prograde or retrograde
- see how stellar orbits have evolved: how do the orbital properties correlate with age and metallicity
- detect kinematic substructure, such as moving stellar groups
- compute stellar orbits to learn about the dynamical structure of the Galaxy.
- measure the properties of the Galactic potential from their spatial and kinematic distributions. In this way, the dark matter content of the Galaxy can be measured from the kinematics and distribution of distant halo stars, and the total density of luminous and dark matter near the sun can be estimated from the properties of nearer disk stars.

Figure 1.33 shows the Lindblad diagram (orbital energy against angular momentum) for a sample of halo and disk stars. This diagram can be constructed for stars with known motions, and is a very useful diagnostic of the dynamical properties of different stellar populations. It is immediately clear if stars are in near-circular orbits, retrograde orbits, or highly energetic orbits. The Toomre diagram (Fig. 1.34) is another diagram derived from stellar kinematics and is much used to identify to which Galactic component (thin disk, thick disk, halo) a star belongs. Here (U, V, W) are components of the star's motion relative to the Local Standard of Rest: U is in the Galactic radial direction (towards $l = 0°$ or $180°$ depending on the convention), V in the direction of rotation (towards $l = 90°$) and W in the vertical direction towards the North Galactic Pole. Stars of the thin disk are clustered towards low velocities, while the stars of the thin disk are hotter kinematically.

If we know the 3D location and velocity for a star, and have a reliable model for the Galactic gravitational potential, then we can compute the Galactic orbit of the star. Figure 1.35 shows the plane and edge-on view of a typical stellar orbit. Stars from the inner and outer Galaxy can pass through the solar neighborhood. Knowing the orbit is not always very useful for Galactic archaeology. There is no guarantee

1 Near Field Cosmology

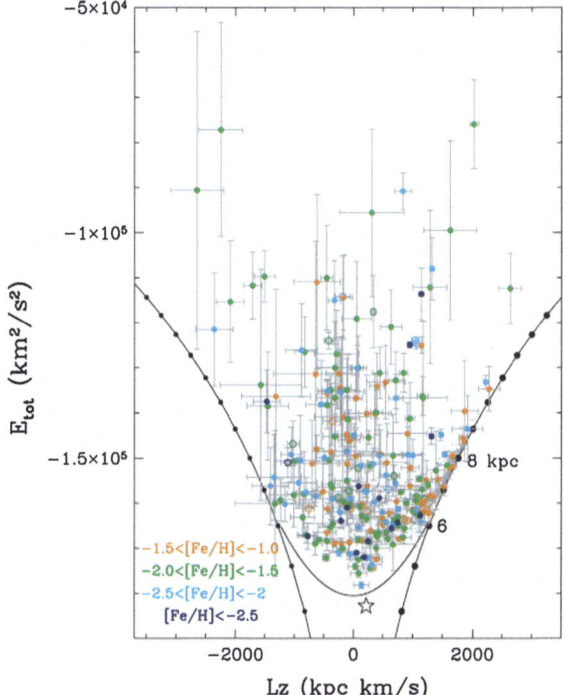

Fig. 1.33 The Lindblad diagram shows orbital energy and angular momentum for Galactic disk and halo stars. Very energetic halo stars lie towards the *top* of the diagram; some halo stars are in retrograde orbits ($L_z < 0$). Disk stars lie near the prograde circular orbit locus (the *black curve* on the *right*). The sun is located near the "8" on the prograde circular orbit locus. The energy E and angular momentum component L_z are both integrals of the motion in a steady-state axisymmetric galaxy (from Morrison et al. 2009)

that the dynamical properties of a star have remained unchanged throughout the life of the star. The Galactic potential has evolved as the Galaxy's mass has gradually increased by accretion of baryonic and dark matter. Stellar orbits can be disturbed as the star interacts with spiral structure and giant molecular clouds, and resonances with the central bar and the spiral structure can flip a star from one near-circular orbit to another.

B.1.4 Stellar Element Abundances

The cosmic abundance distribution in Fig. 1.36 shows the outcome (in the solar neighborhood) of the chemical evolution of our Galaxy. The element abundances of stars come initially from the abundances in the gas from which they formed. This gas has been enriched by previous generations of evolving and dying stars. Different element groups come from different progenitors.

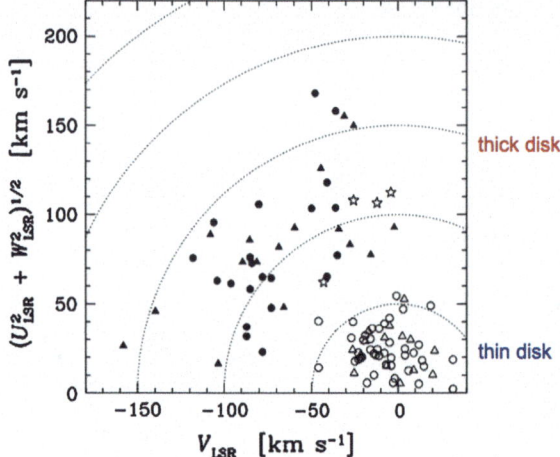

Fig. 1.34 The Toomre diagram for nearby stars of the thin (*open symbols*) and thick (*filled symbols*) disk. The V component of the stellar velocity represents the stellar angular momentum and has an asymmetric distribution with a negative mean: this asymmetric drift increases with velocity dispersion and is a useful diagnostic. The U and W components have more or less symmetric distributions about zero mean, so their combination is a measure of the orbital energy (from Bensby et al. 2005)

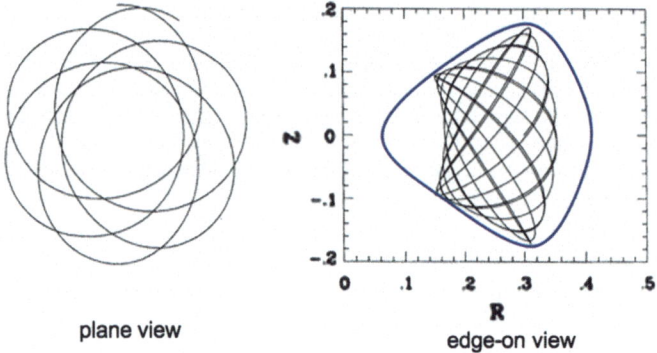

Fig. 1.35 A typical stellar orbit in an axisymmetric potential (adapted from Binney and Tremaine 2008)

- the Fe-peak elements come mainly from type Ia SN
- the α-elements (e.g. Mg, Si, Ca, Ti) and r-process elements come mainly from the more massive type II SN
- the s-process elements come mainly from thermally pulsing AGB stars

For most of the heavier elements, stars remember the abundances with which they are born. The abundances of different element groups in stars can tell us a lot about the star formation history which led to the formation of these particular stars.

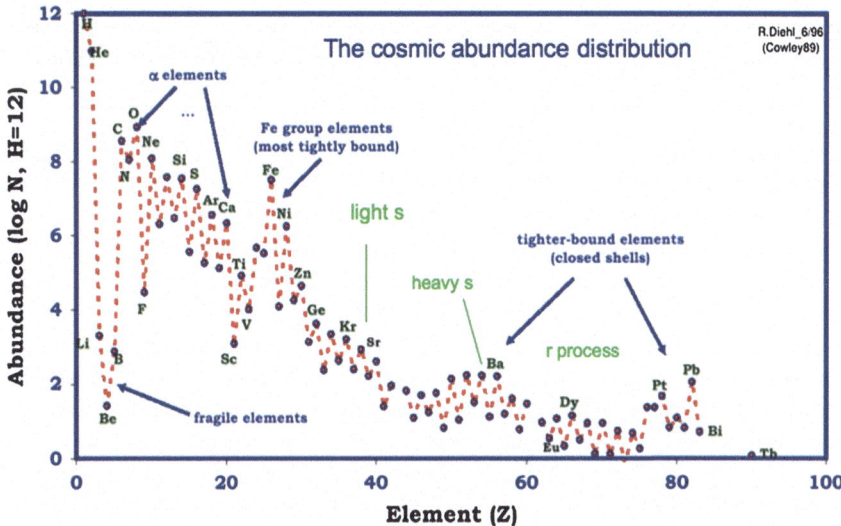

Fig. 1.36 The cosmic abundance distribution, showing structure due to the major chemical evolution processes

For example, α-enrichment relative to Fe indicates that SNII were important for the chemical evolution and the star formation history was fairly rapid, on a timescale of order 1 Gyr: Fe-enrichment from SNIa (which take ∼1 Gyr to evolve) was less prominent.

Different components of the Galaxy (halo, bulge, thick disk, thin disk) each have different characteristic chemical properties. For example, the halo stars are mostly metal poor ($-1 >$ [Fe/H] > -5), while the thick disk stars are more metal rich ($-0.2 >$ [Fe/H] > -2). The halo and thick disk stars are both enriched in α-elements (Mg, Si, Ca, Ti). The [Fe/H] range of the thin and thick disks overlap; the thick disk stars have higher [α/Fe] ratios than thin disk stars of the with similar [Fe/H] abundances, indicating that the chemical evolution of the thick disk proceeded more rapidly.

Groups of stars born together, like open star clusters, usually have almost identical abundances, reflecting the abundances of the gas from which they formed (e.g. De Silva et al. 2009). This is true also for most of the globular clusters: a few have heavy element variations from star to star, and many show correlated variations of lighter elements such as C, N, O, Na, Mg, Al, the origin of which are not full understood yet. Chemical signatures may allow us to recognize groups of stars which were born together but have dispersed and drifted apart . This technique is known as chemical tagging and will be used for analysing the products of large high resolution spectroscopic surveys like the HERMES survey on the AAT and the APOGEE near-IR survey with the Sloan telescope. Although it is readily possible to measure abundances of more than 30 elements, these elements do not all vary independently.

The number of independently varying elements is 8–9 (Ting et al. 2012): this is the dimensionality of the space defined by the abundances of the chemical elements.

B.1.5 Stellar Ages

Stellar ages let us evaluate when events occurred in the evolution of the Galaxy. They are important for measuring the star formation history and for understanding how the metallicity and dynamics of different groups of stars have evolved. For example, how has the star formation rate in the disk of the Galaxy evolved since the disk began to form? The star formation rate is believed to have been roughly constant over time near the sun, with episodic star bursts over the past 10 Gyr (e.g. Rocha-Pinto et al. 2000) but this remains uncertain because of uncertain ages. How have the kinematics and the metallicity of the thin disk near the sun changed from 10 Gyr ago to the present time under the effects of dynamical and chemical evolution? Again, the kinematics and metallicities are not difficult to measure, but the derivation of stellar ages remains problematic. Because stellar ages are still difficult to measure, there remains much uncertainty about the evolution of the Galaxy. Measuring stellar ages is one of the most important goals for Galactic evolution for the future.

B.2 Sources of Data

Now we turn to the sources of the various kinds of data needed for Galactic archaeology.

B.2.1 Photometric Catalogs

Here is an incomplete list of some of the major photometric catalogs.

2MASS is a relatively shallow all-sky near-IR (JHK) survey and includes a point source catalog. This catalog is notable for its excellent astrometry and is an invaluable source of stars for brighter spectroscopic surveys. The UKIDSS survey (*YJHK*) is a deeper survey of 7500 square degrees of northern sky. The VISTA VHS survey (JK) is still in progress and will also go significantly deeper.

The Sloan Digital Sky Survey (SDSS) covers about 8000 square degrees, mainly in the northern sky. It uses a 5-filter system (*ugriz*) which has become a standard system. The Pan-STARRS project will survey about 30,000 square degrees, again mainly in the northern sky, using *grizy* filters. In the future, the LSST will provide a very deep survey of the whole southern sky with *ugrizy* filters. In the meantime, the SkyMapper survey will cover the southern sky several times, using a six-filter system (*uvgriz*), reaching somewhat deeper than the SDSS.

1 Near Field Cosmology

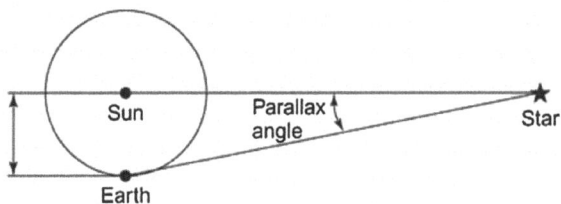

Fig. 1.37 Geometry of parallax measurement

B.2.2 Techniques for Measuring Distances

Trigonometric parallaxes. The stellar distance scale (and the extragalactic distance scale) are based on fundamental trigonometric parallaxes. Stellar positions are measured from the extremes of the earth's orbit around the sun (see Fig. 1.37). Because this technique is used for relatively nearby stars, it is usually necessary to continue the observation for several seasons, in order to separate out the shifts in position due to parallax from the shifts due to the star's transverse or proper motion. From the ground, parallax errors of a few milli-arcseconds (mas) can be achieved, giving distances with 10% errors out to about 30 pc from the sun.

From space, higher precision is possible. The Hipparcos mission provided parallaxes with errors of about 1 mas. The Gaia mission, due for launch in 2013, will give parallax errors of about 10 μ as at a V-magnitude of about 14. Parallaxes with this level of accuracy will be useful out to distances of about 10 kpc. For turnoff stars with $V \sim 14$, the distances will be known to within about 1%. The Gaia parallax errors are larger for fainter stars.

Photometric and spectroscopic parallaxes. Theoretical or empirical isochrones are used to estimate the absolute magnitude and hence the distance of the star. To make this work well, one needs to know the abundance [Fe/H] of the star, and an estimate of its effective temperature and surface gravity or luminosity. These can all be derived from multi-filter photometry or spectroscopy. An assumption of the star's age is needed if the star has evolved from the main sequence.

The more one knows, the better this works. If only broad-band colors are known, the distance errors can be very large. In the best cases, when the errors are $\sim$0.1 in [Fe/H], 100 K in temperature and 0.1 in log g, distance errors of about 15% can be achieved. For stars on the steep giant branch, the distance estimates are usually less accurate.

For some kinds of stars, like the He-core burning RR Lyrae stars, blue horizontal branch stars and red giant branch clump (RGBC) stars, accurate absolute magnitudes are known or can be estimated from periods and colors, and the errors in their photometric distances can be <10%. RR Lyr and BHB stars are usually found in more metal-poor populations like the Galactic halo and thick disk. RGBC stars are particularly useful for studies of galactic structure in the more metal-rich populations because they are so common and it is easy to measure their element abundances.

For star clusters, one can fit theoretical isochrones to their color-magnitude diagrams to derive ages and distances if their metallicities are known. Globular clusters mostly have horizontal branch stars or RR Lyr stars from which accurate distances can be derived.

Interstellar reddening and extinction are a problem for photometric parallaxes. This problem is less significant in the near-infrared. Multicolor photometry can give independent stellar reddening estimates if the wavelength dependence of reddening is known. For individual stars, the diffuse interstellar bands can be used to estimate the reddening directly (e.g. Munari et al. 2008). The reddening can also be derived from reddening maps of the Galaxy, like the Schlegel et al. (1998) maps derived from the COBE/DIRBE near-IR mapping of the Milky Way. These maps give the total reddening along each line of sight. For stars that are located within the reddening layer, correction is needed via models of the reddening distribution along the line of sight.

B.2.3 Techniques for Measuring Stellar Velocities

Radial (line of sight) velocities are measured spectroscopically via the Doppler shift. Proper (transverse) motions are measured astrometrically from the shift of a star's position with time relative to a set of reference stars, and are usually measured from wide field photographic or CCD images.

Some large stellar radial velocity surveys are in progress for Doppler planet searches and for Galactic structure and dynamics. The fiber spectrograph surveys SEGUE, and RAVE surveys are observing several $\times 10^5$ stars for Galactic structure and chemical evolution. The LAMOST survey is one or two orders of magnitude larger. Radial velocities have typical accuracies ranging from about $1\,\mathrm{m\,s^{-1}}$ with special techniques at spectroscopic resolution $R = \lambda/\Delta\lambda \sim 50,000$ to $1\ \mathrm{km\,s^{-1}}$ at $R \sim 7000$ (RAVE) and $5\ \mathrm{km\,s^{-1}}$ at $R \sim 2000$ (SEGUE, LAMOST). For most Galactic programs, $5\ \mathrm{km\,s^{-1}}$ is good enough. For some programs, like finding the kinematically cold debris of tidally disrupting star clusters, an accuracy of $1\ \mathrm{km\,s^{-1}}$ can be very useful. Spectra acquired for radial velocities can also give useful estimates of the stellar parameters T_{eff}, $\log g$ and [Fe/H].

From the ground, the proper motion accuracy can be a few mas yr^{-1} ($\sim 20\ \mathrm{km\,s^{-1}}$ at a distance of 1 kpc). Very large samples of proper motions (10^5 to 10^9 stars) come from many ground-based surveys, such as USNO, UCAC2, SPM, SDSS, 2MASS, GSC, PM2000, PPMXL ...with more to come from the large imaging surveys Pan-STARRS, SkyMapper and LSST.

From space, the Hipparcos/Tycho mission provided proper motions of about 2 million stars, with an accuracy of about 1 mas yr^{-1}. The Gaia mission will give proper motions for about a billion stars: the accuracy depends on the brightness and color of the star, and is about $10\,\mu$ as yr^{-1} at $V = 14$ (i.e. about 0.7 $\mathrm{km\,s^{-1}}$ for a bright giant at a distance of 15 kpc). Gaia will really change Galactic astrophysics, with vast numbers of very precise parallaxes and proper motions. We should be prepared to get the most from this resource (launch is due in 2013). The JASMINE missions are smaller near-IR astrometric space projects, aimed particularly at the Galactic plane and bulge.

B.2.4 Techniques for Measuring Chemical Abundances

Intermediate and broad band photometry, such as the Strömgren photometry and the SDSS/SkyMapper photometry, can give estimates of stellar temperature, gravity and metallicity. The typical metallicity errors are about 0.2 dex but can be smaller. Figure 1.38 shows how the passbands of the $uvgr$ filters of the SkyMapper and SDSS systems align with features in FGK stars. Examples of photometric surveys that have generated large samples of stellar abundances are the Geneva-Copenhagen survey (GCS) of about 14,000 nearby FG dwarfs using Strömgren photometry complemented with Hipparcos astrometry and precise groundbased radial velocities (Nordström et al. 2004), and the large study of the Galactic abundance distribution by Ivezic et al. (2008) using DSS data.

Medium resolution spectroscopy ($R \sim 2000$–$10,000$) uses the strengths of spectral features to estimate stellar parameters, including [Fe/H] and sometimes [α/Fe] and a few other elements. Recent examples are the RAVE and SEGUE surveys of several $\times 10^5$ stars. Various techniques are used to measure the stellar parameters, including empirical calibration of individual line strengths and χ^2 matching of the spectra to grids of synthetic spectra. The [Fe/H] errors are typically about 0.15 dex. Several of the medium resolution spectroscopy facilities can acquire spectra of many stars at once. For example, the AAOmega spectrometer on the AAT is fed by optical fibers and can measure medium resolution spectra of up to about 400 stars simultaneously in a 2° diameter field.

High resolution spectroscopy ($R > 20,000$) allows measurement of abundances for many elements, including the important neutron-capture elements. Many of these elements have relatively weak lines which are difficult to measure at lower resolutions. It is also possible to measure isotopic abundances for some elements, and these can be good diagnostics of nuclear processes. From high resolution spectra with high signal-to-noise ratios, it is possible to measure differential abundances with an error as low as 0.02 dex in the element abundance ratios [X/Fe] when comparing stars of similar temperatures and gravities. Techniques include analysis of equivalent widths of individual spectral lines and matching synthetic spectra in detail to the observed spectra. The newly developed MATISSE method (Recio-Blanco et al. 2006) involves projection of the spectra onto basis vectors to derive the individual parameters.

High resolution spectrographs are usually echelle spectrographs, some with a few hundred fibers for multi-object capability. Existing systems include Hectochelle on the MMT, MIKE and MIKE-fibers on Magellan, and the FLAMES/GIRAFFE/UVES system on the VLT. HERMES on AAT and APOGEE on the Sloan telescope are coming soon and will be used for large high-resolution spectroscopic surveys of 10^5 to 10^6 stars at resolutions of 20,000 (APOGEE, in the NIR H-band) to 30,000 (HERMES in optical bands). HERMES will also have a $R = 48,000$ multi-object capability. The analysis of high resolution spectra is currently laborious, but this will

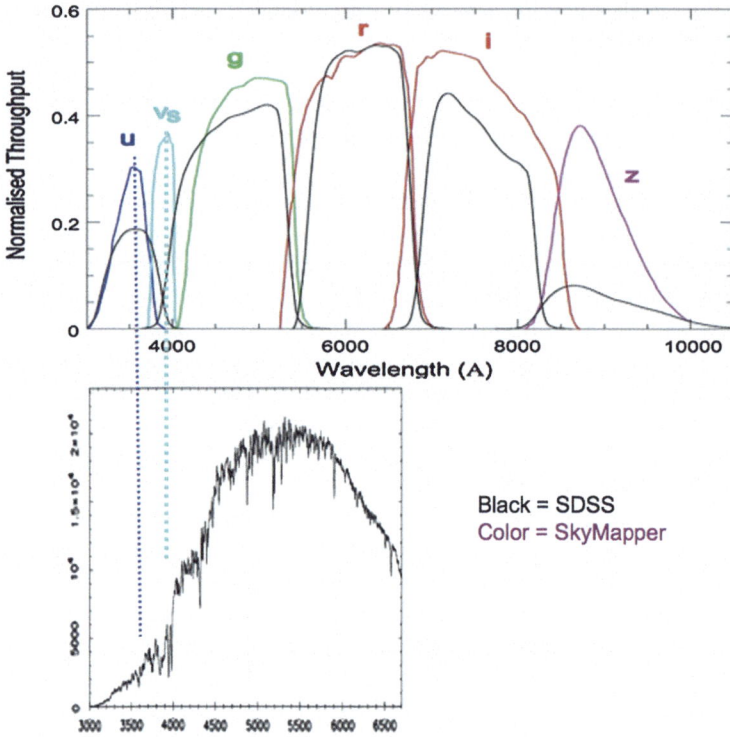

Fig. 1.38 The filter passbands of the SkyMapper and SDSS filter system are aligned with the spectrum of a cooler star to show how the passbands are located relative to features in the spectrum of the star

change as pipelines are developed for the coming high-resolution surveys of large samples of stars.

Currently available compilations of high resolution abundance data for nearby stars include those by Venn et al. (2004: 781 stars) and Soubiran and Girard (2005: 743 stars)

For large optical high-resolution surveys, FGK stars with $T_{\text{eff}} = 4000$–6500 K are usually selected. These stars are cool enough to have plenty of lines and warm enough for analysis to be relatively straightforward. Hotter stars have mostly weaker metallic lines and are often younger and rotating which broadens the lines and makes weak lines difficult to measure. Cooler stars have complex atmospheres with molecules: they are more difficult to analyse from optical spectra, and are better studied in the near-IR, which is also much less affected by interstellar extinction. The infrared is not so good for neutron capture elements.

This is a very brief overview of high resolution spectroscopy and has not attempted to address the many issues involved in abundance analysis related to the physics of energy transfer in stellar atmospheres. The study of hotter and cooler stars are each major specialties in astrophysics.

B.2.5 Techniques for Measuring Stellar Ages

Measuring ages for individual stars remains difficult; see Soderblom (2010) for a comprehensive review.

Nuclear cosmochronology is a fundamental technique for estimating ages: the observed ratios of radioactive and stable species (e.g ratio of Th to U) are compared to the expected production ratios from nucleosynthesis theory. The technique has been used on a few stars. The expected production ratios from theory are somewhat uncertain.

Asteroseismology uses the spectrum of stellar oscillations to estimate the stellar age. The frequency spectrum depends on the structure of the stellar interior, which changes as the star ages. Accurate photometry at the μmag level is needed and is usually done from space. The asteroseismology space missions (CoRoT, Kepler) promise to derive stellar ages with 5–10 % errors. The technique works for main sequence and giant stars for which it is otherwise difficult to determine reliable ages.

Stellar activity and rotation are useful estimators of stellar age. Stars spin down as they age, with rotation periods that typically increase roughly as $(age)^{1/2}$. The rotation periods are measured photometrically or spectroscopically, and the relation between rotation period and age is calibrated empirically on star clusters and the Sun (see Barnes 2007). Chromospheric activity (usually measured from Ca K emission) is associated with rotation and decreases with age. It was much used in the past but is believed to be less accurate for older stars. Figure 1.17 compares rotational (gyro) ages and chromospheric ages for a sample of well studied stars.

If the stellar temperature, metallicity and luminosity (or surface gravity) are known, it is possible to estimate the age of a star from isochrones. In its most direct form, the trigonometric parallax gives the absolute magnitude and hence the stellar luminosity L after bolometric correction. Photometry or spectroscopy gives the temperature T_{eff} and the metallicity. One then compares the location of the star in the $L - T_{\text{eff}}$ plane with theoretical isochrones for the appropriate metallicity. This works if the star is in a region of the $L - T_{\text{eff}}$ plane where the luminosity and temperature depend on age, so it does not work on the unevolved main sequence and not well on the upper giant branch. It can be used for evolved stars which are still close to the main sequence, but accurate temperatures are needed. As seen in Fig. 1.39, one needs to beware of regions near the turnoff where the isochrones cross and the estimated ages can be multi-valued. Subgiants are well suited to isochrone aging because the isochrones are well separated on the subgiant branch: see Fig. 1.39. The Gaia mission will provide a huge increase in the numbers of stars for which accurate luminosities are known, and for which accurate isochrone ages can be derived. Recall that the Gaia distance errors will be about 1 % at $V = 14$.

If parallaxes are not available, isochrone ages can be measured by using the isochrones in the surface gravity—temperature plane. The gravity g is related to the

Fig. 1.39 Isochrones from Bertelli et al. (1994) for stars of near-solar abundance and a range of ages from 1.6 to 16 Gyr. The isochrones can overlap near the turnoff are well separated in the subgiant region

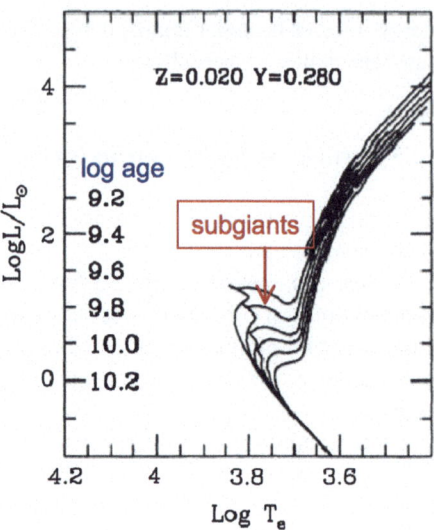

luminosity, mass and temperature by $L = 4\pi G M a T_{\text{eff}}^4/g$ where a is the Stefan-Boltzmann constant and M the stellar mass. The gravity can be measured spectroscopically with a typical error in $\log g$ of about 0.3 for medium resolution spectra and 0.1–0.2 for high resolution spectroscopy. Age errors of about 25 % can be achieved in this way.

Large biases can occur in isochrone ages if the errors in L or $\log g$, temperature and abundance are significant. These occur because the underlying distributions of stellar mass and abundance are not uniform. They reflect the stellar initial mass function, Galactic density distribution and metallicity distribution function. Bayesian techniques can include these underlying mass distributions as priors: see Pont and Eyer (2004).

For populations of stars, such as white dwarfs of the disk, and for clusters of stars, other techniques can be used. The luminosity function of a population of white dwarfs can be used to estimate its age; as the white dwarfs cool and fade, their luminosity function evolves in a predictable way. At the faint end, the white dwarf luminosity function drops rapidly, and the luminosity at which it drops is a measure of the age of the white dwarf population. For example, Leggett et al. (1998) use the luminosity function of white dwarfs in the disk to estimate that the oldest disk stars are about 9 Gyr old. For open and globular clusters, the color-magnitude diagram defines an empirical isochrone which can be dated by comparison with theoretical isochrones of the same metallicity.

B.3 Sources of Models

In this section we will briefly discuss stellar atmosphere models, theoretical isochrones and Galactic models which are important tools for Galactic archaeology.

B.3.1 Stellar Atmosphere Models

Model atmospheres and synthetic spectra are used for deriving element abundances and other stellar parameters from medium and high resolution spectra by comparison with observed spectra. The most widely used synthetic spectra come either from the Kurucz models (e.g. Munari et al. 2005) or the MARCS models (marcs.astro.uu.se). The models available at this time are one-dimensional local thermodynamic equilibrium (LTE) models, giving flux vs wavelength at various spectral resolutions for a wide range of stellar parameters. For example, the RAVE pipeline works by fitting the observed spectra ($R \sim 7000$, SNR ~ 40) in the Ca triplet region (840–880 nm) to a grid of such models. The internal accuracy of these fits is about 0.1 in [M/H], 0.2 in $\log g$ and 135 K in T_{eff}.

B.3.2 Isochrone Models

Isochrones are derived from stellar evolution models. Some of the widely used isochrone libraries come from the Padova, Dartmouth, Victoria-Regina, BaSTI, Geneva and Yale-Yonsei models. They can be found on the www and give isochrones of chosen age metallicity and sometimes [α/Fe], typically tabulating the stellar mass, bolometric luminosity, temperature, gravity, absolute magnitude in various photometric systems, and sometimes useful derived quantities such as the number of stars per solar mass of system at each step along the isochrone.

B.3.3 Galactic Models

Galactic models are constructed to represent the stellar density distribution, kinematics and reddening of the various components of the Galaxy. They are used to simulate observations such as star counts and kinematics and are very useful for planning large surveys. Widely used examples include the Besançon model (Robin et al. 2003), the TRILEGAL model (Girardi et al. 2005) and the Galaxia model (Sharma et al. 2011). Stellar evolution models are part of these Galactic models.

Generating a synthetic catalog of stars in accordance with a given model of galaxy formation has a number of uses. First, it helps to interpret the observational data. Secondly, it can be used to test the theories upon which the models are based. Moreover synthetic catalogs can be used to test the capabilities of different instruments, check for systematics and device strategies to reduce measurement errors. This is well understood by the architects of galaxy redshift surveys who rely heavily on ΛCDM simulations to remove artefacts imposed by the observing strategy (e.g. Colless et al. 2001).

Given the widespread use of synthetic catalogs, a need for faster and accurate methods to generate such synthetic catalogs has recently arisen due to the advent of large scale surveys in astronomy, e.g., future surveys like LSST and GAIA have plans to measure over 1 billion stars. In order to generate a synthetic catalog, one first needs to have a model of the Milky Way. While we are far from a dynamically consistent model, a working framework is fundamental to progress. Inevitably, this will require approximations or assumptions that may not be mutually consistent. Cosmologists already accept such compromises when they relate the observed galaxies to the dark-matter test particles that emerge from cosmological simulations.

There have been various attempts over the past few decades to create a Galaxy model that is constrained by observations. The earliest such attempt was by Bahcall and Soneira (1984) where they assumed an exponential disc with magnitude dependent scale heights. An evolutionary model using population synthesis techniques was presented by Robin and Creze (1986). Given a star formation rate (SFR) and an initial mass function (IMF), one calculates the resulting stellar populations using theoretical evolutionary tracks. Local observations were then used to constrain the SFR and IMF. Bienayme et al. (1987) later introduced dynamical self consistency to constrain the disc scale height (cf. Bertelli et al. 1995).

The present state of the art is described in Robin et al. (2003) and is known as the Besançon model. Here the disc is constructed from a set of isothermal populations that are assumed to be in equilibrium. Analytic functions for density distributions, the age/metallicity relation and the IMF are provided for each population. A similar scheme is also used by the photometric code Trilegal by Girardi et al. (2005).

In spite of its popularity, the current Besançon model has important shortcomings. A web interface exists to generate synthetic catalogs from the model but it has limited applicability for generating wide area surveys and the output is not drawn correctly from a statistical distribution. Discrete step sizes for radial, and angular coordinates-ordinates need to be specified by the user and results might differ depending upon the chosen step size. The scale height and the velocity dispersion of the disc are in reality a function of age but, due to computational complexity, the disc is modeled as a finite set of isothermal discs of different ages. Increasing the number of discs enhances the smoothness of the model but at the price of computational cost (Girardi et al. 2005).

In addition to the disc components, one also needs a model of the stellar halo. Under the hierarchical structure formation paradigm, the stellar halo is thought to have been produced by numerous accretion events and signatures for which should be visible as substructures in the stellar halo. Missions like GAIA, LSST and PanSTARRS are being planned which will enable us to detect substructures in the stellar halo.

A smooth analytic stellar halo as in the Besançon model is inadequate for testing schemes of substructure detection. Furthermore, such a halo does not accomodate known structures like the Sagittarius dwarf stream which may constitute a large fraction of the present halo (Ibata et al. 1995; Chou et al. 2010). Substructures have complex shapes and hence to model them we cannot use the approach of analytic density distributions as discussed earlier. However, N-body models are ideally suited for this task. Brown et al. (2005) attempted to combine a smooth galaxy model with some simulated N-body models of disrupting satellites, but the stellar halo was not simulated in a proper cosmological context.

Recently, using hybrid N-body techniques, Bullock and Johnston (2005) have produced high resolution N-body models of the stellar halo which are simulated within a cosmological context; see also Cooper et al. (2009) and (De Lucia and Helmi 2008) for a similar approach. These can be used to make accurate predictions of the substructures in the stellar halo and also test the ΛCDM paradigm. However, as highlighted by Brown et al. (2005) there are several unresolved issues related to sampling of an N-body model which has prevented their widespread use.

The new Galaxia code (Sharma et al. 2011) allows for fast and accurate methods to convert analytic and N-body models of a galaxy into a synthetic catalog of stars. This relieves the burden of generating catalogs from modelers on one hand and on the other hand allow the testing of models generated by different groups. This is a new scheme for sampling the analytical models which enables the user to generate continuous values of the variables like position and age of stars. Instead of a set of discs at specified ages, our methodology allows us to generate a disc which is continuous in age.

As a concrete example, Sharma et al. (2011) demonstrate the Besançon analytical model for the disc. To model the disc kinematics more accurately, Galaxia employs the Shu (1969) distribution function that describes the non-circular motion in the plane of the disc. This function has now been generalised for a range of rotation curves (Sharma and Bland-Hawthorn 2013). For the stellar halo, Galaxia uses the simulated N-body models of Bullock and Johnston (2005) which can reproduce the substructure in the halo. We show a scheme for sampling the N-body particles such that the sampled stars preserve the underlying phase space density of N-body particles.

Another powerful aspect of Galaxia is the use of Markov Chain Monte (MCMC) Carlo methods that allow the determination of many parameters from model fitting applied to a large stellar survey. For the first time, we find that two very different, large stellar surveys (GCS, RAVE) yield the same kinematic parameters for the local disk (Sharma et al. 2013). Moreover, the more spatially extended RAVE data demand the use of a Shu distribution function to yield meaningful results. This has strong parallels with MCMC analysis of galaxy redshift surveys in the context of ΛCDM simulations. In time, it will be possible to apply MCMC model fitting of the most detailed Galaxy simulations to stellar surveys which of course reinforces the remarkable complementarity between near-field and far-field cosmology.

Finally, we would like to commend the ESA-ESO Working Group Report 4: Galactic Populations, Chemistry and Dynamics (Turon and Primas 2008). This is a very useful compendium of the major problems in Galactic astronomy, ways to attack them, and major surveys past, present and future.

References

Abadi MG, Navarro JF, Steinmetz M & Eke VR 2003, Ap. J. 597, 21.
Abbasi et al.: 2012, Phys. Rev. D. 85, 2002.
Abe, F. et al. 1999, Astron. J. 118, 261–72.
Abel T, Bryan GL & Norman ML 2000, Ap. J. 540, 39.
Abel T, Bryan GL & Norman ML 2002, Science 295, 93.
Abbott BP et al. 2009, Rep. Prog. Phys. 72.
Adams F et al. 2006, Ap. J. 641, 504.
Agertz O, Teyssier R & Moore B, 2009, MNRAS, 397, 64.
Allen L et al. 2007, In Protostars and Planets V, eds. Reipurth, Jewitt & Keil, 361–376.
An FP et al. 2012, Phys. Rev. Lett. 108, 1803.
Ambartsumian VA 1958, IAU Symp. 5, 4.
Anderson ME & Bregman JN 2011, Ap. J. 737, 22.
Antoja T et al. 2012, Mon. Not. R. Astron. Soc. 426, L1–L5.
Argast D, Samland M, Gerhard OE, Thielemann F-K 2000, Astron. Astrophys. 356, 873–87.
Arimoto N & Yoshii Y 1987, Astron. Astrophys. 173, 23–38.
Arnaboldi M, Freeman KC, Hui X, Capaccioli M, Ford H 1994, ESO Messenger 76, 40–44.
Arnould M & Goriely S 2001, In Astrophysical Ages and Time Scales, eds T von Hippel, C Simpson & N Manset, Astron. Soc. Pac. 245, 252–261.
Arp H 1962, Ap. J. 136, 66–74.
Arras P & Wasserman I 1999, MNRAS 306, 257–78.
Ashman KM & Zepf SE 1992, Ap. J. 384, 50–61.
Athanassoula L 2007, MNRAS 377, 1569.
Audouze J & Silk J 1995, Ap. J. 451, L49–52.
Aumer M & Binney J 2009, MNRAS 397, 1286.
Babcock HW 1939, Lick Obs. Bull. 498, 41.
Babich D & Loeb A 2006, Ap. J. 640, 1.
Babul A & Rees M 1992, MNRAS 255, 346–50.
Bagchi J et al. 2006, Science 314, 791.
Bahcall JN & Soneira RM 1984, Ap. J. Suppl. 44, 73.
Bakos J, Trujillo I & Pohlen M 2008, Ap. J. 683, L103.
Baldry IK et al. 2004, Ap. J. 600, 681.
Baldry IK et al. 2008, MNRAS 388, 945.
Barkana R & Loeb A 1999, Ap. J. 523, 54.
Barkana R & Loeb A 2007, Rep. Prog. Phys. 70, 627.
Barnes S 2007, Ap. J. 669, 1167.
Barnes JE & Hernquist LE 1992, Nature 360, 715.
Bastian N et al. 2012, MNRAS 419, 2606.
Battaglia G et al. 2005, MNRAS 364, 433.
Battaglia G et al. 2012, Ap. J. 761, L31.
Baugh C, Cole S, Frenk C 1996, MNRAS 283, 1361–78.
Belenkiy A 2012, Physics Today, 65, 38.
Bell E, de Jong R. 2000. MNRAS 312, 497–520.
Bell EF et al. 2007, Astron. J. 663, 834–843.
Bell EF et al. 2008, Ap. J. 680, 295.
Belokurov V et al. 2009, MNRAS 397, 1748.
Benjamin RA & Danly L 1997, Ap. J. 481, 764.
Benjamin RA et al. 2005, Ap. J. 630, L149.
Bensby T et al. 2005, A&A 433, 185.
Benson AJ 2010, Physics Reports 495, 33.
Benson AJ & Bower R 2011, MNRAS 410, 2653.
Bertelli G et al. 1995, Astron. Astrophys. 301, 381.

Bertelli G et al. 1994, Astron. Astrophys. Suppl. 106, 275.
Bertelli G & Nasi E 2001, Astron. J. 121, 1013–1023.
Bertoldi F & McKee CF 1992, Ap. J. 395, 140.
Bica E, Alloin D & Schmidt A, 1990, MNRAS 242, 241–49.
Bienayme O, Robin A & Creze M 1987, Astron. Astrophys. 180, 94.
Binggeli B, Sandage A & Tammann GA 1985, Astron. J. 90, 1681–771.
Binney JJ 1977, Ap. J. 215, 483.
Binney JJ & Lacey C 1988, MNRAS 230, 597–627.
Binney JJ, Dehnen W & Bertelli G 2000, MNRAS 318, 658.
Binney JJ & Tremaine SD 2008, Galactic Dynamics: Second Edition, (Princeton University Press).
Birnboim Y & Dekel A 2003, MNRAS 345, 349.
Blake C et al. 2011, MNRAS 418, 1707.
Bland-Hawthorn J, Freeman KC & Quinn PJ 1997, Ap. J. 490, 143.
Bland-Hawthorn J et al. 1998, MNRAS 299, 611–624.
Bland-Hawthorn J 1999, Nature 400, 220.
Bland-Hawthorn J & Freeman KC 2000, Science 287, 79–84.
Bland-Hawthorn J & Freeman KC 2004, PASA 21, 110.
Bland-Hawthorn J & Freeman KC 2006, Mem. Soc. della Societa Astron. Ital., 77, 1095.
Bland-Hawthorn J & Maloney PR 2001, In Extragalactic Gas at Low Redshift, eds. J. Mulchaey & J. Stocke, Astron. Soc. Pac. 254, 267–82.
Bland-Hawthorn J et al. 2005, Ap. J. 629, 239.
Bland-Hawthorn J et al. 2007, Ap. J. 670, L109.
Bland-Hawthorn J 2008, In Galaxies in the Local Volume, Astrophys. Sp. Science 1, 259.
Bland-Hawthorn J 2009, In The Galaxy Disk in Cosmological Context, IAU Symp. 254, 241.
Bland-Hawthorn J, Krumholz MR & Freeman KC 2010a, Ap. J. 713, 166.
Bland-Hawthorn J et al. 2010b, Ap. J. 721, 582.
Blanton M & Moustakas J 2009, Annu. Rev. Astron. Astrophys. 47, 159.
Blanton M et al. 2003, Ap. J. 594, 186.
Blitz L, Spergel DN, Teuben PJ, Hartmann D & Burton WB 1999, Ap. J. 514, 818–843.
Blumenthal G, Faber S, Primack J & Rees M 1984, Nature 311, 517–525.
Bondi H 1952, MNRAS 112, 195.
Bondi H & Hoyle F 1944, MNRAS 104, 273.
Bondi H, Hoyle F & Lyttleton RA 1947, MNRAS 107, 184.
Bosma A 1978, PhD, University of Groningen, Netherlands.
Boss L 1908, Astron. J. 26, 31–36.
Bouwens R et al. 2007, Ap. J. 670, 928.
Bower R, Kodama T & Terlevich A 1998, MNRAS 299, 1193–1208.
Bower R et al. 2006, MNRAS 370, 645.
Boylan-Kolchin M et al. 2009, MNRAS 398, 1150.
Braun R & Burton WB 1999, Astron. Astrophys. 341, 437–50.
Bromm V & Yoshida N 2011, Annu. Rev. Astron. Astrophys. 49, 37.
Bromm V, Coppi, PS & Larson RB 1999, Ap. J. 527, L5.
Bromm V, Coppi, PS & Larson RB 2002, Ap. J. 564, 23.
Bromm V, Yoshida N, Hernquist L & McKee CF 2009, Nature 459, 49.
Brook C et al. 2011, MNRAS 415, 1051.
Brooks AM et al. 2009, Ap. J. 694, 396.
Brown RL & van den Bout PA 1991, Astron. J. 102, 1956.
Brown A, Velazquez HM & Aguilar LA 2005, MNRAS 359, 1287.
Bubar E & King JR 2010, Astron. J. 140, 293.
Bullock JS, Kravtsov A & Weinberg D 2000, Ap. J. 539, 517–21.
Bullock JS & Johnston KV 2005 Ap. J. 635, 931.
Bullock JS et al. 2001 Ap. J. 321, 559.
Burbidge EM, Burbidge GR, Fowler WA & Hoyle F 1957, Rev. Mod. Phys. 29, 547–650.

Bureau M, Freeman KC, Pfitzner DW & Meurer GR 1999, Astron. J. 118, 2158–71.
Burris DL et al. 2000, Ap. J. 544, 302–319.
Burrows A et al. 2006, Ap. J. 640, 878.
Burstein D, Faber SM, Gaskell CM & Krumm N 1984, Ap. J. 287, 586–609.
Butcher H 1987, Nature 328, 127–31.
Cabrera-Lavers A et al. 2008 Astron. Astrophys. 491, 781.
Calcaneo-Roldan C, Moore B, Bland-Hawthorn J & Sadler EM 2000, MNRAS 314, 324–33.
Caldwell N & Rose J 1998, Astron. J. 115, 1423–1432.
Caloi V & d'Antona F 2011, MNRAS 417, 228.
Carlberg RG & Sellwood JA 1985, Ap. J. 292, 79–89.
Carlberg RG 1986, Ap. J. 310, 593–596.
Carlberg et al.: 1985, Ap. J. 294, 674.
Carney BW, Aguilar L, Latham DW & Laird JB 1990, Astron. J. 99, 201–20.
Carney BW, Laird JB, Latham DW & Aguilar LA 1996, Astron. J. 112, 668–92.
Carney BW 1993, In The Globular Cluster - Galaxy Connection eds. GH Smith & JP Brodie, Astron. Soc. Pac. 48, 234–45.
Carollo D et al. 2007, Nature 450, 1020.
Carpenter JM 2000, Astron. J. 120, 3139–61.
Carraro G, Ng Y & Portinari L 1998, MNRAS 296, 1045–56.
Carretta E et al. 2002, Astron. J. 124, 481.
Castro S et al. 1997, Astron. J. 114, 376–87.
Castro S, Porto de Mello GF & da Silva L 1999, MNRAS 305, 693.
Cayrel R et al. 2001, Nature 409, 691–92.
Cayrel R et al. 2004, Astron. Astrophys. 416, 1117.
Ceccarelli C et al. 2005, In Protostars and Planets V, eds. Reipurth, Jewitt & Keil, 47–62.
Ceverino D, Dekel A & Bournaud F 2010, Mon. Not. R. Astron. Soc. 404, 2151–2169.
Chaboyer B, Demarque P, Kernan P & Krauss L 1998, Ap. J. 494, 96–110.
Chaboyer B, Demarque P & Sarajedini A 1996, Ap. J. 459, 558–69.
Chaboyer B 1998, Physics Reports, 307, 23–30.
Chandrasekhar S 1961, Hydrodynamic & Hydromagnetic Stability, Dover Books.
Charlot S, Silk J 1994, Ap. J. 432, 453–463.
Chiappini C, Matteucci F & Romano D 2001, Ap. J. 554, 1044–58.
Chiappini C et al. 1999, Ap. J. 515, 226.
Chiba M 2001, Ap. J. 565, 17–23.
Chiba M & Beers T 2000, Astron. J. 119, 2843–65.
Chou M-Y et al. 2010, Ap. J. 708, 1290.
Christensen-Dalsgaard J 1986, IAU Symp. 123, 295.
Christlieb N et al. 2001, In Astrophysical Ages and Time Scales, eds T von Hippel, C Simpson & N Manset, Astron. Soc. Pac. 245, 298–300.
Christlieb N et al. 2002, Nature 419, 904.
Clark PC, Glover SCO & Klessen RS 2008, Ap. J. 672, 757.
Clark PC, Glover SCO, Klessen RS & Bromm V 2011, Ap. J. 727, 110.
Clarke CJ, Bonnell IA, Hillenbrand LA 2000, In Protostars and Planets IV, eds. V Mannings, AP Boss, SS Russell, 151–177 (University of Arizona Press).
Clayton D 1987, Ap. J. 315, 451–459.
Clayton D 1988, MNRAS 234, 1–36.
Codis S et al. 2012, MNRAS 427, 3320.
Cohen JG, Christlieb N, Beers TC, Gratton R & Carretta E 2002, Astron. J. 124, 470.
Cohen JG et al. 2007, Ap. J. 659, L161.
Cole S, Aragon-Salamanca A, Frenk CS, Navarro JF, Zepf SE 1994, MNRAS 271, 781–806.
Cole S et al. 2005, MNRAS 362, 505.
Coleman MG et al. 2004 Astron. J. 127, 832.
Colin P, Avila-Reese V & Valenzuela O 2000, Ap. J. 542, 622–30.

Colless MM et al. 2001, MNRAS 328, 1039.
Combes F et al. 1990, Astron. Astrophys. 233, 82.
Concannon KD, Rose JA & Caldwell N 2000, Ap. J. 536, L19–22.
Connors TW, Kawata D & Gibson BK 2006, MNRAS 371, 108.
Conroy C 2012, Ap. J. 758, 21.
Conti PS, Wallerstein G & Wing RF 1965, Ap. J. 142, 999.
Conroy C & Spergel D 2011, Ap. J. 726, 36.
Cooke R et al. 2011, MNRAS 412, 1047.
Cooper AP et al. 2009, MNRAS 406, 744.
Courteau S, de Jong RS & Broeils AH 1996, Ap. J. 457, L73–76.
Cowan JJ, McWilliam A, Sneden C & Burris DL 1997, Ap. J. 480, 246–54.
Cox TJ & Loeb A 2008, MNRAS 386, 461.
Creze M, Chereul E, Bienayme O & Pichon C 1998, Astron. Astrophys. 329, 920–36.
Croton D et al. 2012, MNRAS 365, 11.
Croton DJ et al. 2006, Mon. Not. R. Astron. Soc. 365, 11–28.
Da Costa G 1999, In The Third Stromlo Symposium, The Galactic Halo, eds. B Gibson, T Axelrod & M Putman, Astron. Soc. Pac. 165:153–166.
Dai X et al. 2010, Astrophys. J. 719, 119–125.
Dalcanton JJ, Spergel DN & Summers FJ 1997, Ap. J. 482, 659–76.
Danforth CW & Shull MJ 2008, Ap. J. 679, 194.
Davis M, Efstathiou G, Frenk CS & White SDM 1985, Ap. J. 292, 371.
Davies RL et al. 2001, Ap. J. 548, L33–L36.
de Bruijne J 1999, MNRAS 306, 381–93.
de Grijs R, Kregel M & Wesson KH 2001, MNRAS 324, 1074–86.
de Jong R & Lacey C 2000, Ap. J. 545, 781–97.
de Jong R et al. 2012, SPIE Conf. 8446, 0.
De Lucia & Helmi 2008, MNRAS 391, 14.
De Silva GM et al. 2006, Astron. J. 131, 455.
De Silva GM et al. 2007a, Astron. J. 133, 694.
De Silva GM et al. 2007b, Astron. J. 133, 1161.
De Silva GM, Freeman KC & Bland-Hawthorn J 2009, PASA 26, 11.
De Silva GM et al. 2013, MNRAS in press.
de Vaucouleurs G 1961, Ap. J. Suppl. 6, 213.
de Vaucouleurs G 1970a, Science 167, 1203.
de Vaucouleurs G 1970b, IAU Symp. 38, 18.
De Young DS 2003, New Astron. Rev. 47, 545.
de Zeeuw PT, Hoogerwerf R, de Bruijne JHJ, Brown AG & Blaauw A 1999, Astron. J. 117, 354–99.
Dehnen WA 2000, Astron. J. 119, 800–12.
Dekel A & Silk J 1986, Ap. J. 303, 39–55.
Dekel A et al. 2009 Nature 457, 451.
Dijkstra M et al. 2004, Ap. J. 601, 666.
Dodd KN 1953 MNRAS 113, 484.
Doroshkevich et al.: 1967, Soviet Astron. 11, 233.
Dressler A 1980, Ap. J. 236, 351–365.
Driver SP et al. 2007, Ap. J. 657, L85.
Driver SP et al. 2011, Mon. Not. R. Astron. Soc. 413, 971–995.
Driver SP et al. 2013, MNRAS in press.
Dubois Y et al. 2012, MNRAS 420, 2662.
Durrell PR, Harris WE & Pritchet CJ 2001, Astron. J. 121, 2557–71.
Edmunds MG 1990, Nature 348, 395–396.
Edvardsson B et al. 1993, Astron. Astrophys. 275, 101–52.
Efremov YN 2011, Astron. Reports 55, 108–122.
Efstathiou GP 1992, MNRAS 256, 43P.

Efstathiou GP et al. 1990, Nature 348, 705.
Efstathiou GP et al. 2002, MNRAS 330, L29–L35.
Eggen OJ, Lynden-Bell D & Sandage AR 1962, Ap. J. 136, 748–66.
Eggen OJ & Sandage AR 1959, MNRAS 119, 255–77.
Eggen OJ & Sandage AR 1969, Ap. J. 158, 669–84.
Eggen OJ 1977, Ap. J. 215, 812–26.
Einasto J, Saar E, Kaasik A & Chernin AD 1974, Nature 252, 111.
Eisenstein DJ et al. 2005, Ap. J. 633, 560.
Ekholm T & Teerikorpi P 1994, Astron. Astrophys. 284, 369.
Ellis RS, Abraham RG, Brinchmann J & Menanteau F 2000, Astron. Geophys. 41, 10–16.
Elmegreen BG 2007, Ap. J. 668, 1064.
Elmegreen BG 2010, Ap. J. 712, L184.
Elmegreen BG, Efremov Y, Pudritz RE & Zinnecker H 2000, In Protostars and Planets IV, eds V Mannings, AP Boss, SS Russell, 179–215 (University of Arizona Press).
Elmegreen D et al. 2005, Ap. J. 631, 85.
Elmegreen BG & Elmegreen DM 2006, Ap. J. 650, 644–660.
Englmaier P & Gerhard OE 2006, Celest. Mech. & Dyn. Astron., 94, 369–379.
Eskridge PB et al. 2000, Astron. J. 119, 536.
Exter K, Barlow MJ, Walton NA & Clegg RES 2001, Astro. Sp. Science 277, 199–99.
Faber S 1973, Ap. J. 179, 731–54.
Fabian AC 1994, Annu. Rev. Astron. Astrophys. 32, 277.
Fabian AC & Nulsen PE 1977 MNRAS 180, 479.
Falcon-Barroso J et al. 2006, MNRAS 369, 529.
Fall SM & Efstathiou GP 1980, MNRAS 193, 189–206.
Fall SM 1983, In Internal Kinematics and Dynamics of Galaxies, ed. E Athanassoula, 391–98 (Reidel).
Fall SM 2006, Ap. J. 652, 1129.
Fall SM, Chandar R & Whitmore BC 2005, Ap. J. Lett. 631, L133.
Fall SM, Chandar R & Whitmore BC 2009, Ap. J. 704, 453.
Fall SM, Krumholz MR & Matzner CD 2010, Ap. J. 710, L142.
Fan X et al. 2002, Astron. J. 123, 1247.
Fardal MA et al. 2001, Ap. J. 562, 605.
Faúndez S et al. 2004, Astron. Astrophys. 426, 97.
Feltzing S, Holmberg J & Hurley JR 2001, Astron. Astrophys. 377, 911–24.
Feltzing S & Holmberg J 2000, Astron. Astrophys. 357, 153–63.
Ferraro F et al. 2009, Nature 462, 483.
Flynn C et al. 2006, MNRAS 372, 1149.
Font A, Navarro J, Stadel J & Quinn T 2001, Ap. J. 563, L1–L4.
Fontani F et al. 2005, Astron. Astrophys. 443, 921.
Fontanot F et al. 2009, MNRAS 397, 1776.
Forero-Romero JE et al. 2011, Mon. Not. R. Astron. Soc. 417, 1434–1443.
Förster-Schreiber NM et al. 2009, Ap. J. 706, 1364.
Fowler WA & Hoyle F 1960, Astron. J. 65, 345–45.
Frayer DT & Brown RL 1997, Ap. J. Suppl. 113, 221–43.
Frebel A et al. 2005, Nature 434, 871.
Frebel A et al. 2010a, Ap. J. 708, 560.
Frebel A et al. 2010b, Nature 464, 72.
Freedman W et al. 2001, Ap. J. 553, 47.
Freeman KC, Illingworth G & Oemler A 1983, Ap. J. 272, 488–508.
Freeman KC 1970, Ap. J. 160, 811–30.
Freeman KC 1987, Annu. Rev. Astron. Astrophys. 25, 603–32.
Freeman KC 1991, In Dynamics of Disk Galaxies, ed. B Sundelius (University of Goteborg), 15.

Freeman KC 1993, In The Globular Cluster - Galaxy Connection, eds. GH Smith & JP Brodie, Astron. Soc. Pac. 48, 608–14.
Freeman KC 1997, In The Nature of Elliptical Galaxies, eds. Arnaboldi, Da Costa & Saha, Astron. Soc. Pac. 116, 1.
Freeman KC 2008, In Formation and Evolution of Galaxy Disks, eds. Funes & Corsini, Astron. Soc. Pac. 396, 3.
Freeman KC & Bland-Hawthorn J 2008, In Panoramic Views of Galaxy Formation and Evolution ASP Conference Series, eds. K Tadayuki, Y Toru & A Kentaro, Astron. Soc. Pacific 399, 439.
Freeman KC & Bland-Hawthorn J 2002, Annu. Rev. Astron. Astrophys. 40, 487.
Freeman KC & Bland-Hawthorn J 2008, In Panoramic Views of Galaxy Formation & Evolution, Astron. Soc. Pac. 399, 439.
Freeman KC et al. 2013, MNRAS 428, 3660.
Friel ED 1995, Annu. Rev. Astron. Astrophys. 33, 381–414.
Friel ED & Janes KA 1993, Astron. Astrophys. 267, 75–91.
Fry AM, Morrison HL, Harding P & Boroson TA 1999, Astron. J. 118, 1209–19.
Fryer CL 1999, Ap. J. 522, 413.
Fryer CL & Heger A 2000, Ap. J. 541, 1033.
Fryer CL, Woosley S & Heger A 2001, Ap. J. 550, 372.
Fulbright JP et al. 2007, Ap. J. 661, 1152.
Fumagalli M et al. 2011, Science 334, 1245.
Furlanetto S & Loeb A 2004, Ap. J. 611, 642.
Gaensler BM et al. 2008, PASA 25, 184.
Gallagher JS et al. 2003, Ap. J. 588, 326.
Gao L & Theuns T 2007, Science 317, 1527.
Garcia-Berro E & Iben I 1994, Ap. J. 434, 306.
Gardner E & Flynn C 2010, MNRAS 405, 545.
Genzel R et al. 2006, Nature 442, 786.
Gilmore G & Reid IN 1983, MNRAS 202, 1025–47.
Gilmore G, Wyse RFG & Jones BJ 1995, Astron. J. 109, 1095–1111.
Gilmore G, Wyse RFG & Kuijken K 1989, Annu. Rev. Astron. Astrophys. 27, 555–627.
Gilroy K, Sneden C, Pilachowski CA & Cowan JJ 1988, Ap. J. 327, 298–320.
Gimenez A & Favata F 2001, In Astrophysical Ages & Time Scales, eds. T von Hippel, C Simpson & N Manset, Astron. Soc. Pac. 245, 304–306.
Girardi L et al. 2005, Astron. Astrophys. 436, 895.
Glazebrook K 2012, astro-ph/1212.3065.
Gnedin NY 2000, Ap. J. 542, 535.
Gnedin NY & Kravtsov A 2006, Ap. J. 645, 1054.
Gnedin OY, Lee HM & Ostriker JP 1999, Ap. J. 522, 935–49.
Götz M & Köppen J 1992, Astron. Astrophys. 262, 455–67.
Gomez A et al. 1997, In Hipparcos - Venice '97, 621–24 (ESA Publ.).
Gonzalez V et al. 2011, Ap. J. 735, L34.
Goriely S & Arnould M 1996, Astron. Astrophys. 312, 327–337.
Goriely S & Arnould M 2001, Astron. Astrophys. 379, 1113–22.
Gottlöber S, Yepes G, Wager C & Sevilla R 2006, astro-ph/0608289.
Gottlöber S, Hoffman Y & Yepes G 2010, astro-ph/1005.2687.
Gough DO 1987, Nature 326, 257–259.
Gough DO 2001, In Astrophysical Ages & Time Scales, eds. T von Hippel, C Simpson & N Manset, Astron. Soc. Pac. 245, 31–43.
Governato F et al. 2012, MNRAS 422, 1231.
Gratton RG et al. 1997, Ap. J. 491, 49–71.
Gratton R, Sneden C & Carretta E 2004, Annu. Rev. Astron. Astrophys. 42, 385.
Grcevich J & Putman ME 2009, Ap. J. 696, 385.
Grebel EK 2001, Astro. Sp. Science 277, 231–39.

Grebel EK & Gallagher JS 2004, Ap. J. 610, L89.
Gregg MD, West MJ. 1998. Nature 396, 549–52.
Grenon M 1999, Astro. Sp. Science 265, 331–36.
Greif T et al. 2011, Ap. J. 736, 147.
Griffin R 1998, Observatory, 118, 223–25.
Guenther DB 1989, Ap. J. 339, 1156–59.
Guenther DB & Demarque P 1997, Ap. J. 484, 937–59.
Gunn JE & Gott JR 1972, Ap. J. 176, 1.
Gunn JE 1977, Ap. J. 218, 592.
Guth A 1997, The Inflationary Universe (Addison-Wesley).
Haiman Z, Thoul AA & Loeb A 1996, Ap. J. 464, 523.
Hamann F & Ferland G 1999, Annu. Rev. Astron. Astrophys. 37, 487–531.
Hansen BMS et al. 2007, Astrophys. J. 671, 380–401.
Harding P et al. 2001, Astron. J. 122, 1397–1419.
Harris WE 1991, Annu. Rev. Astron. Astrophys. 29, 543–79.
Harris WE & Pudritz RE 1994, Ap. J. 429, 177–91.
Harris WE 2002, IAU Symp. 207, 755.
Hartmann L, Ballesteros-Paredes J & Bergin EA 2001, Ap. J. 562, 852.
Hartkopf WI & Yoss KM 1982, Astron. J. 87, 1679–709.
Heckman TM et al. 1989, Ap. J. 342, 735.
Heger A & Woosley S 2002, Ap. J. 567, 532.
Heitsch F & Putman ME 2009 Ap. J. 698, 1485.
Helmi A 2008, Astron. Astrophys. Rev. 15, 145.
Helmi A, Springel V & White SDM 2001, Phys. Rev. D. 66:6, 063502.
Helmi A, White SDM, de Zeeuw PT & Zhao H-S 1999, Nature 402, 53–55.
Helmi A & White SDM 1999, MNRAS 307, 495–517.
Helmi A, Zhao H-S & de Zeeuw PT 1999, In The Third Stromlo Symposium: The Galactic Halo, eds. BK Gibson, TS Axelrod, ME Putman, Astron. Soc. Pac. 165:125–29.
Henry RBC 1998, In Abundance Profiles, Diagnostic Tools for Galaxy History, eds. D Friedli, M Edmunds, C Robert & L Drissen, 47–59.
Hermit S et al. 1996, MNRAS 283, 709.
Hernquist L & Quinn PJ 1988, Ap. J. 331, 682–698.
Heyer MH & Brunt CM 2004, Ap. J. 615, L45.
Hibbard JE & van Gorkom JH 1996, Astron. J. 111, 655.
Hill V et al. 2002, Astron. Astrophys. 387, 560.
Hill A et al. 2008, Ap. J. 686, 363.
Hirshfeld A, McClure R & Twarog BA 1978, In The HR Diagram - the 100th Anniversary of Henry Norris Russell, eds. AG Davis Philip & DS Hayes, 163 (Reidel).
Hoffman Y & Ribak E 1991, Ap. J. 380, L5.
Hogan CJ & Dalcanton JJ 2001 Ap. J. 561, 35–45.
Hohl F 1970, IAU Symp. 38, 368.
Holtzman JA et al. 1992, AJ, 103, 691–702.
Hopkins AM, Irwin MJ & Connolly AJ 2001, Ap. J. 558, L31–33.
Hopkins A & Beacom J 2006, Ap. J. 651, 142.
Hosokawa T, Omukai K, Yoshida N & Yorke HW 2011, Science 334, 1250.
Howard CD et al. 2009, Ap. J. 702, L153.
Hoyle F 1966, Galaxies, Nuclei & Quasars (London: Heinemann).
Hoyle F & Lyttleton RA 1939, Proc. Cam. Phil. Soc. 35, 405.
Hubble E & Humason M 1931, Ap. J. 74, 43–80.
Huff EM & Stahler SW 2006, Ap. J. 644, 355.
Hunt R 1971, MNRAS 154, 141.
Hunt R 1975, MNRAS 173, 465.
Ibata R, Gilmore G & Irwin MJ 1994, Nature 370, 194–196.

Ibata R, Gilmore G & Irwin MJ 1995, MNRAS 277, 781–800.
Ibata R, Irwin M, Lewis G, Ferguson AMN & Tanvir N 2001a, Nature 412, 49–52.
Ibata R, Lewis GF, Irwin MJ, Totten E & Quinn T 2001b, Ap. J. 551, 294–311.
Ibata R, Irwin MJ, Lewis GF & Stolte A 2001c, Ap. J. 547, L133–36.
Ibata R & Lewis G 1998, Ap. J. 500, 575–90.
Ibata R, Wyse RFG, Gilmore G, Irwin MJ & Suntzeff NB 1997, Astron. J. 113, 634–55.
Ibata R et al. 2005, Ap. J. 634, 287–313.
Ikeuchi S 1986, Astro. Sp. Science 118, 509–14.
Irwin MJ et al. 2005, Ap. J. 628, L105–L108.
Ishimaru Y & Wanajo S 2000, In The First Stars, eds. A Weiss, TG Abel & V Hill, 189–193 (Berlin, Springer).
Ivezic Z et al. 2008, Ap. J. 684, 287.
Jablonka P, Martin P & Arimoto N 1996, Astron. J. 112, 1415–22.
Janes KA & Phelps RL 1994, Astron. J. 108, 1773–85.
Janka H et al. 2007, Phys. Rep., 442, 38.
Jeffries RD 2007, MNRAS 381, 1169.
Jenkins A & Binney J 1990, MNRAS 245, 305–17.
Jenkins A et al. 2001, MNRAS 321, 372–84.
Johnson JA, Bolte M 2001, Ap. J. 554, 888–902.
Johnston KV, Hernquist L & Bolte M 1996, Ap. J. 465, 278–287.
Johnston KV, Sackett PD & Bullock JS 2001, Ap. J. 557, 137–149.
Johnston KV, Spergel DN & Haydn C 2002, Ap. J. 570, 656.
Johnston KV 1998, Ap. J. 495, 297–308.
Jones L, Smail I & Couch W 2000, Ap. J. 528, 118–122.
Jones L & Worthey G 1995 Ap. J. 446, L31–L35.
Joung MR, Bryan GL & Putman ME 2012, Ap. J. 745, 148.
Just A 2001, In Disks of Galaxies, Kinematics, Dynamics and Perturbations, eds. E Athanassoula, A Bosma, I Puerari, Astron. Soc. Pac. 275, 117.
Kaiser N 1986, Mon. Not. R. Astron. Soc. 222, 323–345.
Kaiser N 1987, Mon. Not. R. Astron. Soc. 227, 1–21.
Käppeler F, Thielemann FK & Wiescher M 1998, Annu. Rev. Nucl. Part. Sci. 48, 175–251.
Kafle PR, Sharma S, Lewis GF & Bland-Hawthorn J 2012, Ap. J. 761, 98.
Kafle PR, Sharma S, Lewis GF & Bland-Hawthorn J 2013, MNRAS press.
Kahn FD & Woltjer L 1959, Ap. J. 130, 705.
Kaiser N 2002, Elements of Astrophysics (University of Hawaii Press).
Kalirai J 2012, Nature 486, 90.
Kalnajs A 1983, IAU Symp. 100, 109–115 (+comment).
Kamionkowski M & Liddle AR 2000, Phys. Rev. Lett. 84, 4525–8.
Karlsson T 2005, Astron. Astrophys. 439, 93.
Karlsson T & Gustafsson B 2001, Astron. Astrophys. 379, 461–81.
Karlsson T & Gustafsson B 2005, Astron. Astrophys. 436, 879.
Karlsson T, Bland-Hawthorn J, Freeman KC & Silk J 2012, Ap. J. 759, 111.
Karlsson T, Bromm V & Bland-Hawthorn J 2013, Rev. Mod. Phys. in press.
Katz N, Keres D, Dave R & Weinberg DH 2003, In The IGM/Galaxy Connection, 185 (Kluwer).
Kauffmann G 1996, MNRAS 281, 487–492.
Kauffmann G & Charlot S 1998, MNRAS 294, 705–17.
Kauffmann G, White SDM & Guiderdoni B 1993, MNRAS 264, 201–18.
Kauffmann G et al. 2004, MNRAS 353, 713.
Kauffmann G, Li C & Heckman T 2010, MNRAS 409, 491.
Kennicutt RC 1989, Ap. J. 344, 685.
Keres D et al. 2005, MNRAS 363, 2.
Keres D & Hernquist LE 2009, Ap. J. 700, L1.
Keres D et al. 2009, MNRAS 395, 160.

Kimm T et al. 2012, MNRAS 425, L96.
King IR 1958, Astron. J. 63, 109–13.
Kinman TD 1959, MNRAS 119, 538–58.
Kirby E et al. 2008, Ap. J. 685, L43.
Kitayama T & Ikeuchi S 2000, Ap. J. 529, 615.
Klement RJ 2010, Astron. Astrophys. Rev. 18, 567–594.
Klypin A, Kravtsov AV, Valenzuela O & Prada F 1999, Ap. J. 522, 82–92.
Klypin A, Hoffman Y, Kravtsov AV & Gottlöber S 2003, Ap. J. 596, 19.
Klypin A, Trujillo-Gomez S & Primack J 2011, Ap. J. 740, 102.
Kochanek C 1996, Ap. J. 457, 228–43.
Komatsu E et al. 2011, Ap. J. Suppl. 192, 18.
Kormendy J 1993, IAU Symp. 153, 209.
Kormendy J & Kennicutt RC 2004, Annu. Rev. Astron. Astrophys. 42, 603.
Kormendy J, Drory N, Bender R & Cornell ME 2010, Ap. J. 723, 54.
Kravtsov AV, Klypin A & Hoffman Y 2002 Ap. J. 571, 563.
Kroupa P 2001, MNRAS 322, 231.
Kroupa P 2002, MNRAS 330, 707–18.
Kroupa P, Pawlowski M & Milgrom M 2012, Int. J. Mod. Phys., 21:14.
Krumholz MR & Tan JC 2007, Ap. J. 654, 304.
Kuijken K & Gilmore G 1991, Ap. J. 367, L9–L13.
Kuiper E et al. 2011, MNRAS 417, 1088.
Kuntschner H 2000, MNRAS 315, 184–208.
Kuntschner J & Davies RL 1998, MNRAS 295, L29–L33.
Kurucz RL 1991, In The Solar Interior and Atmosphere, eds. AN Cox, WC Livingston, M Matthews, 663 (Tucson, U. Arizona Press).
Kurucz RL 1995, In Laboratory and Astronomical High Resolution Spectra, eds. AJ Sauval, R Blomme, N Grevesse, 17.
Lacey C & Cole S 1993, MNRAS 262, 627.
Lacey C & Fall M 1983, MNRAS 204, 791–810.
Lacey C & Fall M 1985, Ap. J. 290, 154–70.
Lada CJ & Lada EA 2003, Annu. Rev. Astron. Astrophys. 41, 57.
Larsen SS 2009, Astron. Astrophys. 494, 539.
Larson R 1974, MNRAS 169, 229–246.
Law R et al. 2007, Ap. J. 669, 929.
Law DR, Makewski S & Johnston KV 2009, Ap. J. 703, L67.
Law DR et al. 2012, Ap. J. 745, 85.
Lee Y-W 1999, Nature 402, 55–57.
Leggett S et al. 1998, Ap. J. 497, 294.
Lemaître G 1927, Annales de la Societé Scientifique de Bruxelles, Serie A 47:49, 1927.
Letarte B et al. 2006, Astron. Astrophys. 453, 547–554.
Liddle A & Lyth D 2000, Cosmological Inflation and Large-Scale Structure (Cambridge University Press).
Lilly SJ et al. 1996, Ap. J. 406, L1.
Lin D & Pringle J 1987, Ap. J. 320, L87–91.
Lineweaver CH 1999, Science 284, 1503–1507.
Liu W & Chaboyer B 2000, Ap. J. 544, 818–29.
Lockman FJ et al. 2008, Ap. J. 679, L21.
Loeb A 2010, How Did the First Stars and Galaxies Form? (Princeton University Press, Princeton).
López-Corredoira M, Cabrera-Lavers A & Gerhard OE 2005, Astron. Astrophys. 439, 107.
López-Corredoira M et al. 2007, Astron. J. 133, 154.
Luck RE & Bond HE 1985, Ap. J. 292, 559–77.
Lynden-Bell D & Kalnajs A 1972, MNRAS 157, 1–30.
Lynden-Bell D & Lynden-Bell RM 1995, MNRAS 275, 429–42.

Mac Low M & Ferrara A 1999, Ap. J. 513, 142–55.
Madau P, Meiksin A & Rees MJ 1997, Astrophys. J. 475, 429.
Madau P, Pozetti L & Dickinson M 1998, Ap. J. 498, 106.
Maeda K et al. 2002, Ap. J. 565, 405.
Majewski SR, Hawley SL & Munn JA 1996, In Formation of the Halo ...Inside & Out eds. H Morrison, A Sarajedini, Astron. Soc. Pac. 92, 119–129.
Majewski SR, Ostheimer JC, Kunkel WE & Patterson RJ 2000, Astron. J. 120, 2550–68.
Majewski SR 1993, Annu. Rev. Astron. Astrophys. 31, 575–638.
Malin DF & Carter D 1980, Nature 285, 643–45.
Malin DF & Hadley B 1997, PASA 14, 52–58.
Maller AH & Bullock JS 2004, MNRAS 355, 694.
Maloney PR 1993, Ap. J. 414, 41–56.
Maloney PR & Bland-Hawthorn J 2001, Ap. J. 553, L129.
Mandelbaum R et al. 2006, MNRAS 368, 715.
Manuel O 2000, Origins of Elements in the Solar System, (New York: Kluwer).
Marquez A & Schuster WJ 1994, Astron. Astrophys. Suppl. 108, 341–58.
Marinacci F et al. 2011, MNRAS 415, 1534.
Martinez-Delgado D et al. 2010, Astron. J. 140, 962–967.
Mastropietro C et al. 2005, MNRAS 363, 509.
Mateo M. 1998. Annu. Rev. Astron. Astrophys. 36, 435–506.
Mathews GJ, Bazan G & Cowan JJ 1992, Ap. J. 391, 719–735.
Matsuda Y et al. 2005, Ap. J. 634, L125.
Matteucci F & Francois P 1989, MNRAS 239, 885–904.
Matthews LD, Gallagher JS & van Driel W 1999, Astron. J. 118, 2751–2766.
Matzner CD 2007, Ap. J. 659, 1394.
McCarthy IG et al. 2008, MNRAS 383, 593.
McConnachie A 2012, Astron. J. 144:4, 1.
McConnachie A et al. 2009, Nature 461, 66.
McGaugh SS, Schombert JM, Bothun GD & de Blok WJG 2000, Ap. J. 533, L99–L102.
McGaugh SS, Schombert JM, de Blok WJG & Zagursky MJ 2010, Ap. J. 708, L14–L17.
McKee CF & Ostriker EC 2007, Annu. Rev. Astron. Astrophys. 45, 565.
McKee CF & Tan JC 2002, Nature 416, 59–61.
McKee CF & Williams JP 1997, Astrophys. J. 476, 144.
McKellar A 1940, PASP 52, 187.
McWilliam A, Preston GW, Sneden C & Searle L 1995, Astron. J. 109, 2736–2756.
McWilliam A & Rich RM 1994, Ap. J. Suppl. 91, 749–91.
Melendez L et al. 2008, Astron. Astrophys. 484, L21.
Metcalf RB 2001, astro-ph/0109347.
Meusinger H, Stecklum B & Reimann H-G 1991, Astron. Astrophys. 245, 57–74.
Mitschang AW et al. 2013, MNRAS 428, 2321.
Minchev I et al. 2012, Astron. Astrophys. 548, 24.
Miyamoto M & Nagai R 1975, PASJ 27, 533.
Mo H, van den Bosch FC & White SDM 2010, Galaxy Formation & Evolution (Cambridge University Press).
Moore CE, Minnaert MGJ & Houtgast J 1966, The Solar Spectrum 2935 Å to 8770 Å (NBS Monogr. 61) (Washington: NBS).
Molla M, Ferrini F & Diaz A 1996, Ap. J. 466, 668–85.
Moore B, Katz N & Lake G 1996, Ap. J. 457, 455–59.
Moore B et al. 1999, Ap. J. 524, L19–22.
Moore B et al. 2001, Phys. Rev. D. 64, 063508–19.
Moore B et al. 2006, Ap. J. 368, 563.
Morgan WW 1959, Astron. J. 64, 432.
Morrison HL 1993, Astron. J. 106, 578–90.

Morrison HL et al. 2009, Ap. J. 694, 130.
Mortlock D et al. 2011, Nature 474, 616.
Mueller KE et al. 2002 Ap. J. Suppl. 143, 469.
Muldrew S et al. 2012, MNRAS 419, 2670.
Munari U et al. 2005, Astron. Astrophys. 442, 1127.
Munari U et al. 2008, Astron. Astrophys. 488, 969.
Murray SD & Lin DNC 1990, Ap. J. 363, 50.
Murray SD & Lin DNC 2004, Ap. J. 615, 586.
Mushotzky R 1999, In The Hy-Redshift Universe, eds. AJ Bunker & WJM van Breugel, Astron. Soc. Pac. 193, 323–335.
Nakamura F & Li Z-Y 2007, Ap. J. 662, 395.
Navarro JF, Frenk CS & White SDM 1997 Ap. J. 490, 493.
Ness M et al. 2012, Ap. J. 756, 22.
Ness M et al. 2013, MNRAS 430, 836.
Newberg HJ et al. 2001, Ap. J. 569, 245.
Neyman J, Page T & Scott E 1961, Astron. J. 66, 633.
Nichols M & Bland-Hawthorn J 2009, Ap. J. 707, 1642.
Nichols M & Bland-Hawthorn J 2011, Ap. J. 732, 17.
Nichols M & Bland-Hawthorn J 2013, Ap. J. 775, 10.
Nissen PE & Schuster WJ 2012, Astron. Astrophys. 543, 28.
Noguchi M 1998, Nature 392, 253–55.
Nomoto K 1987, Ap. J. 322 206.
Nomoto K et al. 2005, Nucl. Phys. A 758, 263.
Norberg P et al. 2001, MNRAS 328, 64–70.
Nordström B et al. 2004, A&A, 418 919.
Norman ML 2011, IAU Symp. 270, 7.
Norris JE, Ryan SG, Beers TC 1996, Ap. J. Suppl. 107, 391–421.
Norris JE, Ryan SG, Beers TC 1997, Ap. J. 488, 350–63.
Norris JE, Ryan SG 1989, Ap. J. 336, L17–19.
Norris JE et al. 2007, Ap. J. 670, 774.
Norris JE, Yong D, Gilmore G & Wyse RFG 2010a, Ap. J. 711, 150.
Norris JE et al. 2010b, Ap. J. 723, 1632.
Okamoto T, Gao L & Theuns T 2008, Mon. Not. R. Astron. Soc. 390, 920–928.
O'Shea BW & Norman ML 2007, Ap. J. 654, 66.
O'Shea BW & Norman ML 2008, Ap. J. 673, 14.
Odenkirchen M et al. 2001, Ap. J. 548, L165–69.
Okrochov M & Tumlinson J 2010, Ap. J., 716, L41.
Oort J 1965, In Galactic Structure, eds. A. Blaauw, M. Schmidt (University of Chicago Press), 455–511.
Oort J 1966, Bull. Astron. Inst. Neth. 18, 421.
Oosterloo T, Fraternali F & Sancisi R 2007, Astron. J. 134, 1019.
Ortolani S et al. 1995, Nature 377, 701–703.
Ostriker JP & Peebles PJE, 1973 Ap. J. 186, 467.
Ostriker JP, Peebles PJE & Yahil A 1974, Ap. J. 193, L1.
Oswalt TD, Smith JA, Wood MA & Hintzen P 1996, Nature 382, 692.
Pagel BEJ 1965, Roy. Obs. Bull. 104, 127–51.
Pagel BEJ 1989, In Evolutionary Phenomena in Galaxies (Cambridge University Press), 201–223.
Pagel BEJ & Patchett BE 1975, MNRAS 172, 13–40.
Pagel BEJ & Tautvaisiene G 1997, MNRAS 288, 108–116.
Panter B, Jimenez R, Heavens AF & Charlot S 2007, MNRAS 378, 1550.
Peacock JA 1999, Cosmological Physics (Cambridge University Press).
Peacock JA et al. 2001, Nature 410, 169.
Peacock JA 2013, astro-ph/1301.7286.

Peebles PJE & Dicke RH 1968, Ap. J. 154, 891–908.
Peebles PJE 1970, Astron. J. 75, 13–20.
Peebles PJE 1971, Physical Cosmology (Princeton University Press).
Peebles PJE 1974, Ap. J. 189, L51–53.
Peebles PJE 1993, Principles of Physical Cosmology (Princeton University Press).
Peebles PJE, Seager S & Hu W 2000, Ap. J. 539, L1–4.
Peletier R & de Grijs R 1998, MNRAS 300, L3–6.
Penton S, Stocke JT & Shull JM 2004, Ap. J. Suppl. 152, 29.
Penzias A & Wilson R 1965, Ap. J. 142, 419.
Percival W et al. 2001, MNRAS 327, 1297.
Perlmutter S et al. 1998, Nature 391, 51.
Peterson JR et al. 2003, Ap. J. 590, 207.
Pettini M et al. 2003, Ap. J. 594, 695.
Pfenniger D, Combes F & Martinet L 1994, Astron. Astrophys. 285, 79–93.
Pflamm-Altenburg J & Kroupa P 2007, MNRAS 375, 855.
Phelps R & Janes K 1996, Astron. J. 111, 1604–08.
Pichon C et al. 2011, MNRAS 418, 2493.
Pitts E & Tayler RJ 1989, MNRAS 240, 373–395.
Plume R et al. 1997, Ap. J. 476, 730.
Podsiadlowski P et al. 2002, Ap. J. 567, 491.
Pohlen M, Dettmar R-J & Lütticke R 2000, Astron. Astrophys. 357, L1–L4.
Pohlen M, Dettmar R-J & Lütticke R 2002, Astron. Astrophys. 392, 807–816.
Pohlen M et al. 2008, Astron. Soc. Pac. 390, 247.
Pont F & Eyer L 2004, MNRAS 351, 487.
Portnoy D, Pistinner S & Shaviv G 1993, Ap. J. Suppl. 86, 95.
Prada F et al. 2012, MNRAS 423, 3018.
Prescott MKM et al. 2012, Ap. J. 752, 25.
Prescott MKM et al. 2013, Ap. J. in press.
Press WH & Schecter P 1974, Ap. J. 187, 425–438.
Preston GW, Beers TC & Shectman SA 1994, Astron. J. 108, 538–554.
Pritchet C & van den Bergh S 1994, Astron. J. 107, 1730–1736.
Prochaska JX, Naumov SO, Carney BW, McWilliam A & Wolfe AM 2000, Astron. J. 120, 2513–2549.
Purcell S, Bullock JS & Kaplinghat M 2009, Ap. J. 703, 2275.
Putman ME et al. 2003, Ap. J. 597, 948.
Putman ME, Peek JEG & Joung MR 2012, Annu. Rev. Astron. Astrophys. 50, 491.
Quillen AC & Garnett D. 2001 In Galaxy Disks and Disk Galaxies, eds. G Jose, SJ Funes and EM Corsini, Astron. Soc. Pac. 230, 87–88.
Quillen AC 2002, astro-ph/0202253.
Quinn PJ 1984, Ap. J. 279, 596–609.
Quinn PJ & Goodman J 1986, Ap. J. 309, 472–495.
Quinn PJ, Hernquist L & Fullagar D 1993, Ap. J. 403, 74–93.
Quinn PJ & Zurek W 1988, Ap. J. 331, 1–18.
Quinn T, Katz N & Efstathiou GP 1996, MNRAS 278, 49–54.
Rafelski M et al. 2012, Ap. J. 755, 89.
Rakic O et al. 2011, Mon. Not. R. Astron. Soc. 414, 3265–3271.
Ramirez I et al. 2010, Astron. Astrophys. 521, 33.
Rauch M et al. 2005, Ap. J. 632, 58.
Rauch M et al. 2011, MNRAS 418, 1115.
Rauscher T, Heger A, Hoffman RD & Woosley SE 2002, Ap. J. 576, 323.
Recio-Blanco, A et al. 2006, MNRAS 370, 141.
Rees MJ 1986, MNRAS 218, 25–30.
Rees MJ & Ostriker JP 1977, MNRAS 179, 541.

Reid IN 1998, In Highlights of Astronomy, ed. J Andersen, 11A, 562 (Kluwer).
Reid IN et al. 2007, Ap. J. 665, 767.
Renzini A 2000, In From Extrasolar Planets to Cosmology, eds. J Bergeron, A Renzini (Berlin: Springer).
Renzini A 2002 In Chemical Enrichment of Intracluster and Intergalactic Medium, eds. F Matteucci and R Fusco-Femiano, Astron. Soc. Pac. 253, 331.
Ribaudo J et al. 2011 Ap. J. 743, 207.
Rich RM 2001, In Astrophysical Ages and Time Scales, eds. T von Hippel, C Simpson & N Manset, Astron. Soc. Pac. 245, 216–25.
Richardson J et al. 2011, Ap. J. 732, 76.
Riess A et al. 1998, Astron. J. 116, 1009.
Rix H-W & Bovy J 2012, Astron. Astrophys. Rev. in press.
Rix H-W & Zaritsky D 1995, Ap. J. 447, 82–102.
Roberts MS 1967, IAU Symp. 31, 189.
Roberts MS 1975, Ap. J. 201, 327.
Roberts MS 1976, Comm. Astrophys. 6, 105.
Roberts MS & Rots AH 1973, Astron. Astrophys. 26, 483.
Robertson B et al. 2010, Nature 468, 49.
Robin A & Creze M 1986, Astron. Astrophys. 157, 71.
Robin A et al. 2003, A&A 409 523.
Robotham A et al. 2012, MNRAS 424, 1448.
Rocha-Pinto HJ et al. 2000, Ap. J. 531, L115.
Rocha-Pinto HJ, Scalo J, Maciel WJ & Flynn C 2000, Astron. Astrophys. 358, 869–85.
Rodgers AW, Harding P & Sadler EM 1981, Ap. J. 244, 912–18.
Rood HJ, Page TL, Kintner EC & King IR 1972, Ap. J. 175, 627.
Rosdahl J & Blaizot J 2012 MNRAS 423, 344.
Rose J 1994, Astron. J. 107, 206–29.
Roskar R et al. 2008a, Ap. J. 675, L65.
Roskar R et al. 2008b, Ap. J. 684, L79.
Roth K et al. 1993, Ap. J. 413, 67.
Rubin V & Ford K 1970, Ap. J. 159, 379.
Rubin V, Thonnard N & Ford WK 1978, Ap. J. 225, L107.
Rubin KHR et al. 2012, Ap. J. 747, L26.
Ryan SG, Norris JE & Beers TC 1996, Ap. J. 471, 254–78.
Ryan Weber E, Pettini M & Madau P 2006, MNRAS 371, L78.
Ryan Weber E, Pettini M, Madau P & Zych BJ 2009, MNRAS 395, 1476.
Salvadori S & Ferrara A 2009, MNRAS 395, L6.
Sanchez-Gil, C et al. 2011, MNRAS 415, 753.
Sancisi R et al. 2008, Astron. Astrophys. Rev. 15, 189.
Sandage A 1990, JRASC 84, 70–88.
Sandage A & Cacciari C 1990, Ap. J. 350, 645–661.
Sandage A & Visvanathan N 1978, Ap. J. 225, 742–750.
Sarajedini A et al. 2007, Astron. J. 133, 1658.
Schaye J 2004, Ap. J. 609, 667.
Schlegel DJ et al. 1998, Ap. J. 500, 525.
Schmidt M 1963, Ap. J. 137, 758–69.
Schneider R & Omukai K 2010, MNRAS 402, 429.
Schönmakers RHM, Franx M & de Zeeuw PT 1997, MNRAS 292, 349–64.
Schommer RA, Suntzeff NB, Olszewski EW & Harris HC 1992, Astron. J. 103, 447–59.
Schönrich RA, Binney JJ & Asplund MA 2012, MNRAS 420, 1281.
Schuster WJ et al. 2012, Astron. Astrophys. 538, 21.
Schweizer F 1987, In Nearly Normal Galaxies - From the Planck Time to the Present, ed. S Faber, 18–25 (New York: Springer-Verlag).

Scott D & Rees MJ 1990, MNRAS 247, 510.
Searle L & Zinn R 1978, Ap. J. 225, 357–79.
Searle L 1977, In The Evolution of Galaxies & Stellar Populations, eds. BM Tinsley, RB Larson, 219 (New Haven, Yale).
Sedgewick R 1992, Algorithms in C^{++} (Addison-Wesley).
Sellwood JA 1999, In Astrophysical Disks, eds. J Sellwood & J Goodman, 327.
Sellwood JA & Binney JJ 2002, MNRAS 336, 785.
Sembach KR et al. 2003, Ap. J. Suppl. 146, 165.
Shang Z et al. 1998, Ap. J. 504, L23–26.
Shapiro PR & Kang H 1987, Ap. J. 318 32.
Shapiro PR et al. 2006, Ap. J. 646, 681.
Sharma S, Bland-Hawthorn J, Johnston K & Binney JJ 2011, Ap. J. 730, 3.
Sharma S, Steinmetz M & Bland-Hawthorn J 2012, Ap. J. 750, 107.
Sharma S & Bland-Hawthorn J 2013, Ap. J. in press.
Sharma S et al. 2013, Ap. J. in press.
Sheth RK & Tormen G 1999, MNRAS 308, 119–126.
Sheth RK & Tormen G 2001, MNRAS 323, 1.
Shigeyama T & Tsujimoto T 1998, Ap. J., 507, L135–39.
Shima E et al. 1985, MNRAS 217, 367.
Shirley YL et al. 2003, Ap. J. Suppl. 149, 375.
Shu FH 1969, Ap. J. 158, 505.
Shull JM, Smith BD & Danforth CW 2012, Ap. J. 759:23.
Siebert A et al. 2011, Astron. J. 141, 187.
Silk J 1977, Ap. J. 211, 638.
Silk J & Mamon GA 2012, Res. Astron. Astrophys. 12, 917–946.
Simon JD & Geha M 2007, Ap. J. 670, 313.
Slipher VM 1913, Low. Obs. Bull., 2, 56.
Slipher VM 1914, Low. Obs. Bull., 2, 66.
Smith S 1936, Ap. J. 83, 23.
Smith MC et al. 2007, MNRAS 379, 755.
Sneden C et al. 2000, Ap. J. 533, L139–L142.
Soderblom D 2010, Annu. Rev. Astron. Astrophys. 48, 581.
Soubiran C & Girard P 2005, Astron. Astrophys. 438, 139.
Spergel D & Steinhardt P 2000, Phys. Rev. Lett. 84, 3760–63.
Spiegel EA 1970, IAU Symp. 39, 201.
Spite M & Spite F 1978, Astron. Astrophys. 67, 23–31.
Spitzer L & Schwarzschild M 1953, Ap. J. 118, 106–12.
Springel V, Hernquist L 2003, Mon. Not. R. Astron. Soc. 339, 289–311.
Springel V et al. 2005, Nature 435, 629.
Springel V et al. 2008, MNRAS 391, 1685.
Stacy A, Greif TH & Bromm V 2010, MNRAS 403, 45.
Stacy A, Greif TH & Bromm V 2012, MNRAS 422, 290.
Stadel J et al. 2009, MNRAS 398, L21.
Stahler SW, Palla F & Ho PTP 2000, In Protostars and Planets IV, eds. V Mannings, AP Boss & SS Russell, 327–351 (University of Arizona Press).
Starkenburg E et al. 2010, Astron. Astrophys. 513, 34.
Steidel CC et al. 2011, Ap. J. 736, 160.
Sternberg A, McKee CF & Wolfire MG 2002, Ap. J. Suppl. 143, 419.
Sternberg A & Soker N 2008, MNRAS 389, L13.
Strateva I et al. 2001, Astron. J. 122, 1861.
Strömgren B 1987, In The Galaxy: Proceedings of the NATO ASI, eds. G Gilmore, R Carswell, 229–246 (Reidel).
Sun XH et al. 2008 Astron. Astrophys. 477, 573.

Sutherland RS & Dopita MA 1993, Ap. J. Suppl. 88, 253.
Tabor G & Binney JJ 1993, MNRAS 263, 323.
Talbot RJ & Arnett WD 1971, Ap. J. 170, 409–422.
Tamburro D et al. 2008, Astron. J. 136, 2872.
Tamm A et al. 2012, Astron. Astrophys. 546, 4.
Tan JC, Krumholz MR & McKee CF 2006, Ap. J. 641, L121.
Tanvir N et al. 2009, Nature 461, 1254.
Taylor BJ 2000, Astron. Astrophys. 362, 563–579.
Thoul A & Weinberg D 1996, Ap. J. 465, 608–16.
Ting YS et al. 2012, MNRAS 421, 1231.
Tinsley B 1980, Fund. Cos. Phys. 5, 287–388.
Tolstoy E, Hill V & Tosi M 2009, Annu. Rev. Astron. Astrophys. 47, 371.
Trager S, Faber SM, Worthey G & Gonzalez J 2000, Astron. J. 120, 1645–76.
Trager S, Worthey G, Faber SM, Bustein D & Gonzales J 1998, Ap. J. Suppl. 116 1–28.
Travaglio C, Gallino R, Zinner E, Amari S & Woosley S 1998, Meteoritics & Planetary Science 33, A155–156.
Travaglio C et al. 1999, Ap. J. 521, 691–702.
Tremaine SD 1993, In Back to the Galaxy, eds. SS Holt, F. Verter (New York: AIP), 599–609.
Tremaine SD 1999, MNRAS 307, 877–83.
Tremaine SD 1999, Ap. J. 525, 1223 (Comment).
Tremaine SD & Gunn JE 1979, Phys. Rev. Lett. 42, 407–410.
Tripicco MJ & Bell RA 1995, Astron. J. 110, 3035–3049.
Trimble V 1987, In Annual review of astronomy and astrophysics 25, 425–472.
Truran JW, Burles S, Cowan J & Sneden C 2001, In Astrophysical Ages and Time Scales, eds. T von Hippel, C Simpson & N Manset, Astron. Soc. Pac. 245, 226–234.
Truran JW 1981, Astron. Astrophys. 97, 391–93.
Tsikoudi V 1980, Ap. J. Suppl. 43, 365–77.
Tsujimoto T & Shigeyama T 1998, Ap. J. 508, L151.
Tsujimoto T, Shigeyama T & Yoshii Y 2000, Ap. J. 531, L33–36.
Tsuribe T & Omukai K 2006, Ap. J. 642, L61.
Tsuribe T & Omukai K 2008, Ap. J. 676, L45.
Tully RB 1987, Catalogue of Nearby Galaxies (University of Hawaii Press).
Tully RB & Fisher JR 1977, Astron. Astrophys. 54, 661–673.
Tully RB, Verheijen MA, Pierce MJ, Huang J-S & Wainscoat RJ 1996, Astron. J. 112, 2471–2499.
Tully RB et al. 2008, Ap. J. 676, 184.
Tumlinson J 2010, Ap. J. 708, 1398.
Tumlinson J et al. 2011, Sci. 334, 948.
Turk MJ, Abel T & OShea B 2009, Science 325, 601.
Turner E 1976a, Ap. J. 208, 20.
Turner E 1976b, Ap. J. 208, 304.
Turon C & Primas F 2008, ESA-ESO WG Report No. 4: Galactic Populations, Chemistry & Dynamics (ESO: Garching).
Twarog BA, Ashman KM & Anthony-Twarog BJ 1997, Astron. J. 114, 2556–2585.
Twarog BA 1980, Ap. J. 242, 242–259.
Tyson JA et al. 1998, Astron. J. 116, 102–110.
Ulrich RK 1986, Ap. J. 306, L37–40.
Umeda H & Nomoto K 2003, Nature 422, 871.
Unavane M, Wyse RFG & Gilmore G 1996, MNRAS 278, 727–36.
van Albada TS, Bahcall JN, Begeman K & Sancisi R 1985, Ap. J. 295, 305.
van den Bergh S 1962, Astron. J. 67, 486–90.
van den Bergh S 2000, The Galaxies of the Local Group (Cambridge University Press).
van den Bergh S & McClure RD 1980, Astron. Astrophys. 88, 360–62.
van der Kruit P 1979, Astron. Astrophys. Suppl. 38, 15.

van der Kruit P 2002, In The Dynamics, Structure & History of Galaxies - Workshop in Honour of KC Freeman, eds. G Da Costa & H Jerjen, 7.
van der Kruit PC 2007, Astron. Astrophys. 466, 883–893.
Vanzella E et al. 2011, Ap. J. 730, 35.
Vazdekis A 1999, Ap. J. 513, 224–41.
Veilleux S, Cecil G & Bland-Hawthorn J 2005, Annu. Rev. Astron. Astrophys. 43, 769.
Venn K et al. 2004 Astron. J. 128, 1177.
Vivas AK et al. 2001, Ap. J. 554, L33–36.
Wakker BP 2001, Ap. J. Suppl. 136, 463.
Walker IR, Mihos JC & Hernquist L 1996, Ap. J. 460, 121–35.
Wallace L, Hinkle K & Livingston W 1998, An atlas of the spectrum of the solar photosphere (3570–7405Å) (Tucson: NOAO).
Wallerstein G, Greenstein JL, Parker R, Helfer HL & Aller LH 1963, Ap. J. 137, 280–300.
Wallerstein G et al. 1997, Rev. Mod. Phys. 69, 995–1084.
Walsh SM, Jerjen H & Willman B 2007, Ap. J. 662, L83.
Weil M, Bland-Hawthorn J & Malin DF 1997, Ap. J. 490, 664–81.
Weinberg S 1977, The First Three Minutes: A Modern View of the Origin of the Universe (London: Andre Deutsch).
Weiner BJ, Vogel SN & Williams TB 2001, In Extragalactic Gas at Low Redshift, eds. J. Mulchaey & J. Stocke, Astron. Soc. Pac. 254, 256–266.
Weinmann SM et al. 2012, Mon. Not. R. Astron. Soc. 426, 2797–2812.
Weinmann SM et al. 2013, MNRAS in press (1204.4184).
Westin J, Sneden C, Gustafsson B & Cowan J 2000, Ap. J. 530, 783–99.
White M & Croft RAC 2000, Ap. J. 539, 497–504.
White SDM 1976, MNRAS 177, 717–33.
White SDM 1981, MNRAS 195, 1037.
White SDM 1983, Ap. J. 274, 53.
White SDM 1984, Ap. J. 286, 38.
White SDM 1996, In Cosmology & Large Scale Structure, eds. R Schaeffer, J Silk, M Spiro & J Zinn-Justin, 349 (Amsterdam: Elsevier).
White SDM & Rees MJ 1978, MNRAS 183, 341–58.
White SDM et al. 1993, Nature 366, 429.
Whitmore BC, Schweizer F, Leitherer C, Borne K & Robert C 1993, Astron. J. 106, 1354–70.
Whitmore BC & Schweizer F 1995, Astron. J. 109, 960–80.
Whitmore B, Chandar R & Fall SM 2007, Astron. J. 133, 1067.
Widrow L & Dubinski J 1998, Ap. J. 504, 12–26.
Williams JP & McKee CF 1997, Ap. J. 476, 166.
Williams MEK et al. 2011, Ap. J. 728, 102.
Williams BF et al. 2013, Ap. J. in press.
Wolfe AM, Gawiser E & Prochaska JX 2006, Annu. Rev. Astron. Astrophys. 43, 861.
Woosley SE & Weaver TA 1995, Ap. J. Suppl. 101, 181.
Worthey G, Dorman B & Jones L 1996, Astron. J. 112, 948–53.
Worthey G, Faber SM, Gonzales J & Burstein D 1994, Ap. J. Suppl. 94, 687–722.
Worthey G 1994, Ap. J. Suppl. 95, 107–149.
Wright EL 2003, New Astron. Rev. 47, 877–881.
Wright EL 2006, PASP 118, 1711.
Wylie de Boer L et al. 2012, Ap. J. 755, 35.
Wyse RFG 2001, In Galaxy Disks and Disk Galaxies, eds. JG Funes, SJ & EM Corsini, Astron. Soc. Pac. 230, 71.
Wyse RFG & Silk J 1989, Ap. J. 339, 700–711.
Xie T, Allen M & Langer WD 1995, Ap. J. 440, 674.
Yepes G, Gottlöber S, Martinez-Vaquero LA & Hoffman Y 2009, AIP Conf. 1115, 80–91.
Yokoi K, Takahashi K & Arnould M 1983, Astron. Astrophys. 117, 65–82.

Yoshida N, Bromm V & Hernquist L 2004, Ap. J. 605, 579.
Yoshida N, Omukai K, Hernquist L & Abel T 2006, Astrophys. J. 652, 6–25.
Yoshida N, Oh SP, Kitayama T & Hernquist L 2007, Ap. J. 663, 687.
Yoshida N, Omukai K & Hernquist L 2008, Science 321, 669.
Yoshii Y & Peterson BA 1995, Ap. J. 444, 15.
Younger JD et al. 2007, Ap. J. 671, 1531.
Yun MS, Carilli CL, Kawabe R, Tutui Y, Kohno K & Ohta K 2000, Ap. J. 528, 171–178.
Zheng Z et al. 1999, Astron. J. 117, 2757–2780.
Ziegler BL et al. 2005, å433, 519.
Zinn R 1985, Ap. J. 293, 424–444.
Zinnecker H & Cannon RD 1986, In Star Forming Galaxies and Related Objects, eds. D Kunth, TX Thuan, J Thanh Van, 155 (Paris: Editions Frontiéres).
Zwicky F 1933, Helvetica Physica Acta, 6, 110–127.
Zucker DB et al. 2004, Ap. J. 612, L117.
Zucker DB et al. 2006, Ap. J. 650, L41.
Zucker DB et al. 2007, Ap. J. 659, L21.
Zurek W, Quinn PJ & Salmon J 1988, Ap. J. 330, 519–34.

Chapter 2
Chemical Evolution of the Milky Way and Its Satellites

Francesca Matteucci

In recent years a great deal of high resolution spectroscopic data, relating to chemical abundances in the stars of the Milky Way and its satellites, has appeared. Through the analysis of chemical abundances we can reconstruct the star formation histories of galaxies in terms of an astro-archaeological approach. In these lectures I describe how to interpret abundances and abundance ratios in galaxies by means of detailed galactic chemical evolutionary models. After comparing model results and observational data we can put constraints on the star formation history, stellar nucleosynthesis and time-scales for the formation of galaxies. First I describe the chemical evolution of the Milky Way and try to reconstruct the history of its formation, then I describe and interpret the chemical evolution of dwarf and ultra faint dwarf galaxies, the satellites of our Galaxy. A comparison between the abundance patterns observed in these objects and in the Milky Way allows us to discuss the possibility that these satellites were part of the building blocks of the Milky Way. The chemical evolution of some spiral galaxies in the Local Group is also presented. Finally, I discuss cosmic chemical evolution, namely the chemical evolution of a unitary comoving volume of the Universe where different galaxies contribute to the chemical enrichment process. Cosmic supernova (Types II and Ia) rates are also discussed.

2.1 How to Model Galactic Chemical Evolution

Before going into the detailed chemical evolution history of the Milky Way and its satellites, it is necessary to understand how to model, in general, galactic chemical evolution. The basic ingredients to build a model of galactic chemical evolution can be summarized as:

F. Matteucci (✉)
Astronomy Department, Trieste University, Trieste, Italy
e-mail: matteucci@ts.astro.it

Osservatorio Astronomico (INAF), Trieste, Italy

- Initial conditions;
- Stellar birthrate function (the rate at which stars are formed from the gas and their mass spectrum);
- Stellar yields (how elements are produced in stars and restored into the interstellar medium);
- Gas flows (infall, outflow, radial flow).

When all these ingredients are ready, we need to write a set of equations describing the evolution of the gas and its chemical abundances which include all of them. These equations will describe the temporal variation of the gas content and its abundances by mass (see next sections). The chemical abundance of a generic chemical species i is defined as:

$$X_i = \frac{M_i}{M_{gas}}. \tag{2.1}$$

According to this definition it holds:

$$\sum_{i=1,n} X_i = 1, \tag{2.2}$$

where n represents the total number of chemical species. Generally, in theoretical studies of stellar evolution it is common to adopt X, Y and Z as indicative of the abundances by mass of hydrogen (H), helium (He) and metals (Z), respectively. The baryonic universe is madeup mainly of H and some He while only a very small fraction resides in metals (all the elements heavier than He), roughly 2%. However, the history of the growth of this small fraction of metals is crucial for understanding how stars and galaxies were formed and subsequently evolved; and last but not least, because human beings exist only because of this small amount of metals! We will focus then our attention is studying how the metals were formed and evolved in galaxies, with particular attention to our own Galaxy.

2.1.1 The Initial Conditions

The initial conditions for a model of galactic chemical evolution consist in establishing whether: (a) the chemical composition of the initial gas is primordial or pre-enriched by a pre-galactic stellar generation; (b) the studied system is a closed box or an open system (infall and/or outflow).

2.1.2 Birthrate Function

The birthrate function, namely:

$$B(M, t) = \psi(t)\varphi(m) \tag{2.3}$$

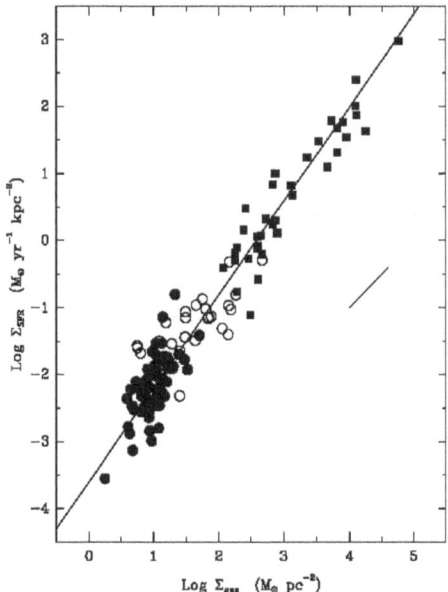

Fig. 2.1 The SFR as measured by Kennicutt (1998a) in star forming galaxies. The continuous line represents the best fit to the data and it can be achieved either with the SF law in Eq. (2.6) with $k = 1.4$ or with the SF law in Eq. (2.9). The short, *diagonal line* shows the effect of changing the scaling radius by a factor of 2. Figure from Kennicutt (1998a)

where the quantity:

$$\psi(t) = SFR \qquad (2.4)$$

is called the star formation rate (SFR), namely the rate at which the gas is turned into stars, and the quantity:

$$\varphi(m) = IMF \qquad (2.5)$$

is the initial mass function (IMF), namely the mass distribution of the stars at birth.

2.1.2.1 The Star Formation Rate

The most common parametrization of the SFR is the Schmidt (1959) law:

$$SFR = \nu \sigma_{gas}^k, \qquad (2.6)$$

where $k = 1$–2 with a preference for $k = 1.4 \pm 0.15$, as suggested by Kennicutt (1998a) for spiral disks (see Fig. 2.1), and ν is a parameter describing the star formation efficiency, in other words, the SFR per unit mass of gas, and it has the dimensions of the inverse of a time. Other physical quantities such as gas temperature, viscosity and magnetic field are usually ignored.

Other common parametrizations of the SFR include a dependence on the total surface mass density besides the surface gas density:

$$\psi(t) = \nu \sigma_{tot}^{k_1} \sigma_{gas}^{k_2}, \qquad (2.7)$$

as suggested by observational results of Dopita and Ryder (1994) and taking into account the influence of the potential well in the star formation process (i.e. feedback between SN energy input and star formation, see also Talbot and Arnett (1975)). Other suggestions concern the star formation induced by spiral density waves (Wyse and Silk 1989) with expressions like:

$$\psi(t) = \nu V(R) R^{-1} \sigma_{gas}^{1.5}, \qquad (2.8)$$

or

$$\psi(t) = 0.017 \Omega_{gas} \sigma_{gas} \propto R^{-1} \sigma_{gas} \qquad (2.9)$$

with Ω_{gas} being the angular rotation speed of gas (Kennicutt 1998a). Also this law provides a good fit to the data of Fig. 2.1.

2.1.2.2 The Initial Mass Function

The most common parametrization of the IMF is a one-slope (Salpeter 1955) or multi-slope (Scalo 1986, 1998; Kroupa et al. 1993; Chabrier 2003) power law. The most simple example of a one-slope power law is:

$$\varphi(m) = am^{-(1+x)}, \qquad (2.10)$$

generally defined in a mass range of 0.1–100 $M_\odot$, where a is the normalization constant derived by imposing that $\int_{0.1}^{100} m\varphi(m) dm = 1$.

The Scalo and Kroupa IMFs were derived from stellar counts in the solar vicinity and suggest a three-slope function. Unfortunately, the same analysis cannot be done in other galaxies and we cannot test if the IMF is the same everywhere. Kroupa (2001) suggested that the IMF in stellar clusters is a universal one, very similar to the Salpeter IMF for stars with masses larger than $0.5 M_\odot$. In particular, this universal IMF is:

$$x_1 = 0.3 \quad for \quad 0.08 \le M/M_\odot \le 0.50$$

$$x_2 = 1.3 \quad for \quad M/M_\odot > 0.5 \qquad (2.11)$$

However, Weidner and Kroupa (2005) suggested that the IMF integrated over galaxies, which controls the distribution of stellar remnants, the number of SNe and the chemical enrichment of a galaxy, is generally different from the IMF in stellar clusters. This galaxial IMF is given by the integral of the stellar IMF over the embedded star cluster mass function which varies from galaxy to galaxy. Therefore, we should expect that the chemical enrichment histories of different galaxies cannot be reproduced by an unique invariant Salpeter-like IMF. In any case, this galaxial IMF is always steeper than the universal IMF in the range of massive stars.

2.1.2.3 How to Derive the IMF

We define the current mass distribution of local Main Sequence (MS) stars as the present day mass function (PDMF), $n(m)$. Let us suppose that we know $n(m)$ from observations. Then, the quantity $n(m)$ can be expressed as follows: for stars with initial masses in the range 0.1–1.0 $M_\odot$ which have lifetimes larger than a Hubble time we can write:

$$n(m) = \int_0^{t_G} \varphi(m)\psi(t)dt \tag{2.12}$$

where $t_G = 14$ Gyr (the age of the Universe). The IMF, $\varphi(m)$, can be taken out of the integral if assumed to be constant in time, and the PDMF becomes:

$$n(m) = \varphi(m) <\psi> t_G \tag{2.13}$$

where $<\psi>$ is the average SFR in the past. For stars with lifetimes negligible relative to the age of the Universe, namely for all the stars with $m > 2M_\odot$, we can write:

$$n(m) = \int_{t_G - \tau_m}^{t_G} \varphi(m)\psi(t)dt, \tag{2.14}$$

where τ_m is the lifetime of a star of mass m. Again, if we assume that the IMF is constant in time we can write:

$$n(m) = \varphi(m)\psi(t_G)\tau_m \tag{2.15}$$

having assumed that the SFR did not change during the time interval between $(t_G - \tau_m)$ and t_G. The quantity $\psi(t_G)$ is the SFR at the present time. We cannot derive the IMF betwen 1 and 2 $M_\odot$ because none of the previous simplifying hypotheses can be applied. Therefore, the IMF in this mass range will depend on a quantity, $b(t_G)$:

$$b(t_G) = \frac{\psi(t_G)}{<\psi>} \tag{2.16}$$

Scalo (1986) assumed:

$$0.5 \leq b(t_G) \leq 1.5 \tag{2.17}$$

in order to fit the two branches of the IMF in the solar vicinity. In Fig. 2.2 we show the differences between a single-slope IMF and multi-slope IMFs, which are preferred according to the last studies.

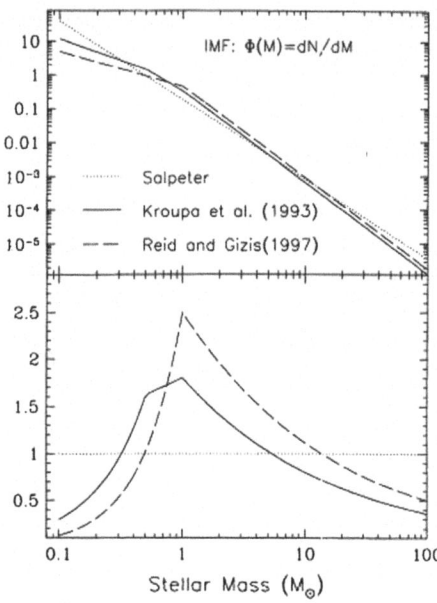

Fig. 2.2 *Upper panel* different IMFs. *Lower panel* normalization of the multi-slope IMFs to the Salpeter IMF. Figure from Boissier and Prantzos (1999)

2.1.3 Stellar Yields

The stellar yields, namely the amount of newly formed and pre-existing elements ejected by stars of all masses at their death, represent a fundamental ingredient to compute galactic chemical evolution. They can be calculated by knowing stellar evolution and nucleosynthesis.

I recall here the various stellar mass ranges and their nucleosynthesis products. In particular:

- Brown dwarfs: are stars with masses $M < 0.1 M_\odot$ which never ignite H. They do not enrich the ISM in chemical elements but only lock up gas.
- Low and Intermediate mass stars ($0.8 \leq M/M_\odot \leq 8.0$). Calculations are available from Marigo et al. (1996); van den Hoeck and Groenewegen (1997), HG97 (Forestini and Charbonnel 1997; Marigo 2001; Meynet and Maeder 2002a; Ventura et al. 2002; Siess et al. 2002; Karakas and Lattanzio 2007; Karakas 2010). These stars produce mainly ^4He, ^{12}C, ^{14}N plus some CNO isotopes and s-process ($A > 90$) elements. In Fig. 2.3 we show an example of integrated yields from stars in this mass range.
- Massive stars ($8 < M/M_\odot \leq 40$). In the mass range 10–40 $M_\odot$, available calculations are from Woosley and Weaver (1995, hereafter WW95), Langer and Henkel (1995); Thielemann et al. (1996); Nomoto et al. (1997); Limongi and Chieffi (2003); Rauscher et al. (2002); Meynet and Maeder (2002a); Nomoto et al. (2006), among others. These stars end their life as Type II SNe and explode by core-collapse; they produce mainly α-elements (O, Ne, Mg, Si, S, Ca), some

2 Chemical Evolution of the Milky Way and Its Satellites

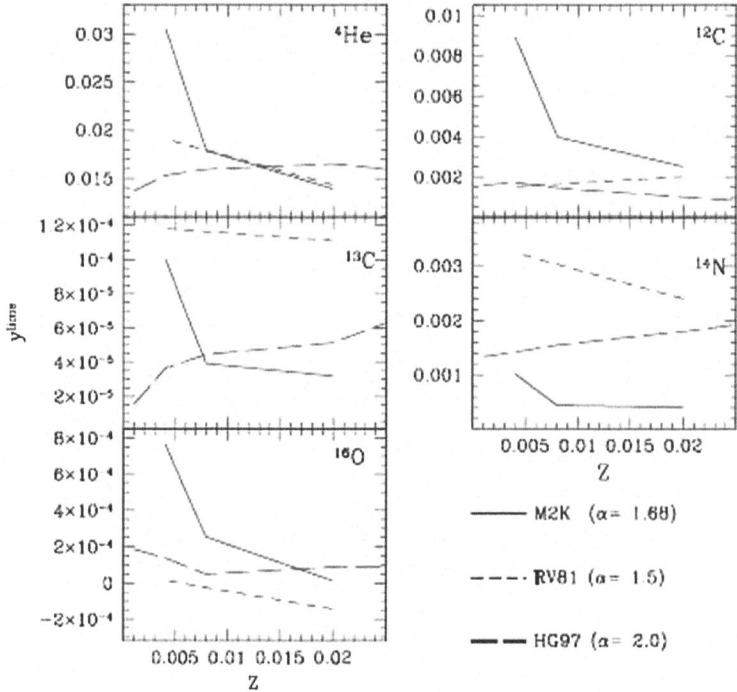

Fig. 2.3 The yields integrated over the Salpeter (1955) IMF of He, C and N produced by low and intermediate mass stars as functions of the initial stellar metallicity. Different results are compared here: those of RV81 (Renzini and Voli 1981), those of HG97 (van den Hoeck and Groenewegen 1997) and those of M2K (Marigo 2001). The mixing length parameters (α) adopted by the authors are indicated. Figure from Marigo (2001)

Fe-peak elements, s-process elements ($A < 90$) and r-process elements. Stars more massive than $40\,M_\odot$ can end up as Type Ib/c SNe. They are also core-collapse SNe and are linked to γ-ray bursts (GRB).

- Type Ia SNe (white dwarfs in binary systems, see later). Calculations are available from Nomoto et al. (1997); Iwamoto et al. (1999). They produce mainly Fe-peak elements.
- Very massive objects ($M > 100\,M_\odot$). Calculations are available from e.g. Portinari et al. (1998); Umeda and Nomoto (2001). They should produce mainly oxygen although many uncertainties are still present.

All the elements with mass number A from 12 to 60 have been formed in stars during the quiescent burnings. Stars transform H into He and then He into heaviers until the Fe-peak elements, where the binding energy per nucleon reaches a maximum and the nuclear fusion reactions stop. H is transformed into He through the proton-proton chain or the CNO-cycle, then ^4He is transformed into ^{12}C through the triple-α reaction.

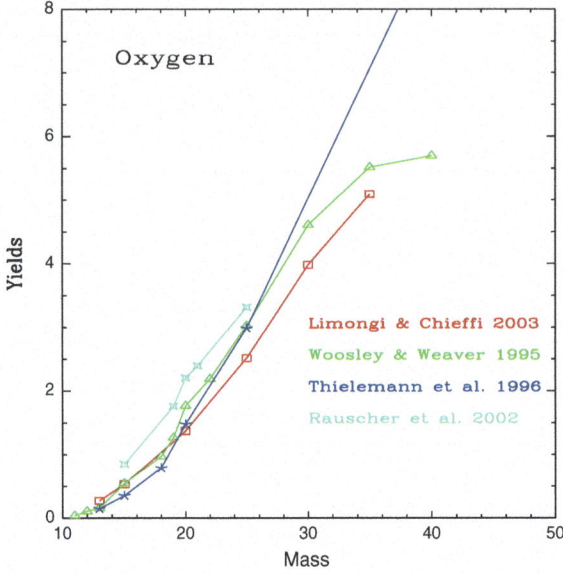

Fig. 2.4 The yields of oxygen for massive stars as computed by several authors, as indicated in the Figure. None of these calculations takes into account mass loss by stellar wind

Elements heavier than ^{12}C are then produced by synthesis of α-particles: they are called α-elements (O, Ne, Mg, Si and others).

The last main burning in stars is the ^{28}Si-burning which produces ^{56}Ni, which then decays into ^{56}Co and ^{56}Fe. Si-burning can be quiescent or explosive (depending on the temperature).

Explosive nucleosynthesis occurring during SN explosions mainly produces Fe-peak elements. Elements originating from s- and r-processes (with A > 60 up to Th and U) are formed by means of slow or rapid (relative to the β- decay) neutron capture by Fe seed nuclei; s-processing occurs during quiescent He-burning, whereas r-processing occurs during SN explosions.

In Figs. 2.4, 2.5, 2.6, 2.7 and 2.8 we show a comparison between stellar yields for massive stars computed for different initial stellar metallicities and with different assumptions concerning the mass loss. In particular, some yields are obtained by assuming mass loss by stellar winds with a strong dependence on metallicity (e.g. Maeder 1992; Langer and Henkel 1995), whereas others (e.g. WW95) are computed by means of conservative models without mass loss. One important difference arises for oxygen in massive stars for solar metallicity and mass loss: in this case, the O yield is strongly depressed as a consequence of mass loss. In fact, the stars with masses $> 25 M_\odot$ and solar metallicity lose a large amount of matter rich of He and C, thus subtracting these elements to further processing which would lead to O and heavier elements. So the net effect of mass loss is to increase the production of He and C and to depress that of oxygen (see Fig. 2.9). More recently,

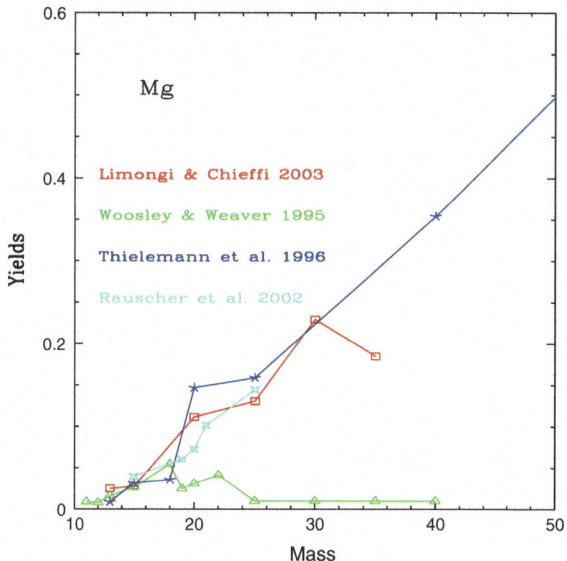

Fig. 2.5 The same as Fig. 2.4 for magnesium

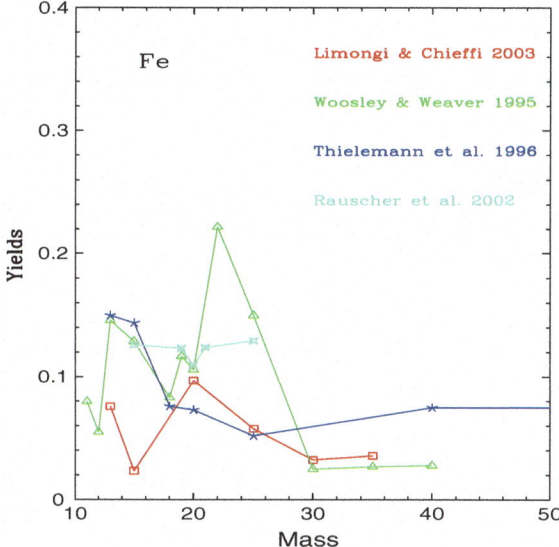

Fig. 2.6 The same as Fig. 2.4 for Fe

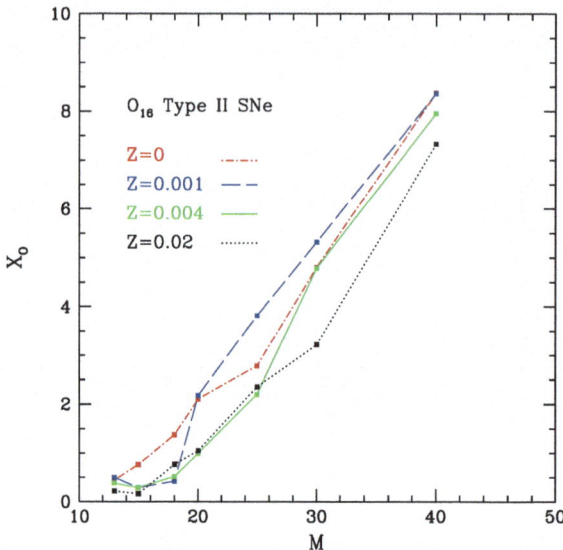

Fig. 2.7 The O yields as calculated by Nomoto et al. (2006) for different metallicities. These calculations do not take into account mass loss by stellar wind

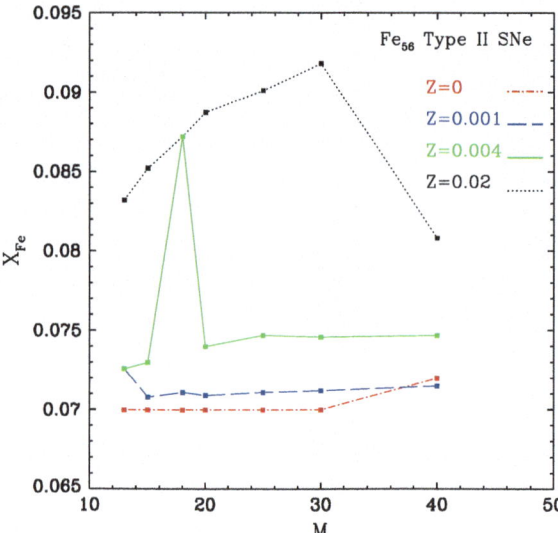

Fig. 2.8 The same as Fig. 2.7 for Fe

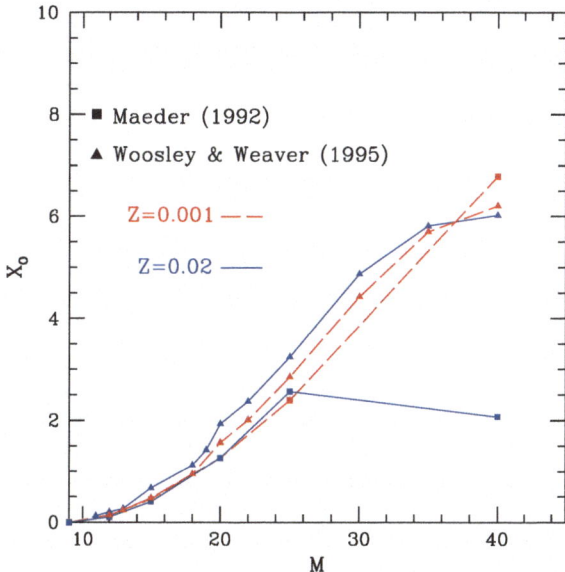

Fig. 2.9 The effect of metallicity dependent mass loss on the oxygen yield. The comparison is between the conservative yields of WW95 for $Z = 0.001$ and $Z = 0.02$ and the yields with mass loss of Maeder (1992) for the same metallicity. As one can see the effect of mass loss for a solar metallicity is a quite important one

Meynet and Maeder (2002a, 2003, 2005) have computed a grid of models for stars with $M > 20 M_\odot$ including rotation and metallicity dependent mass loss. The effect of metallicity dependent mass loss in decreasing the O production in massive stars was confirmed, although they employed significantly lower mass loss rates compared with Maeder (1992). With these models they were able to reproduce the frequency of WR stars and the observed WN/WC ratio, as was the case for the previous Maeder results. Therefore, it appears that the earlier mass loss rates made-up for the omission of rotation in the stellar models. On the other hand, the dependence upon metallicities of the yields computed with conservative stellar models is not very strong except perhaps for the yields computed with zero intial stellar metallicity (Pop III stars).

In Figs. 2.7 and 2.8 we show the more recent results of Nomoto et al. (2006) for conservative stellar models of massive stars at different metallicities. While the O yields are not much dependent upon the initial stellar metallicity, as in WW95, the Fe yields seem to change dramatically with the stellar metallicity.

2.1.3.1 Type Ia SN Progenitors

There is a general consensus about the fact that SNeIa originate from C-deflagration in C–O white dwarfs (WD) in binary systems, but several evolutionary paths can lead to such an event. The C-deflagration produces ~ 0.6–$0.7\,M_\odot$ of Fe plus traces of other elements from C to Si, as observed in the spectra of Type Ia SNe.

Two main evolutionary scenarios for the progenitors of Type Ia SNe have been proposed:

- Single Degenerate (SD) scenario (see Fig. 2.10): the classical scenario of Whelan and Iben (1973), revised by Han and Podsiadlowski (2004), namely C-deflagration in a C–O WD reaching the Chandrasekhar mass $M_{Ch} \sim 1.44\,M_\odot$ after accreting material from a red giant companion. One of the limitations of this scenario is that the accretion rate should be defined in a quite narrow range of values. To avoid this problem, Kobayashi et al. (1998) had proposed a similar scenario, where the companion can be either a red giant or a main sequence star, including a metallicity effect which suggests that no Type Ia systems can form for [Fe/H] < -1.0 dex. This is due to the development of a strong radiative wind from the C–O WD which stabilizes the accretion from the companion, allowing for larger mass accretion rates than the previous scenario. The clock to the explosion is given by the lifetime of the secondary star in the binary system where the WD is the primary star (the originally more massive one). Therefore, the largest mass for a secondary is $8\,M_\odot$, which is the maximum mass for the formation of a C–O WD. As a consequence, the minimum timescale for the occurrence of Type Ia SNe is ~ 30 Myr (i.e. the lifetime of a $8\,M_\odot$) after the beginning of star formation. Observations in radio-galaxies by Mannucci et al. (2005, 2006) seem to confirm the existence of such prompt Type Ia SNe.

  The minimum mass for the secondary is $0.8\,M_\odot$ which is a star with a lifetime equal to the age of the universe. Stars with masses below this limit are obviously not considered. In summary, the mass range for both primary and secondary stars is, in principle, between 0.8 and $8\,M_\odot$, although two stars of $0.8\,M_\odot$ are too small to give rise to a WD with a Chandrasekhar mass, and therefore the mass of the primary star should be assumed to be high enough to ensure that, even after accretion from a $0.8\,M_\odot$ star secondary, it will reach the Chandrasekhar mass. The clock to the explosion here is provided by the lifetime of the secondary star.

- Double Degenerate (DD) scenario: the merging of two C–O white dwarfs, due to loss of angular momentum caused by gravitational wave radiation, which explode by C-deflagration when M_{Ch} is reached (Iben and Tutukov 1984). In this scenario, the two C–O WDs should be of $\sim 0.7\,M_\odot$ in order to give rise to a Chandrasekhar mass after they merge, therefore their progenitors should be in the range 5–$8\,M_\odot$. The clock to the explosion here is given by the lifetime of the secondary star plus the gravitational time delay which depends on the original separation of the two WDs. The minimum timescale for the appearance of the first Type Ia SNe in this scenario is as low as ~ 40 Myr (see Tornambé and Matteucci 1986). For more recent results on the DD scenario see Greggio (2005).

Within any scenario the explosion can occur either when the C–O WD reaches the Chandrasekhar mass and carbon deflagrates at the center or when a massive enough helium layer is accumulated on top of the C–O WD. In this last case there is

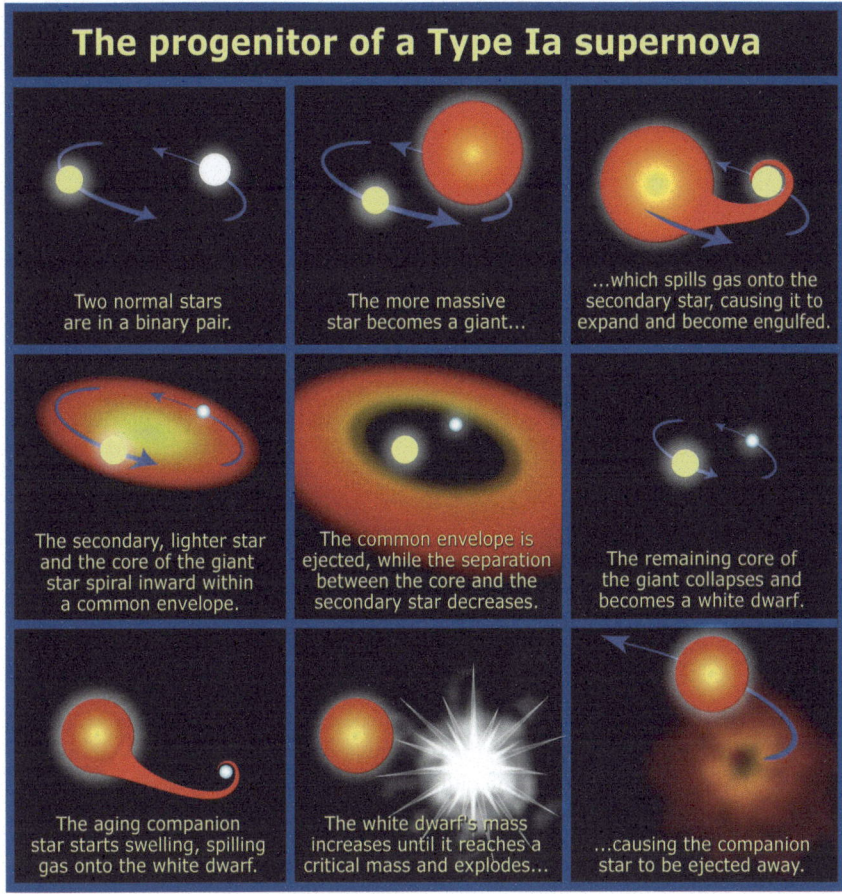

Fig. 2.10 The progenitor of a Type Ia SN in the context of the single-degenerate model (Illustration credit: NASA, ESA, and A. Field (STSci))

He-detonation which induces an off-center carbon deflagration before the Chandrasekhar mass is reached (sub-chandra exploders, e.g. Woosley and Weaver 1994).

While the chandra-exploders are supposed to produce the same nucleosynthesis (C-deflagration of a Chandrasekhar mass), they predict a different evolution of the Type Ia SN rate and different typical timescales for the SNe Ia enrichment. A way of defining the typical Type Ia SN timescale is to assume it as the time when the maximum in the Type Ia SN rate is reached (Matteucci and Recchi 2001). This timescale varies according to the chosen progenitor model and to the assumed star formation history, which varies from galaxy to galaxy. For the solar vicinity, this timescale is at least 1 Gyr, if the SD scenario is assumed, whereas for elliptical galaxies, where the stars formed much more quickly, this timescale is only 0.5 Gyr (Matteucci and Greggio 1986; Matteucci and Recchi 2001).

2.1.4 Gas Flows

Various parametrizations have been suggested for gas flows and the most common is an exponential law for the gas infall rate:

$$IR \propto e^{-t/\tau} \tag{2.18}$$

with the timescale τ being a free parameter, whereas for the galactic outflows the wind rate is generally assumed to be proportional to the SFR:

$$WR = -\lambda\, SFR \tag{2.19}$$

where λ is again a free parameter. Both τ and λ should be fixed by reproducing the majority of observational constraints.

2.2 Basic Equations for Chemical Evolution

2.2.1 Yields per Stellar Generation

Under the assumption of Instantaneous Recycling Approximation (IRA) which states that all stars more massive than $1\,M_\odot$ die immediately, whereas all stars with masses lower than $1\,M_\odot$ live forever, one can define the yield per stellar generation (Tinsley 1980):

$$y_i = \frac{1}{1-R} \int_1^\infty m p_{im} \varphi(m) dm \tag{2.20}$$

where p_{im} is the stellar *new* yield of the element i, namely the newly formed and ejected element i by a star of mass m, and $\varphi(m)$ is the IMF.

The quantity R is the so-called Returned Fraction:

$$R = \int_1^\infty (m - M_{rem})\varphi(m) dm \tag{2.21}$$

and is the mass fraction of gas restored into the ISM by an entire stellar generation. The term fraction derives from the fact that in its definition R is divided by $\int_0^\infty m\varphi(m)dm = 1$, which is the normalization of the IMF.

2.2.2 Analytical Models

2.2.2.1 Simple Model

The Simple Model for the chemical evolution of the solar neighbourhood is the simplest approach to model chemical evolution. The solar neighbourhood is assumed to be a cylinder of 1 Kpc radius centered around the Sun.

The basic assumptions of the Simple Model are:

- the system is one-zone and closed, no inflows or outflows are considered,
- the initial gas is primordial (no metals),
- IRA holds,
- the IMF, $\varphi(m)$, is assumed to be constant in time,
- the gas is well mixed at any time (instantaneous mixing approximation, IMA).

The Simple Model fails in describing the evolution of the Milky Way (G-dwarf metallicity distribution, elements produced on long timescales and abundance ratios) and the reason is that at least two of the above assumptions are manifestly wrong, epecially if one intends to model the evolution of the abundance of elements produced on long timescales, such as Fe. In particular, the assumptions of the closed box and the IRA.

However, it is interesting to know the solution of the Simple Model and its implications. Let Z be the abundance by mass of metals, if $Z \ll 1$, which is generally true, we obtain the solution of the Simple Model for metals. This solution is obtained analytically by ignoring the stellar lifetimes:

$$Z = y_Z \ln(\frac{1}{G}) \tag{2.22}$$

where $G = M_{gas}/M_{tot}$ is the gas mass fraction of the system and y_Z is the yield per stellar generation, as defined above, otherwise called *effective yield*.

In particular, the effective yield is defined as:

$$y_{Z_{eff}} = \frac{Z}{\ln(1/G)} \tag{2.23}$$

namely the yield that the system would have if behaving as the simple closed-box model. This means that if $y_{Z_{eff}} > y_Z$, then the actual system has attained a higher metallicity at a given gas fraction G. Generally, given two chemical elements i and j, the solution of the Simple Model for primary elements (Eq. 2.22) implies:

$$\frac{X_i}{X_j} = \frac{y_i}{y_j} \tag{2.24}$$

which means that the ratio of two element abundances is always equal to the ratio of their yields. This is no more true when IRA is relaxed. In fact, relaxing IRA is necessary to study in detail the evolution of the abundances of single elements produced on long timescales (e.g. Fe, N).

2.2.2.2 Analytical Models in the Presence of Outflows

One can obtain analytical solutions also in the presence of infall and/or outflow but the necessary condition is to assume IRA, as well as precise forms for the infall and outflow rates.

Matteucci and Chiosi (1983) found solutions for models with outflow and infall and Matteucci (2001) found it for a model with infall and outflow acting at the same time. The main assumption in the model with outflow but no infall is that the outflow rate is:

$$W(t) = -\lambda(1-R)\psi(t) \tag{2.25}$$

where $\lambda > 0$ is the wind parameter.

The solution of this model is:

$$Z = \frac{yz}{(1+\lambda)} ln[(1+\lambda)G^{-1} - \lambda] \tag{2.26}$$

for $\lambda = 0$ the equation becomes the one of the Simple Model (2.22).

As one can see from Eq. (2.26), the presence of an outflow decreases the effective yield, in the sense that the true yield of a system is lower than the effective yield. Models with galactic winds or outflows in general are suitable for ellipticals, irregulars and for the Galactic halo. A popular analytical model with outflow is that suggested by Hartwick (1976) for the evolution of the Galactic halo, under the assumption that during the halo collapse stars were forming while the gas was dissipating energy and falling into the bulge and disk, thus producing a net gas loss from the halo. This hypothesis was suggested by the fact that the stellar metallicity distribution of the halo can be reproduced only with an effective yield lower than that of the disk. In Hartwick's model the ouflow rate is assumed to be simply proportional to the SFR:

$$W(t) = -\lambda\psi(t) \tag{2.27}$$

which is similar to Eq. (2.25). Hartwick used this model to reproduce the metallicity distribution of halo stars and also to alleviate the G-dwarf problem in the disk, namely the fact that the Simple Model of chemical evolution predicts too many disk stars than observed (see Tinsley 1980 for a review on the subject). However, the gas lost from the halo cannot have contributed to form the whole disk since the distribution of the specific angular momentum of halo and disk stars are quite different, thus indicating that only a negligible amount of halo gas can have formed the disk. On the other hand, the similarity of the distributions for the halo and bulge indicates that the bulge must have formed out of gas lost from the halo (see Wyse and Gilmore 1992). The G-dwarf problem is instead easily solved if one assumes that the Galactic disk

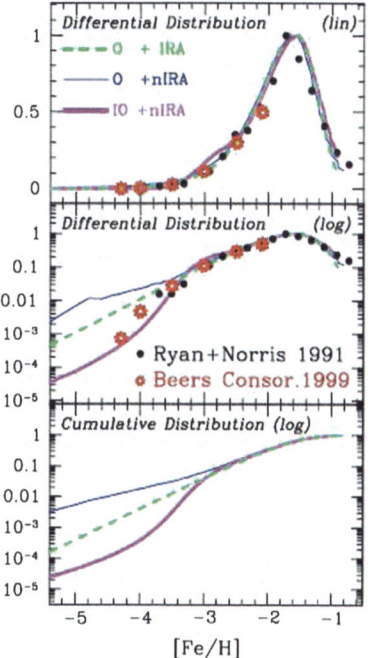

Fig. 2.11 Metallicity distribution for the halo stars. *Upper panel* observed and predicted metallicity distributions. The models are: pure outflow with IRA (*dashed curve*), pure outflow without IRA (*thin solid curve*) and early infall +outflow without IRA (*thick solid curve*). The distribution is on a linear scale. *Middle panel* the same as above but the distribution is on a logarithmic scale. *Lower panel* predicted cumulative distributions. Figure from Prantzos (2003)

has formed by means of slow infall of extragalactic material, as we will see in the next sections. Hartwick's model has been revisited by Prantzos (2003) to interpret the more recent metallicity distribution of halo stars, which is quite different with respect to the G-dwarf metallicity distribution in the local disk. In particular, the halo metallicity distribution is peaked at around [Fe/H] $= -1.6$ dex, whereas the G-dwarf distribution is peaked at around ~ -0.2 dex. Prantzos (2003) suggested that an outflow with $\lambda = 8$ as well as a formation of the halo by early infall are necessary to reproduce the observed halo metallicity distribution. In Fig. 2.11 we show the results of Prantzos (2003) compared with observations.

2.2.2.3 Analytical Models in Presence of Infall

The solution of the equation of metals for a model without a wind but with a primordial infalling material ($Z_A = 0$) at a rate:

$$A(t) = \Lambda(1-R)\psi(t) \tag{2.28}$$

and $\Lambda \neq 1$ is:

$$Z = \frac{y_Z}{\Lambda}[1 - (\Lambda - (\Lambda - 1)G^{-1})^{-\Lambda/(1-\Lambda)}] \tag{2.29}$$

For $\Lambda = 1$ one obtains the well known case of *extreme infall* studied by Larson (1972) whose solution is:

$$Z = y_Z[1 - e^{-(G^{-1}-1)}] \tag{2.30}$$

This extreme infall solution shows that when $G \to 0$ then $Z \to y_Z$. The infall can solve the G-dwarf problem for disk stars except for the extreme infall solution which predicts too few low metallicity stars below [Fe/H] $= -1.0$ (see Tinsley 1980). Moreover, the infall is very important for explaining both the halo and the disk formation.

2.2.2.4 Analytical Models in Presence of Infall and Outflow

Matteucci (2001) presented an analytical solution for infall and outflow present at the same time. The solution refers to the outflow and infall rates of Eqs. (2.25) and (2.28), respectively.

In particular:

$$Z = \frac{y_Z}{\Lambda}\{1 - [(\Lambda - \lambda) - (\Lambda - \lambda - 1)G^{-1}]^{\frac{\Lambda}{\Lambda-\lambda-1}}\}, \tag{2.31}$$

for a primordial infalling gas ($Z_A = 0$).

2.2.3 Detailed Numerical Models

Detailed models of galactic chemical evolution require consideration of the stellar lifetimes, namely they should relax IRA. However, the majority of them still retain the instantaneous mixing approximation (IMA), which assumes that the material ejected by stars at their death is instantaneously mixed with the surrounding interstellar medium (ISM). This approximation seems to be good in the majority of the cases with perhaps the exception of the very early phases of galactic evolution.

The basic equations of chemical evolution follow the evolution of the abundances of single chemical species and the gas as a whole.

If σ_i is the surface mass density of an element i, with $\sigma_{gas} = \sum_{i=1,n} \sigma_i$, being the total surface gas density, we can write:

$$\dot{\sigma}_i(t) = -\psi(t)X_i(t)$$
$$+ \int_{M_L}^{M_{Bm}} \psi(t - \tau_m)Q_{mi}(t - \tau_m)\varphi(m)dm$$
$$+ A \int_{M_{Bm}}^{M_{BM}} \phi(m)$$

2 Chemical Evolution of the Milky Way and Its Satellites

$$\cdot \left[\int_{\gamma_{min}}^{0.5} f(\gamma)\psi(t-\tau_{m2})Q_{mi}(t-\tau_{m2})d\gamma \right] dm$$
$$+ B \int_{M_{Bm}}^{M_{BM}} \psi(t-\tau_m)Q_{mi}(t-\tau_m)\varphi(m)dm$$
$$+ \int_{M_{BM}}^{M_U} \psi(t-\tau_m)Q_{mi}(t-\tau_m)\varphi(m)dm$$
$$+ X_{A_i}A(t) - X_i(t)W(t) \qquad (2.32)$$

The variable here is the the quantity σ_i which represents the surface gas density in the form of a chemical element i. The quantities $X_i(t)$ are the abundances as defined in Eq. (2.1). The quantity Q_{mi} contains all the information about stellar evolution and nucleosynthesis: in practice it gives the mass of gas produced and ejected in the form of an element i by a star of initial mass m, together with the mass of that element which was already present in the star at birth. The various integrals represent the rates at which the mass of a given element is restored into the ISM by stars of different masses which can evolve into WDs or supernovae (II, Ia, Ib). The integral representing the rate of matter restoration by Type Ia SNe is the second one on the right hand side. The quantity A is a constant: it is the fraction, in the IMF, of binary systems with those specific features required to give rise to Type Ia SNe, whereas B = 1 − A is the fraction of all the single stars and binary systems in the same mass range of definition of the progenitors of Type Ia SNe (third integral). The parameter A is obtained by imposing that the predicted Type Ia SN rate reproduces the observed rate at the present time (14 Gyr). Values of A = 0.05–0.09 are found for the evolution of the solar vicinity when an IMF of (Scalo 1986, 1998) or Kroupa et al. (1993) is adopted. If one adopts a flatter IMF such as the Salpeter (1955) one, then A is different. The integral of the Type Ia SN contribution is made over a range of mass going from $M_{Bm} = 3 M_\odot$ to $M_{BM} = 16 M_\odot$, which represents the total masses of binary systems able to produce Type Ia SNe in the framework of the single degenerate scenario. There is also an integration over the mass distribution of binary systems; in particular, one considers the function $f(\gamma)$ where $\gamma = \frac{M_2}{M_1+M_2}$, with M_1 and M_2 being the primary and secondary mass of the binary system, respectively (for more details see Matteucci and Greggio 1986 and Matteucci 2001). The third and fourth integrals represent the rates of Type II and Type Ib/c SNe, respectively. The occurrence of Type Ib SNe seems to be partly related to Wolf-Rayet stars which have original masses larger than $25 M_\odot$ and depends on the mass loss rate which is more active at high metallicities. However, it has been proposed that Type Ib SNe can also originate from massive stars in binary systems. Finally, the functions A(t) and W(t) are the infall and wind rate, respectively.

2.3 The Milky Way

We will first analyze the chemical evolution of our Galaxy, the Milky Way.

2.3.1 The Formation of the Milky Way

2.3.1.1 Observational Evidence

The Milky Way galaxy has four main stellar populations: (1) the halo stars with low metallicities (the most common metallicity indicator in stars is [Fe/H] = $log(Fe/H)_* - log(Fe/H)_\odot$ and eccentric orbits, (2) the bulge population with a large range of metallicities and dominated by random motions, (3) the thin disk stars with an average metallicity $< [Fe/H] > = -0.5$ dex and circular orbits, and finally (4) the thick disk stars which possess chemical and kinematical properties intermediate between those of the halo and those of the thin disk. The halo stars have average metallicities of $< [Fe/H] > = -1.5$ dex and a maximum metallicity of ~ -1.0 dex, although stars with [Fe/H] as high as -0.6 dex and halo kinematics are observed. The average metallicity of thin disk stars is ~ -0.6 dex, whereas the one of bulge stars is ~ -0.2 dex.

The kinematical and chemical properties of the different Galactic stellar populations can be interpreted in terms of the Galaxy formation mechanism. Eggen et al. (1962), in a cornerstone paper suggested a rapid monolithic collapse for the formation of the Galaxy lasting $\sim 2 \times 10^8$ years. This suggestion was based on a kinematical and chemical study of solar neighbourhood stars and the value of the suggested timescale was chosen to allow for the orbital eccentricities to vary in a potential not yet in equilibrium but sufficiently long so that massive stars forming in the collapsing gas could have time to die and enrich the gas with heavy elements (Fig. 2.12).

Later on, Searle and Zinn (1978) measured Fe abundances and horizontal branch morphologies of 50 globular clusters and studied their properties as a function of the galactocentric distance. As a result of this, they proposed a central collapse like the one envisaged by Eggen et al., but also that the outer halo formed by merging of large fragments taking place over a considerable timescale > 1 Gyr. The Searle and Zinn scenario is close to what is predicted by modern cosmological theories of galaxy formation. In particular, in the framework of the hierarchical galaxy formation scenario, galaxies form by accretion of smaller building blocks (e.g. White and Rees 1978; Navarro et al. 1997). Obvious candidates for these building blocks are either dwarf spheroidal (dSph) or dwarf irregular (dIrr) galaxies. However, as we will see in detail later, the chemical composition and in particular the chemical abundance patterns in dSphs or dIrrs are not compatible with the same abundance patterns in the Milky Way (see Geisler et al. 2007), thus arguing against the identification of the building blocks with these galaxies. On the other hand, Carollo et al. (2007) have

Fig. 2.12 Schematic edge-on view of the major components of the Milky Way. Illustration credit from R. Buser, http://www.astro.unibas.ch/forschung/rb/structure.shtml

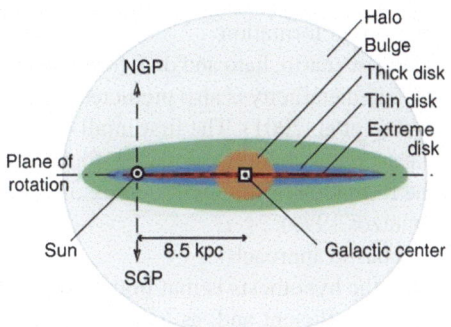

obtained medium resolution spectroscopy of 20,336 stars from the Sloan Digital Sky Survey (SDSS). They showed that the Galactic halo is divisible into two broadly overlapping structural components. In particular, they find that the inner halo is dominated by stars with very eccentric orbits, exhibits a peak at [Fe/H] $= -1.6$ dex and has a flattened density distribution with a modest net prograde rotation. The outer halo includes stars with a wide range of eccentricities, exhibits a peak at [Fe/H] $= -2.2$ dex and a spherical density distribution with highly statistically significant net retrograde rotation. They conclude that most of the Galactic halo should have formed by accrection of multiple distinct sub-systems. However, an analysis of the abundance ratios of these stars is still missing.

2.3.1.2 Theoretical Models

From an historical point of view, the modelization of the Galactic chemical evolution has passed through different phases that I summarize in the following:

- Serial formation
  The Galaxy is modeled by means of one accretion episode lasting for the entire Galactic lifetime, where halo, thick and thin disk form in sequence as a continuous process. The obvious limit of this approach is that it does not allow us to predict the observed overlapping in metallicity between halo and thick disk stars and between thick and thin disk stars, but it gives a fair representation of our Galaxy (e.g. Matteucci and François 1989).
- Parallel formation
  In this formulation, the various Galactic components start at the same time and from the same gas but evolve at different rates (e.g. Pardi et al. 1995). It predicts overlapping of stars belonging to the different components but implies that the thick disk formed out of gas shed by the halo and that the thin disk formed out of gas shed by the thick disk, and this is at variance with the distribution of the stellar angular momentum per unit mass (Wyse and Gilmore 1992), which indicates that the disk did not form out of gas shed by the halo.

- Two-infall formation
  In this scenario, halo and disk formed out of two separate infall episodes (overlapping in metallicity is also predicted) (e.g. Chiappini et al. 1997; Chang et al. 1999; Alibés et al. 2001). The first infall episode lasted no more than 1–2 Gyr whereas the second, where the thin disk formed, lasted much longer with a timescale for the formation of the solar vicinity of 6–8 Gyr (Chiappini et al. 1997; Boissier and Prantzos 1999).
- Stochastic approach
  Here the hypothesis is that in the early halo phases ([Fe/H] < −3.0 dex), mixing was not efficient and, as a consequence, one should observe, in low metallicity halo stars, the effects of pollution from single SNe (e.g. Tsujimoto et al. 1999; Argast et al. 2000; Oey 2000). These models predict a large spread for [Fe/H] < −3.0 dex for all the α-elements, which is not observed, as shown by recent data with metallicities down to −4.0 dex (Cayrel et al. 2004). However, inhomogeneities could explain the observed spread of s- and r-elements at low metallicities (see later).

2.3.2 The Two-Infall Model

The two-infall model of Chiappini et al. (1997) predicts two main episodes of gas accretion: during the first one, the halo the bulge and thick disk formed, while the second gave rise to the thin disk. In Fig. 2.13 we show an artistic representation of the formation of the Milky Way in the two-infall scenario. In the upper panel we see the sequence of the formation of the stellar halo, in particular the inner halo, following a monolithic-like collapse of gas (first infall episode) but with a longer timescale than originally suggested by Eggen et al. (1962): here the time scale is 1–2 Gyr. During the halo formation also the bulge is formed on a very short timescale in the range 0.1–0.5 Gyr. During this phase also the thick disk assembles or at least part of it, since part of the thick disk, like the outer halo, could have been accreted. The second panel from left to right shows the beginning of the disk formation, namely the assembly of the innermost disk regions just around the bulge. This is due to the second infall episode which gives rise to the thin disk. The thin-disk assembles inside-out, in the sense that the outermost regions take a much longer time to form. This is shown in the third panel. In Fig. 2.13 each panel is connected to temporal phases where the Type II and then the Type Ia SN rates are present. So, it is clear that the early phases of the halo and bulge formation are dominated by Type II SNe (and also by Type Ib/c SNe) producing mostly α-elements such as O and Mg. On the other hand, Type Ia SNe start to be non negligible only after 1 Gyr and they pollute the gas during the thick and thin disk phases. The minimum shown in the Type II SN rate is due to a gap in the star formation rate occurring as a consequence of the adoption of a threshold density in the star formation process, as we will see next (Fig. 2.14).

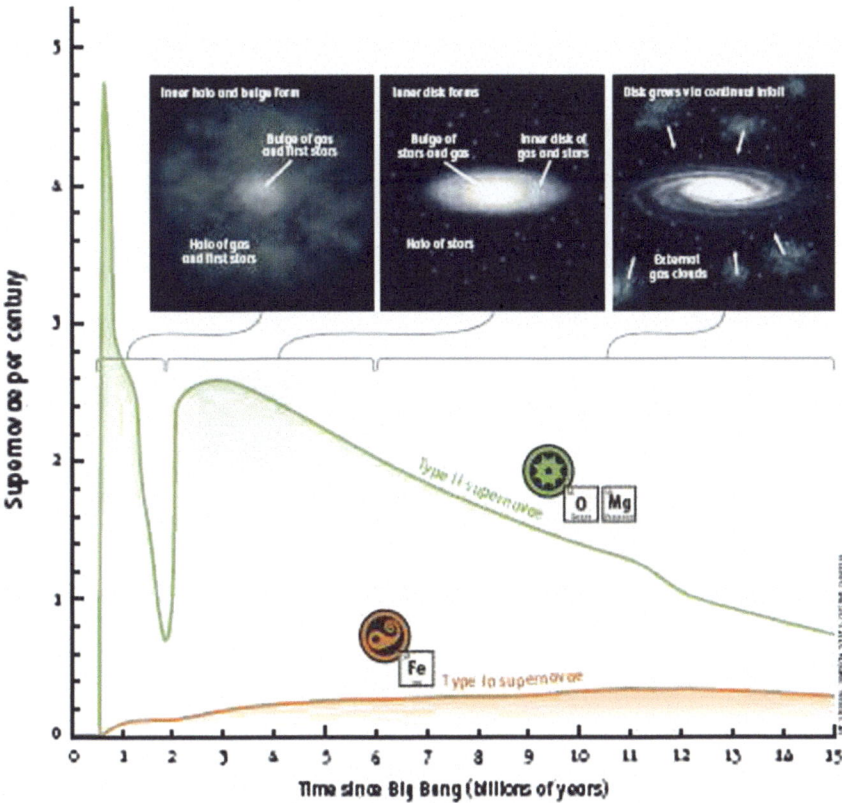

Fig. 2.13 Artistic view of the two-infall model by Chiappini et al. (1997). The predicted SN II and Ia rates per century are also sketched, together with the fact that Type II SNe produce mostly α-elements (e.g. O, Mg), whereas Type Ia SNe produce mostly Fe. (Illustration credit: C. Chiappini, Sky and Telescope 2004, Vol. 108, No. 4, p.32)

2.3.3 Detailed Recipes for the Two-Infall Model

The main assumption of this model are:

- The IMF is that of Scalo (1986) normalized over a mass range of 0.1–100 $M_\odot$.
- The infall law is:

$$A(r,t) = a(r)e^{-t/\tau_H(r)} + b(r)e^{-(t-t_{max})/\tau_D(r)} \tag{2.33}$$

where $A(r,t) = (\frac{d\sigma(r,t)}{dt})_{infall}$ is the rate at which the total surface mass density changes because of the infalling gas. The quantities $a(r)$ and $b(r)$ are two parameters fixed by reproducing the total present time surface mass density in the solar vicinity ($\sigma_{tot} = 51 \pm 6 M_\odot$ pc^{-2}, see Boissier and Prantzos 1999), $t_{max} = 1$ Gyr is the time for the maximum infall on the thin disk, $\tau_H = 2.0$ Gyr is the time scale

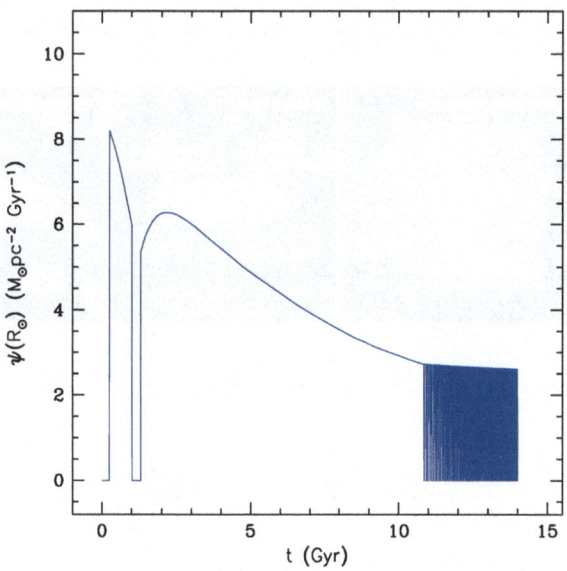

Fig. 2.14 The SFR in the solar vicinity as predicted by the two-infall model. Figure from Chiappini et al. (1997). The oscillating behaviour in the Type II SN rate at late times is due to the assumed threshold density for SF. The threshold gas density is also responsible for the gap in the SFR seen at around 1 Gyr

for the formation of the halo thick-disk and $\tau_D(r)$ is the timescale for the formation of the thin disk and it is a function of the galactocentric distance (formation inside-out, Matteucci and François 1989; Chiappini et al. 2001).
In particular, it is assumed that:

$$\tau_D = 1.033 r(\text{Kpc}) - 1.267 \,\text{Gyr} \tag{2.34}$$

where r is the galocentric distance.
- The SFR is the Kennicutt law with a dependence on the surface gas density and also on the total surface mass density (see Dopita and Ryder 1994). In particular, the SFR is based on the law originally suggested by Talbot and Arnett (1975) and then adopted by Chiosi (1980):

$$\psi(r,t) = \nu \left(\frac{\sigma(r,t)\sigma_{gas}(r,t)}{\sigma(r_\odot,t)^2} \right)^{(k-1)} \sigma_{gas}(r,t)^k. \tag{2.35}$$

where the constant ν is the efficiency of the SF process, as defined in Eq. (2.6), and is expressed in Gyr^{-1}: in particular, $\nu = 2\,\text{Gyr}^{-1}$ for the the halo and $1\,\text{Gyr}^{-1}$ for the disk ($t \geq 1\,\text{Gyr}$). The total surface mass density is represented by $\sigma(r,t)$, whereas $\sigma(r_\odot,t)$ is the total surface mass density at the solar position, assumed to be $r_\odot = 8\,\text{Kpc}$ (Reid 1993). The quantity $\sigma_{gas}(r,t)$ represents the surface gas

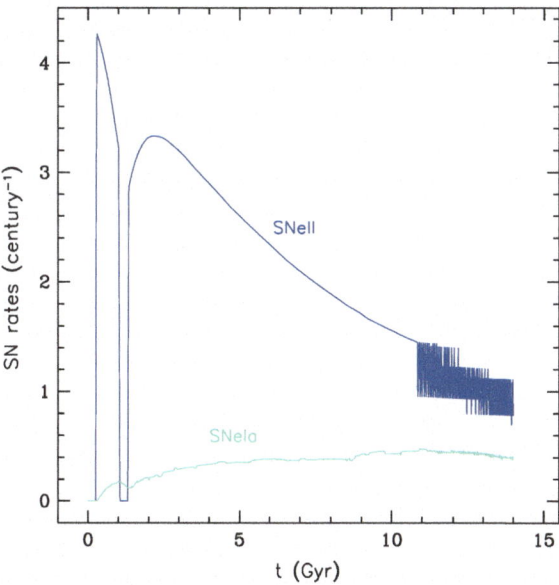

Fig. 2.15 The Type II and Ia rate in the solar vicinity as predicted by the two-infall model. Figure from Chiappini et al. (1997). The oscillating behaviour at late times is due to the assumed threshold density for SF. The threshold gas density is also responsible for the gap in the SFR seen at around 1 Gyr

density. The exponent of the surface gas density, k, is set equal to 1.5, similar to what suggested by Kennicutt (1998a). These choices for the parameters allow the model to fit very well the observational constraints, in particular in the solar vicinity. We recall that below a critical threshold for the surface gas density (7 $M_\odot pc^{-2}$ for the thin disk and 4 $M_\odot\, pc^{-2}$ for the halo phase) we assume that the star formation is halted. The existence of a threshold for the star formation has been suggested by Kennicutt (1998a, b) and Martin and Kennicutt (2001).

The predicted behaviour of the SFR, obtained by adopting Eq. (2.35) with the threshold is shown in Fig. 2.14.

- The Type Ia SN model is the single-degenerate one with the recipe first adopted in Greggio and Renzini (1983a) and Matteucci and Greggio (1986) and more recently in Matteucci and Recchi (2001). The minimum time for the explosion is 30 Myr, whereas the the timescale for restoring the bulk of Fe is 1 Gyr, for the SFR adopted in the solar vicinity. It is worth recalling that this timescale is not universal since it depends on the assumed SNIa progenitor model but also on the assumed star formation history. The SN rates in the solar vicinity are shown in Fig. 2.15.

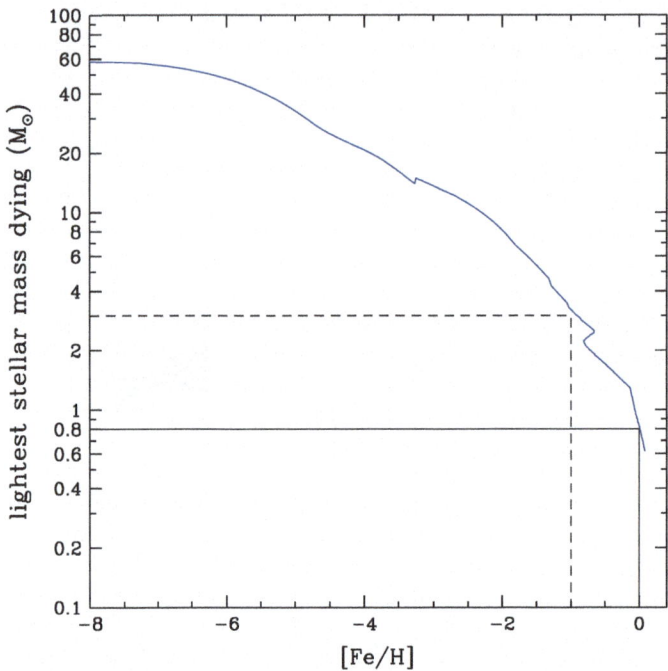

Fig. 2.16 In this figure we show the smallest stellar mass which dies at any given [Fe/H] achieved by the ISM as a consequence of chemical evolution. Thus, it is clear that in the early phases of the halo only massive stars are dying and contributing to the chemical enrichment process. Clearly this graph depends upon the assumed stellar lifetimes and upon the age-[Fe/H] relation. It is worth noting that the Fe production from Type Ia SNe appears before the gas has reached [Fe/H] = −1.0, therefore during the halo and thick disk phase. This clearly depends upon the assumed Type Ia SN progenitors (in this case the single degenerate model)

2.3.4 The Chemical Enrichment History of the Solar Vicinity

We study first the solar vicinity, namely the local ring at 8 Kpc from the Galactic center. By integrating Eq. (2.32) without the wind term we obtain the evolution of the abundances of several chemical species (H, D, He, Li, C, N, O, α-elements, Fe, Fe-peak elements, s-and r- process elements). In Fig. 2.16 we show the smallest mass dying at any cosmic time corresponding to a given abundance of [Fe/H] in the ISM. This is because there is an age-metallicity relation and the [Fe/H] abundance increases with time. We recall that, for a generic chemical element i, with abundance X_i, one defines:

$$[X_i/H] = log(X_i/H)_* - log(X_i/H)_\odot, \quad (2.36)$$

where $log(X_i/H)_\odot$ refers to the solar abundance of the element i.

2.3.4.1 The Observational Constraints

A good model of chemical evolution should be able to reproduce a minimum number of observational constraints and the number of observational constraints should be larger than the number of free parameters which are: τ_H, τ_D, k_1, k_2, ν and A (the fraction of binary systems which can give rise to Type Ia SNe).

The main observational constraints in the solar vicinity that a good model should reproduce (see Chiappini et al. 2001; Boissier and Prantzos 1999 and references therein) are:

- The present time surface gas density: $\sigma_{gas} = 13 \pm 3 M_\odot \, \text{pc}^{-2}$
- The present time surface star density $\sigma_* = 43 \pm 5 M_\odot \, \text{pc}^{-2}$
- The present time total surface mass density: $\sigma_{tot} = 51 \pm 6 M_\odot \, \text{pc}^{-2}$
- The present time SFR: $\psi_o = 2\text{--}5 \, M_\odot \, \text{pc}^{-2} \, \text{Gyr}^{-1}$
- The present time infall rate: $0.3\text{--}1.5 \, M_\odot \, \text{pc}^{-2} \, \text{Gyr}^{-1}$
- The present day mass function (PDMF)
- The solar abundances, namely the chemical abundances of the ISM at the time of birth of the solar system 4.5 Gyr ago as well as the present time abundances
- The observed $[X_i/\text{Fe}]$ versus $[\text{Fe}/\text{H}]$ relations
- The G-dwarf metallicity distribution
- The age-metallicity relation

And finally, a good model of chemical evolution of the Milky Way should reproduce the distributions of abundances, gas and star formation rate along the disk as well as the average SNII and Ia rates (SNII $= 1.2 \pm 0.8 \, 100 \, \text{year}^{-1}$ and SNIa $= 0.3 \pm 0.2 \, 100 \, \text{year}^{-1}$).

2.3.4.2 The Time-Delay Model

What we call time-delay model is the interpretation of the behaviour of abundance ratios such $[\alpha/\text{Fe}]$ (where α-elements are O, Mg, Ne, Si, S, Ca and Ti) versus $[\text{Fe}/\text{H}]$, a typical way of plotting the abundances measured in the stars. The time-delay refers to the delay with which Fe is ejected into the ISM by SNe Ia relative to the fast production of α-elements by core-collapse SNe. Tinsley (1979) first suggested that this time delay would have produced a typical signature in the $[\alpha/\text{Fe}]$ versus $[\text{Fe}/\text{H}]$ diagram. In the following years, Greggio and Renzini (1983b), by means of simple models (star formation burst or constant star formation) studied the effects of the delayed Fe production by Type Ia SNe on the $[\text{O}/\text{Fe}]$ versus $[\text{Fe}/\text{H}]$ diagram. Matteucci and Greggio (1986) included for the first time the Type Ia SN rate formulated by Greggio and Renzini (1983a) in a detailed model for the chemical evolution of the Milky Way. The effect of the delayed Fe production is to create an overabundance of O relative to Fe ($[\text{O}/\text{Fe}] > 0$) at low $[\text{Fe}/\text{H}]$ values, and a continuous decline of the $[\text{O}/\text{Fe}]$ ratio until the solar value ($[O/Fe]_\odot = 0.0$) is reached for $[\text{Fe}/\text{H}] > -1.0 \, \text{dex}$. This is what is observed and indicates that during the halo phase the $[\text{O}/\text{Fe}]$ ratio is due only to the production of O and Fe by SNe II. However, since the bulk of Fe is produced by Type

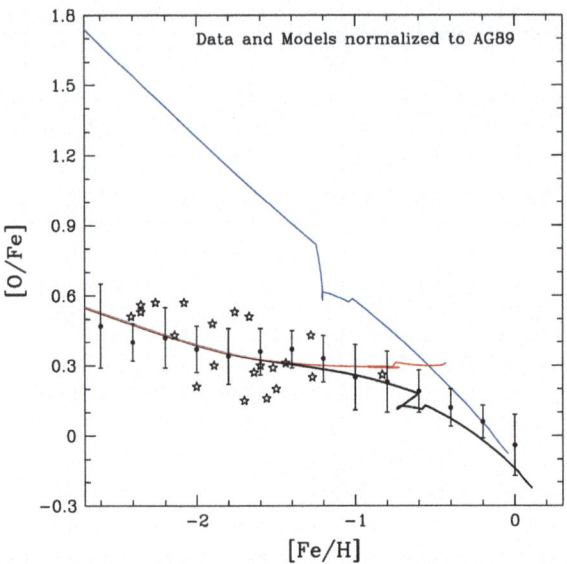

Fig. 2.17 The relation between [O/Fe] versus [Fe/H] for Galactic stars in the solar vicinity. The models and the data are normalized to the solar meteoritic abundances of Anders and Grevesse (1989). The *thick curve* represents the predictions of the two-infall model where Type Ia SNe produce ∼70% of Fe and Type II SNe the remaining ∼30%. The *upper thin curve* represents the case where all the Fe is assumed to be produced by Type Ia SNe, whereas the *thin lower line* refers to the case where all the Fe is assumed to be produced in Type II SNe. The data are from Meléndez and Barbuy (2002)

Ia SNe, when these latter start to be important then the [O/Fe] ratio begins to decline. This effect was predicted by Matteucci and Greggio (1986) to occur also for other α-elements (e.g. Mg, Si). At the present time, a great amount of stellar abundances is available and the trend of the α-elements has been confirmed. Before showing some of the most recent data, it is worth showing better the time-delay model. In Fig. 2.17 it is shown that a good fit of the [O/Fe] ratio as a function of [Fe/H] is obtained only if the α-elements are mainly produced by Type II SNe and the Fe by Type Ia SNe. If one assumes that only SNe Ia produce Fe as well as if one assumes that only Type II SNe produce Fe, the agreement with observations is lost. Therefore, the conclusion is that both Types of SNe should produce Fe in the proportions of 1/3 for Type II SNe and 2/3 for Type Ia SNe. The IMF also plays a role in this game and these proportions are obtained for "normal" Salpeter-like IMFs, which includes both Salpeter (1955) and Scalo (1986) or Kroupa et al. (1993) IMFs.

As an illustration of the time-delay model we show in Figs. 2.18, 2.19 and 2.20 the [X/Fe] versus [Fe/H] relations both observed and predicted for stars in the solar vicinity belonging to halo, thick- and thin-disk. The adopted yields for massive stars are those suggested by François et al. (2004) in order to best fit these relations and the solar abundances (namely the abundances in the ISM 4.5 Gyr ago). These yields are

2 Chemical Evolution of the Milky Way and Its Satellites 173

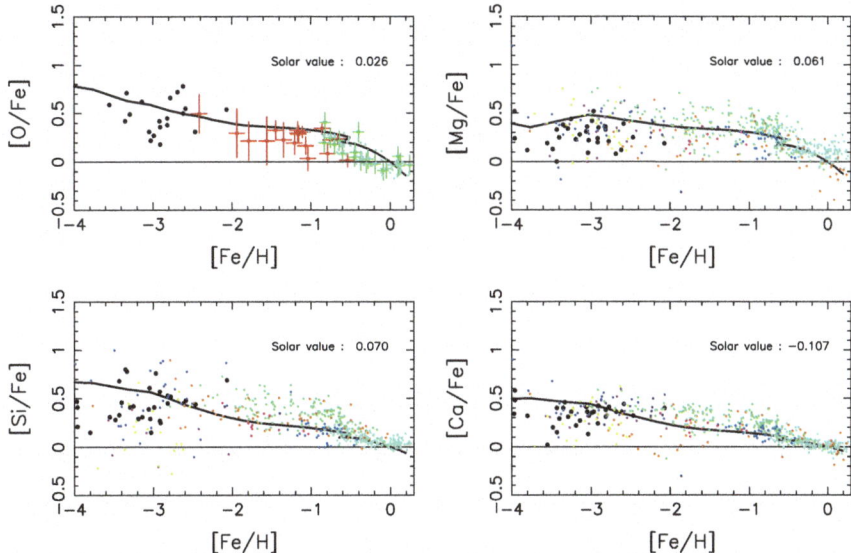

Fig. 2.18 Predicted and observed [α/Fe] versus [Fe/H] in the solar neighbourhood. The models and the data are from François et al. (2004). The models are normalized to the predicted solar abundances. The predicted abundance ratios at the time of the Sun formation (Solar value) are shown in each panel and indicate a good fit (all the values are close to zero)

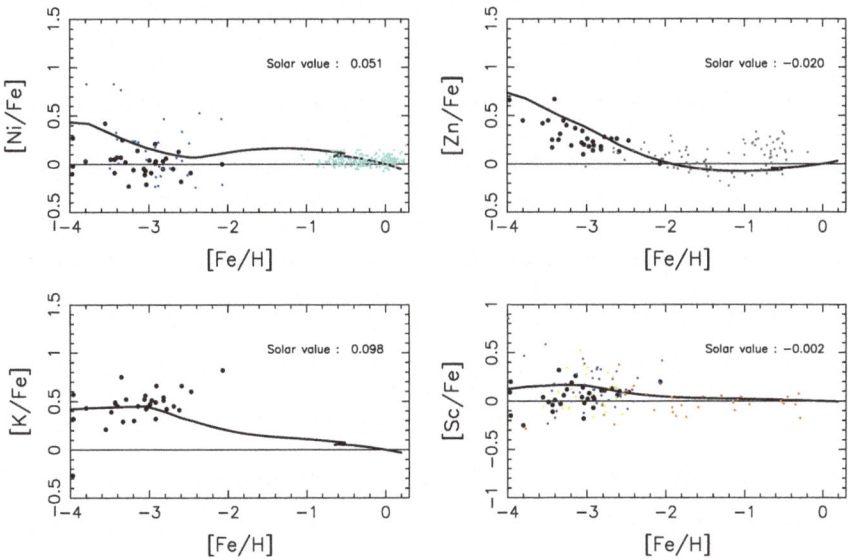

Fig. 2.19 The same as Fig. 2.18 for Ni, Zn, K and Sc. The models and the data are from François et al. (2004). The models are normalized to the predicted solar abundances. The predicted abundance ratios at the time of the Sun formation are shown in each panel and indicate a good fit

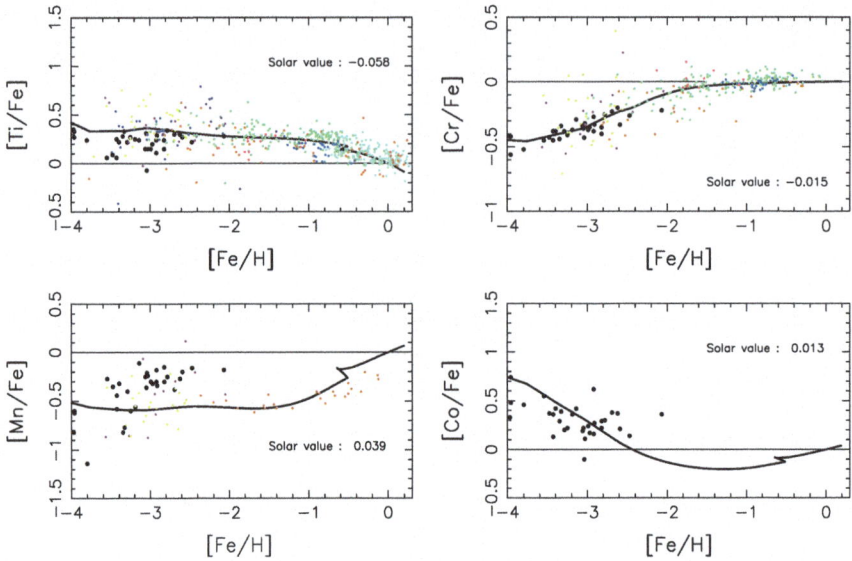

Fig. 2.20 The same as in Fig. 2.18 for Ti, Cr, Mn and Co. The models and the data are from François et al. (2004). The models are normalized to the predicted solar abundances. The predicted abundance ratios at the time of the Sun formation are shown in each panel and indicate a good fit

obtained by applying some corrections to the yields of WW95, as shown in Fig. 2.21, where the ratios between the suggested and WW95 yields are reported.

In Fig. 2.22 we show the predictions of a chemical evolution model for the solar vicinity where the yields from massive stars of Nomoto et al. (2006) have been adopted, except for C, N and O whose yields are taken from the models with mass loss and rotation of the Geneva group. For the low and intermediate mass stars the adopted yields are from Karakas (2010). In the same Figure are shown the predictions obtained with the yields from massive stars by WW95 and those from low and intermediate mass stars from HG97. As one can see, the best yields seem to be the combination of Nomoto et al. (2006), Geneva and Karakas (2010) yields. In general, this combination of yields shows that some of the problems present in the previous yields have been alleviated, whereas for other elements the disagreement still persists.

2.3.4.3 The G-dwarf Metallicity Distribution and Constraints on the Thin Disk Formation

The G-dwarf metallicity distribution is a quite important constraint for the chemical evolution of the solar vicinity. It is the fossil record of the star formation history of the thin disk. If one is able to reproduce such a distribution, then can have an idea of the SFR and the IMF and, as a consequence, of the gas accretion history. Therefore, to fit the G-dwarf metallicity distribution means to obtain constraints on

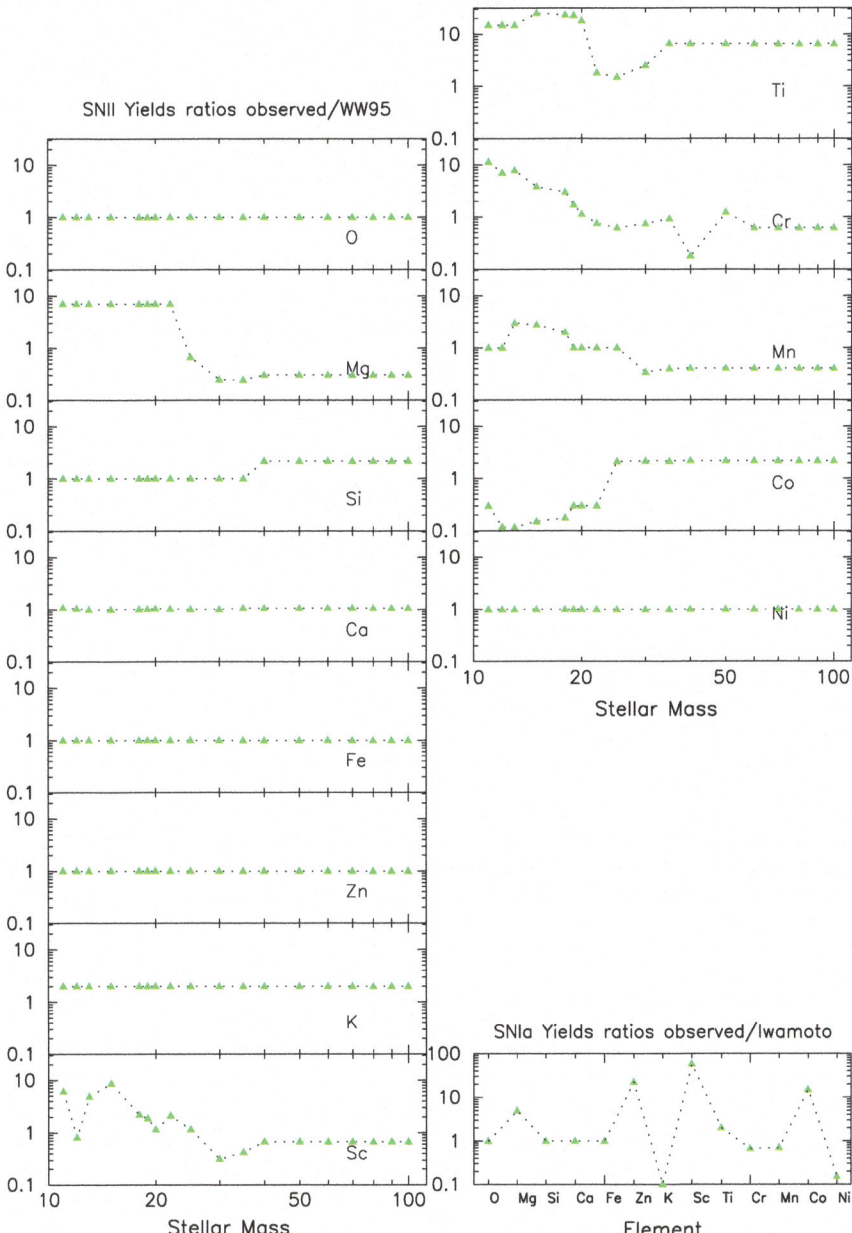

Fig. 2.21 Ratios between the empirical yields derived by François et al. (2004) and the yields of WW95 for massive stars. In the small panel at the *bottom right* we show the same ratios for SNe Ia and the comparison is with the yields of Iwamoto et al. (1999)

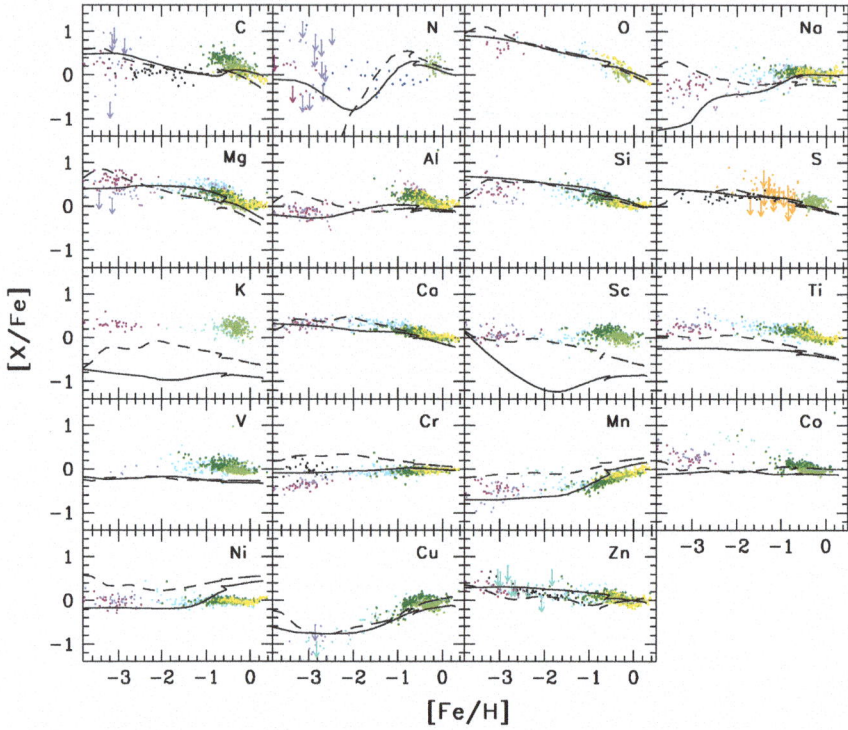

Fig. 2.22 Predicted and observed [α/Fe] versus [Fe/H] in the solar neighbourhood. Two sets of yields have adopted. *Dashed lines* the yields from massive stars are from Nomoto et al. (2006) and from the Geneva group, the yields from low and intermediate mass stars are from Karakas (2010) and those for Type Ia SNe from Iwamoto et al. (1999). *Continuous lines* yields from massive stars from Woosley and Weaver (1995), yields from low and intermediate mass stars from van den Hoeck and Groenewegen (1997), yields from SNe Ia from Iwamoto et al. (1999). Figure from Romano et al. (2010)

the mechanism of formation of the thin disk. Originally, there was the "G-dwarf problem" which means that the Simple Model of galactic chemical evolution could not reproduce the distribution of the G-dwarfs. It has been since long demonstrated that relaxing the closed-box assumption and allowing for the solar region to form gradually by accretion of gas can solve the problem (Tinsley 1980; Pagel 1997). Also a variable IMF could solve the problem but it would create other problems (see Martinelli and Matteucci 2000). Assuming that the disk forms from pre-enriched gas can also solve the problem but still the gas infall is necessary to have a realistic picture of the disk formation. The two-infall model can reproduce very well the G-dwarf distribution and also that of K-dwarfs (see Fig. 2.23), as long as a timescale for the formation of the disk in the solar vicinity of 7–8 Gyr is assumed. This conclusion is shared by other authors (Alibés et al. 2001; Boissier and Prantzos 1999). More recently, Casagrande et al. (2011) reanalysed the Geneva-Copenhagen

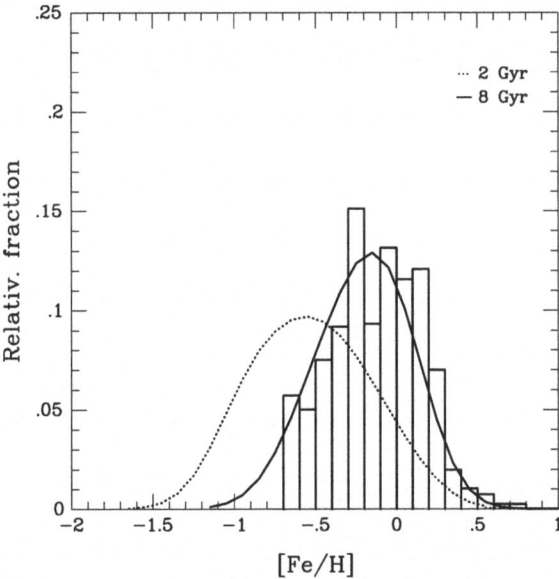

Fig. 2.23 The figure is from Kotoneva et al. (2002) and shows the comparison between a sample of K-dwarfs and model predictions in the solar neighbourhood. The *dotted curve* refers to the two-infall model with a timescale $\tau = 2\,\text{Gyr}$, whereas the continuous line refers to $\tau = 8\,\text{Gyr}$

stellar survey and derived a metallicity distribution function where the stars are separated according to different age intervals (Fig. 2.24). The interesting aspect of this study is that it shows that young stars (ages $<$ 1 Gyr) show a quite narrow distribution, whereas intermediate (between 1 and 5 Gyr) and old age (ages $>$ 5 Gyr) stars present a broader distribution. In particular, old stars seem to have born with the largest range of metallicities ([Fe/H]). The interpretation given by Casagrande et al. (2011) is that this broad distribution can be the sign of stellar migration (Sellwood and Binney 2002). In this picture, the solar neighbourhood could have been not only assembled from local stars, following a local age metallicity relation, but also from stars originating from the inner (more metal-rich) and outer (more metal-poor) Galactic disk which have migrated to the present position.

2.3.4.4 Carbon and Nitrogen Evolution

Carbon and nitrogen deserve a separate discussion from the other elements, in particular ^{14}N whose observational behaviour is difficult to reconcile with the theory. First of all, we should distinguish between *primary* and *secondary* elements: primary elements are those synthesized directly from H and He, whereas secondary elements are those deriving from metals already present in the star at birth. In the framework of the Simple Model of galactic chemical evolution, the abundance of a secondary

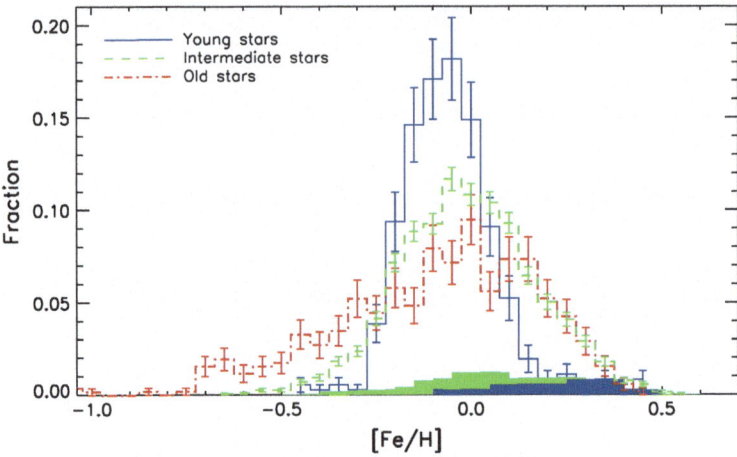

Fig. 2.24 The G-dwarf metallicity distribution obtained by Casagrande et al. (2011) by a reanalysis of the Geneva-Copenhagen survey. They have divided the stars according to their ages: stars having age <1 Gyr are shown with a *continuous line* (*blue*), $1 \leq$ age < 5 Gyr with a *dashed line* (*green*) and age ≥ 5 Gyr with a *dot-dashed line* (*red*). Shaded areas identify the subgroup of stars in the same age intervals as above, but with absolute magnitudes (< 2); no such bright stars are present in the old sample. Only stars with well determined ages were used. Figure from Casagrande et al. (2011)

element evolves like the square of the abundance of the progenitor metal, whereas the evolution of the abundance of a primary element does not depend on the metallicity (Fig. 2.25).

In Fig. 2.25 we show the predictions of the Simple Model for the ratio N/O, together with data for extragalactic HII regions and Damped Lyman-α systems (DLAs).

It is worth noting that the solutions of the Simple Model for a primary and a secondary element are over-simplifications since the Simple Model does not take into account stellar lifetimes which are very important in ^{14}N production, which arises mainly from low and intermediate mass stars, both as a secondary and primary element (e.g. Renzini and Voli 1981; van den Hoeck and Groenewegen 1997). Also ^{12}C originates mainly from low and intermediate mass stars. The contribution to ^{12}C from massive stars becomes very important only for metallicities oversolar, if the metallicity dependent mass loss is adopted (e.g. Maeder 1992). The interpretation of the diagram of Fig. 2.25 is not so straightforward since extragalactic HII regions and DLAs are galaxies, and not necessarily that diagram is an evolutionary one, in the sense that O/H does not trace the time unlike [Fe/H] in the Galactic stars. Galaxies, in fact, may have started forming stars at different cosmic epochs and with different SF histories. However, if we interpret the diagram of Fig. 2.25 as an evolutionary one, then the DLAs and the extragalactic HII regions of low metallicity should be younger and reflect the nucleosynthesis in massive stars and perhaps in intermediate mass stars. The observed plateau for N/O at low metallicity then would indicate

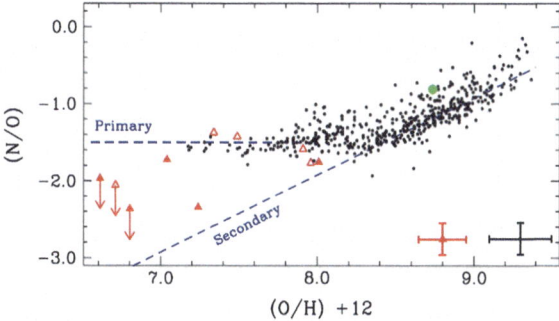

Fig. 2.25 The plot of log (N/O) versus log (O/H) + 12: small dots represent extragalactic HII regions, *red triangles* are Damped-Lyman α systems (DLA), which are high redshift objects. *Dashed lines* mark the solution of the Simple Model for a primary and a secondary element. Figure from Pettini et al. (2002)

a primary production of N in massive stars. Nitrogen, in fact, is also produced in massive stars: until a few years ago, the N production in massive stars was considered only a secondary process, until Meynet and Maeder (2002b, 2003, 2005) showed that stellar rotation in massive stars can produce primary N. A better test for the primary/secondary nature of N are the Galactic stars, since they really represent an evolutionary sequence. In Figs. 2.26 and 2.27 we show data on C and N compared with chemical evolution models including N from rotating massive stars.

As one can see in Figs. 2.26 and 2.27, the fit with data is good when primary N from massive stars is included. However, there are a few warnings, first of all the measurements of N abundance in stars of low metallicity are still uncertain and then the fact that the N measurement in the gas in DLAs at high redshift show that at low O abundances there are systems with a log (N/O) < -2.0, below the plateau shown by Galactic stars. A plateau in [N/Fe] is also observed in Galactic stars for [Fe/H] < -3.0 dex, as shown in Fig. 2.27. In Fig. 2.27 we show also the [C/Fe] values for Galactic stars but only for low metallicity stars: they indicate a roughly solar ratio like the stars with higher metallicities. Therefore, both [N/Fe] and [C/Fe] seem to show roughly constant solar values over the total [Fe/H] range. In the framework of the time-delay model, this means that C, N and Fe are all formed in the same stars and that N is mainly a primary element. However, more data are necessary to assess this point and to reconcile the Galactc star data with high redshift DLAs.

2.3.4.5 S- and R- Process Elements

The s- and r- process elements are generally produced by neutron capture on Fe seed nuclei. The former are formed during the He-burning phase both in low and massive stars, whereas the latter occur in explosive events such as Type II SNe. François et al. (2006) have measured the abundances of several very heavy elements (e.g. Ba and

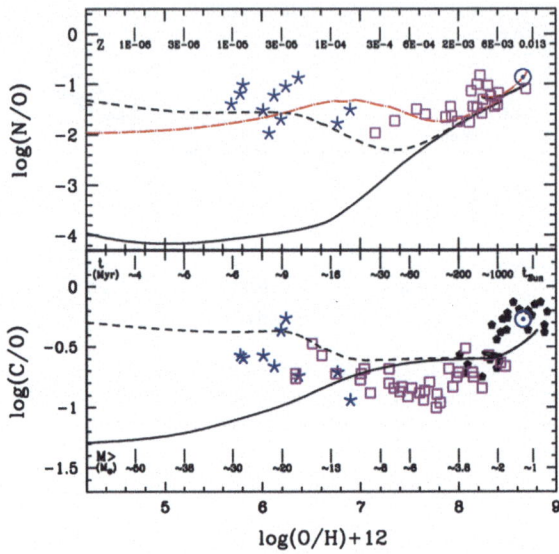

Fig. 2.26 *Upper panel* solar vicinity diagram log (N/O) versus log (O/H) + 12. The data points are from Israelian et al. (2004) (*large squares*) and Spite et al. (2005) (*asterisks*). Models: the *dashed line* represents a model with substantial primary N production from massive stars. This was obtained by means of stellar models (Meynet et al. 2006; Hirschi 2007) with faster rotation relative to the work of Meynet and Maeder (2002a) for $Z = 10^{-8}$. *Lower panel* solar vicinity diagram log (C/O) versus log (O/H) + 12. The data are from Spite et al. (2005) (*asterisks*), Israelian et al. (2004) (*squares*) and Nissen (2003) (*filled pentagons*). Solar abundances (Asplund et al. 2005, and references therein) are also shown. Figure from Chiappini et al. (2006)

Eu) in extremely metal poor stars of the Milky Way. Previous work on the subject had shown a large spread in the abundance ratios of these elements to iron, especially at low metallicities. This spread is confirmed by this more recent study although is less than before, and is at variance with the lack of spread observed in the other elements shown before (e.g. α-elements). Apart from this problem, not yet solved, these diagrams can be very useful to place constraints on the nucleosynthetic origin of these elements. In particular, Cescutti et al. (2006) by adopting the two-infall model predicted the evolution of [Ba/Fe] and [Eu/Fe] versus [Fe/H], as shown in Figs. 2.28 and 2.29. They can well fit the average trend but not the spread at very low metallicities since the model assumes instantaneous mixing. In order to fit the Ba evolution, they assumed that Ba is mainly produced as s-process element in low mass stars ($1-3 M_\odot$) but that a fraction of Ba is also produced as an r-process element in stars with masses $12-30 M_\odot$. Europium is assumed to be only an r-process element produced in the range $12-30 M_\odot$.

In order to explain why the s- and r- process elements show a large and probably real spread at very low metallicities, whereas elements such as the α-elements show only a little spread, one could think of a moderately inhomogeneous model coupled with differences in the nucleosynthesis between s- and r- process elements on one

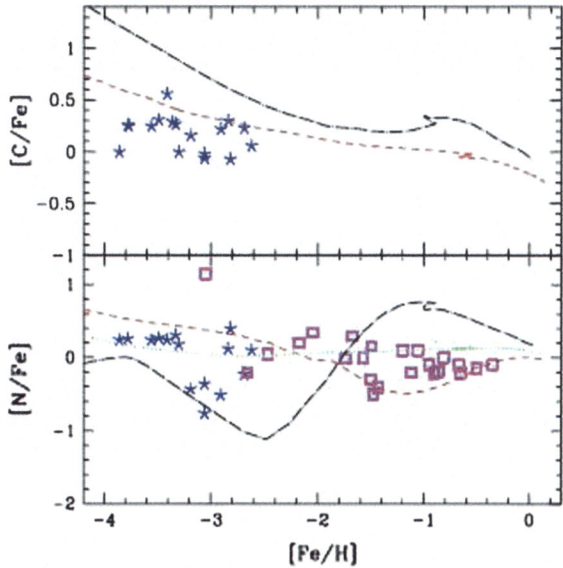

Fig. 2.27 Observed and predicted [C/Fe] versus [Fe/H] (*upper panel*) and [N/Fe] versus [Fe/H] (*bottom panel*) in the solar neighbourhood. The data points are from Cayrel et al. (2004); Spite et al. (2005) (*asterisks*) and Israelian et al. (2004) (*squares*). The *dot-dashed line* represents a model with yields from Chieffi and Limongi (2002, 2004) for a metallicity $Z = 10^{-6}$ connected to the Pop III stars (only massive stars for that metallicity). The *dashed line* and the *dotted lines* represent heuristic models where the yields of C and N have been assumed "ad hoc". In particular, the fraction of primary N from massive stars is obtained by the fit to the data at low metallicity. Figure from Chiappini et al. (2005)

side and α-elements on the other side. Highly inhomogeneous models for the halo evolution, in fact, predict a too large spread for the α-elements at low metallicity (e.g. Argast et al. 2000). It is worth noting the typical secondary behaviour of Ba, whose main production is by means of the s-process, which needs Fe seed nuclei already present in the star, and neutrons which are accreted on these nuclei. The production of neutrons is also dependent on the original metal content, therefore it would be even more precise to speak of Ba as a *tertiary* element.

2.3.5 The Galactic Disk

A good model of chemical evolution for the Milky Way should reproduce also the features of the Galactic disk. In particular: abundance gradients, gas and SFR distribution with the galactocentric distance.

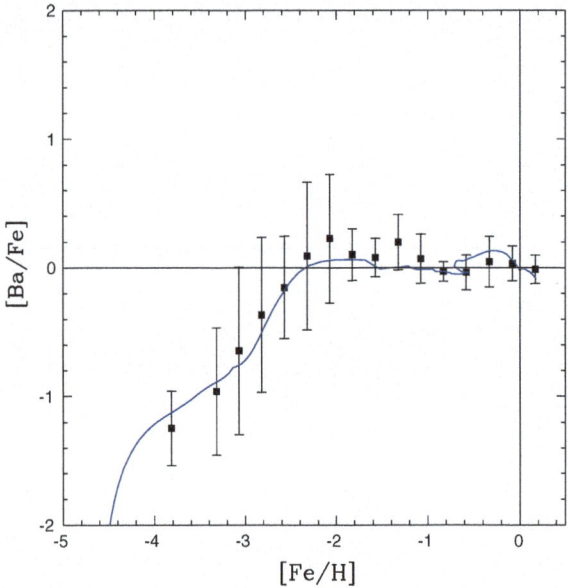

Fig. 2.28 The evolution of Barium in the solar vicinity as predicted by the two-infall model (Cescutti et al. 2006). Data are from François et al. (2006)

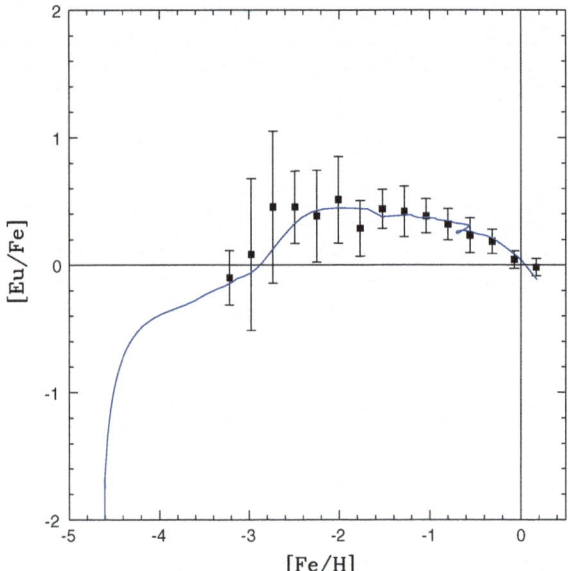

Fig. 2.29 The evolution of Europium in the solar vicinity (Cescutti et al. 2006). Data are from François et al. (2006)

2.3.5.1 Abundance Gradients

The chemical abundances measured along the disk of the Galaxy suggest that the metal content decreases from the innermost to the outermost regions, in other words there is a negative gradient in metals. Abundance gradients can be derived from HII regions, planetary nebulae (PNe), open clusters and stars (O, B stars and Cepheids). There are two types of abundance determinations in HII regions: one is based on recombination lines which should have a weak dependence on the temperature of the nebula (He, C, N, O), the other is based on collisionally excited lines where a strong dependence is intrinsic to the method (C, N, O, Ne, Si, S, Cl, Ar, Fe and Ni). This second method has predominated until now. A direct determination of the abundance gradients from HII regions in the Galaxy from optical lines is difficult because of extinction, so usually the abundances for distances larger than 3 Kpc from the Sun are obtained from radio and infrared emission lines.

Abundance gradients can also be derived from optical emission lines in PNe. However, the abundances of He, C and N in PNe are giving only information on the internal nucleosynthesis of the star. So, to derive gradients one should look at the abundances of O, S and Ne, unaffected by stellar processes. Abundance gradients are derived also from measuring the Fe abundance in open clusters (e.g. Carraro et al. 2004; Yong et al. 2005) or from abundances in Cepheids (e.g.Andrievsky et al. 2002a, b, c, 2004; Luck et al. 2003; Yong et al. 2006; Luck and Lambert 2011) or from abundances in O, B stars (e.g. Daflon and Cunha 2004).

In Fig. 2.30 we show theoretical predictions of abundance gradients along the disk of the Milky Way compared with data from HII regions, B stars and PNe. The adopted model is from Chiappini et al. (2001) and is based on an inside-out formation of the thin disk. The assumed model does not allow for exchange of gas between different regions of the disk. The disk is, in fact, divided in several concentric shells 2 Kpc wide with no interaction between them.

As already mentioned, most of the current models agree on the inside-out scenario for the disk formation, however not all models agree on the evolution of the gradients with time. In fact, some models, although assuming an inside-out formation of the disk, predict a flattening with time (Boissier and Prantzos 1999; Alibés et al. 2001), whereas others such as that of Chiappini et al. (2001) predict a steepening, as shown in Fig. 2.30. The reason for the steepening is that in the model of Chiappini et al. there is included a threshold density for SF, which induces the SF to stop when the density decreases below the threshold. This effect is particularly strong in the external regions of the Galactic disk, thus contributing to a slower evolution in those regions and therefore to a steepening of the gradients with time. In Fig. 2.31 we show models and some more recent data including Cepheids.

In the Chiappini et al. model, the fit to the gradients is obtained by means of the inside-out formation of the Galactic disk. Numerical simulations of abundance gradients show that no gradient arises if one assumes the same timescale of disk formation at any galactocentric distance. The different timescale of accretion influences the SFR, thus creating a gradient in the star formation rate and therefore in the

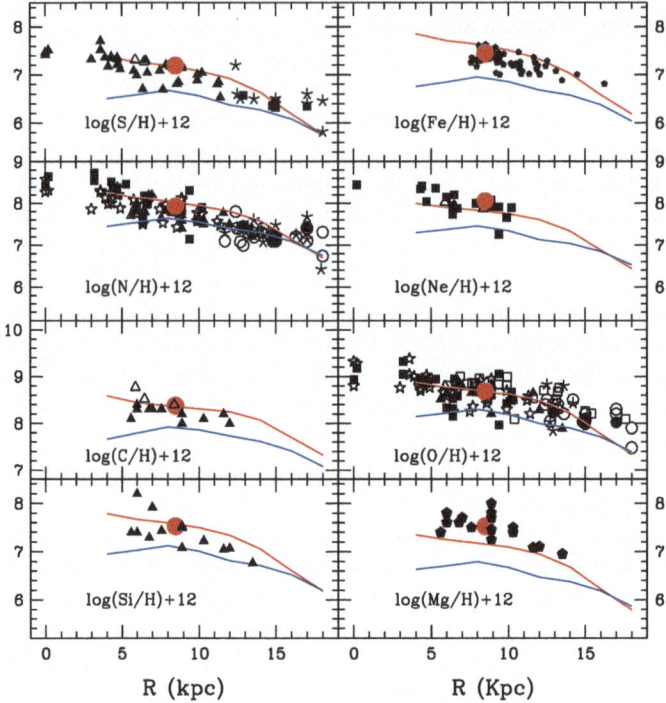

Fig. 2.30 Spatial and temporal behaviour of abundance gradients along the Galactic disk as predicted by the best model of Chiappini et al. (2001). The *upper lines* in each panel represent the present time gradient, whereas the *lower ones* represent the gradient a few Gyr ago. It is clear that the gradients tend to stepeen in time, a still controversial result. The data are from HII regions, B stars and PNe (see Chiappini et al. 2001)

resulting metal content. However, it should be said that the effect of the threshold is also important and tends to steepen the gradients in the outer regions.

In Fig. 2.32 we show the results of Boissier and Prantzos (1999) for abundance gradients and also for the gas and SFR distribution along the disk. Here, one can see that this model predicts a flattening of gradients with time.

The effect of radial gas flows along the thin disk on the abundance gradients has been studied by several authors such as Portinari and Chiosi (2000); Schoenrich and Binney (2009); Spitoni and Matteucci (2011): these latter included radial flows into the two-infall model. In Fig. 2.33 we show the result of a model assuming inside-out formation of the thin disk, a threshold in the gas density for star formation and radial gas flows with variable speed as a function of Galactocentric distance. As one can see, this model seems to reproduce at best the most recent data on Cepheids (Luck and Lambert 2011). It is worth noting that these data on Cepheids suggest an O gradient in excellent agreement with that derived from HII regions and PNe.

2 Chemical Evolution of the Milky Way and Its Satellites

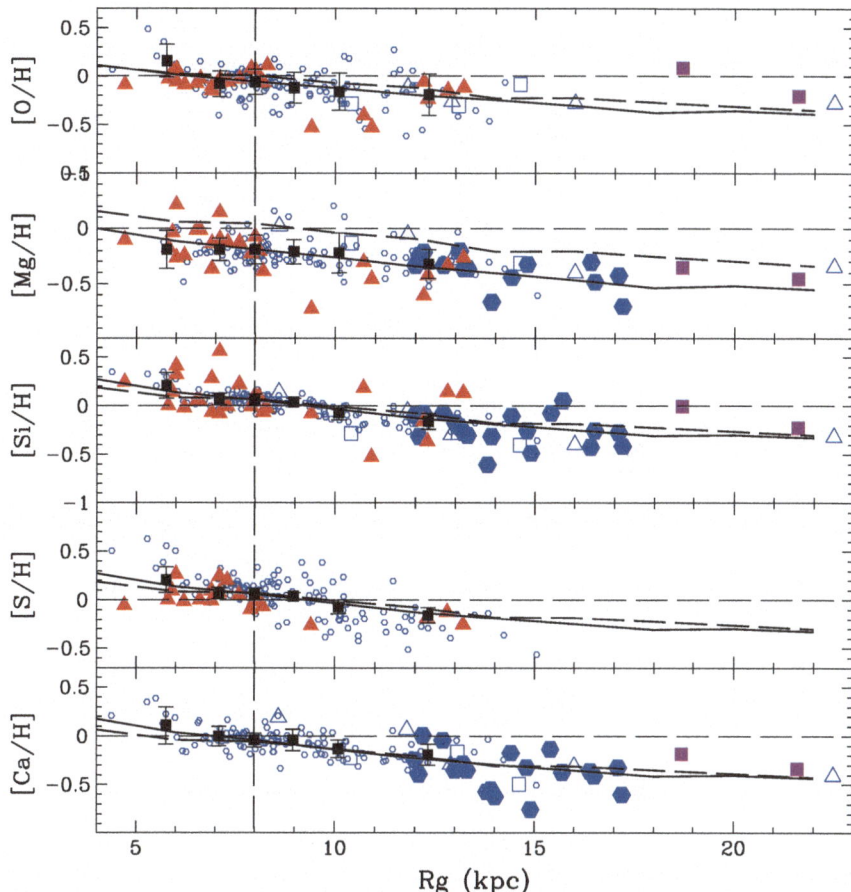

Fig. 2.31 Gradients of the α-elements along the disk. The predicted gradients for O, Mg, Si, S and Ca are compared with different sets of data. The *small open circles* are the data of the Cepheids by Andrievsky et al. (2002a, b, c, 2004) and Luck et al. (2003). The *solid triangles* are the data by Daflon and Cunha (2004) (OB stars), the *open squares* are the data by Carney et al. (2005) (*red giants*), the *solid hexagons* are the data by Yong et al. (2006) (Cepheids), the *open triangles* are the data by Yong et al. (2005) (*open clusters*) and the *solid squares* are the data by Carraro et al. (2004) (*open clusters*). The most distant value for Carraro et al. (2004) and Yong et al. (2005) refers to the same object: the *open cluster* Berkeley 29. The *thin solid line* represents the model predictions at the present time normalized to the mean value of the Cepheids at 8 Kpc; the *dashed line* represents the predictions of the model at the epoch of the formation of the solar system normalized to the observed solar abundances by Asplund et al. (2005). This prediction should be compared with the data for *red giant stars* and *open clusters* (Carraro et al. 2004; Carney et al. 2005; Yong et al. 2005). The models and the Figure are from Cescutti et al. (2007)

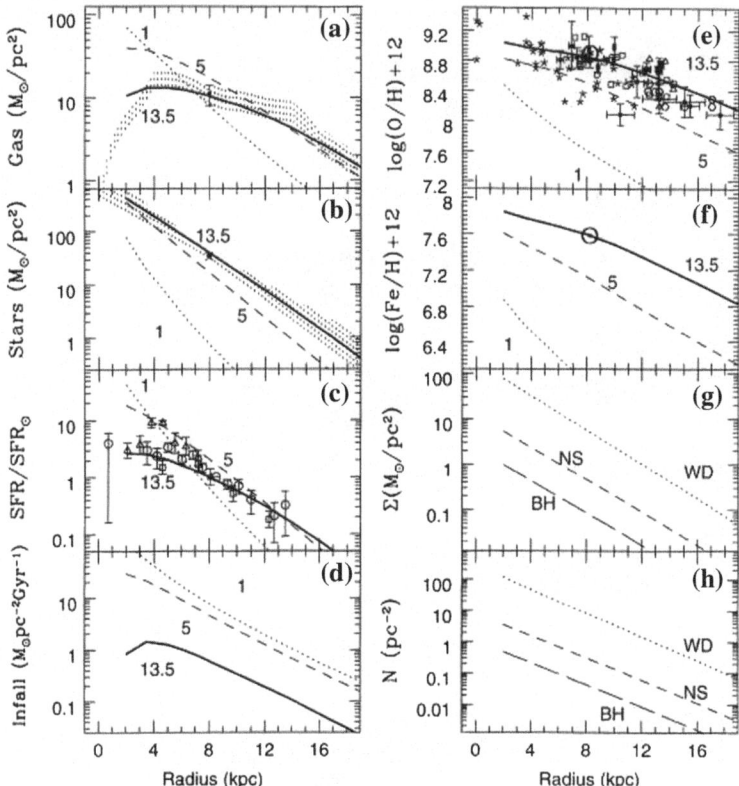

Fig. 2.32 Comparison between model predictions and observations for the disk of the Milky Way. The figure is from Boissier and Prantzos (1999). *Top left panel* gas distribution along the disk. *Top right panel* the O gradient at the present time (curve with label 13.5) and at two other different cosmic epochs (5 and 1 Gyr from the beginning). Second *left panel* the surface mass density of living stars. Second *right panel* the Fe gradient. Third *left panel* the gradient of the SFR normalized to the value at the solar ring. Third *right panel* the predicted distribution of the current surface mass densities of stellar remnants (WDs), black holes (BH) and neutron stars (NS). Fourth *left panel* the predicted infall rate along the disk at three different cosmic epochs. Fourth *right panel* the predicted distributions of surface densities by number of the stellar remnants

2.3.6 The Galactic Bulge

2.3.6.1 Bulge Formation

The bulges of spiral galaxies are generally distinguished in true bulges, hosted by S0-Sb galaxies and "pseudobulges" hosted in later type galaxies (see Renzini (2006) for references). Generally, the properties (luminosity, colors, line strenghts) of true bulges are very similar to elliptical galaxies. In the following, we will refer only to true bulges and in particular to the bulge of the Milky Way. The bulge of the

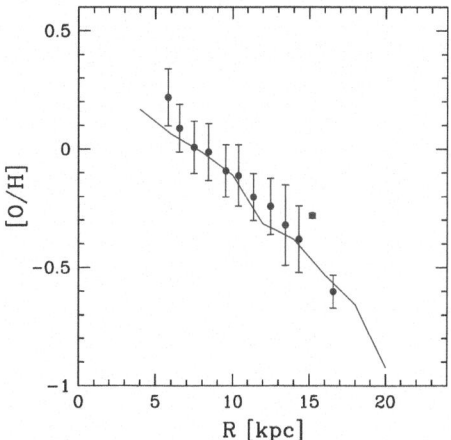

Fig. 2.33 Observed and predicted present time radial abundance gradient for oxygen along the the thin disk of the Milky Way. The model includes inside-out formation of the thin disk, a threshold in the gas density for star formation and radial flows with variable speed as a function of the galactocentric distance (Mott et al. 2013). The data are the most recent ones about Cepheids (Luck and Lambert 2011)

Milky Way is, in fact, the best studied bulge at the moment: recent studies have suggested that a bar is present in the Galactic bulge (e.g. Ness and Freeman 2012) so the scenarios for the bulge formation appear rather complex. Here, we will recall the history of the interpretation of the bulge stellar populations and the models which have been suggested to explain them.

Before doing that, we summarize the proposed scenarios for the Galactic bulge formation:

- **Accretion of stellar satellites**: the bulge formed by accretion of extant stellar systems which eventually settled in the center of the Galaxy.
- **In situ star formation**: the bulge was formed by accumulation of gas at the center of the Galaxy and subsequent evolution with either fast or slow star formation. The gas could have been primordial or enriched and originating from the halo, thick or thin disk.
- **Secular evolution**: the bulge formed as a result of secular evolution through a bar, thus forming a pseudo-bulge.
- **Mixed scenario**: the bulge contains two stellar populations, one formed by an early fast collapse of gas and the other formed later in the bar.

2.3.6.2 The History of Bulge Chemical Evolution

In the context of chemical evolution, the Galactic bulge was first modeled by Matteucci and Brocato (1990) who predicted that the [α/Fe] ratio for some elements (O, Si and Mg) should be supersolar over almost the whole metallicity range, in analogy with the halo stars, as a consequence of assuming a fast bulge evolution which involved rapid gas enrichment in Fe mainly by Type II SNe. At that time, no data were available for chemical abundances; the predictions of Matteucci and Brocato (1990) were confirmed for a few α-elements (Mg, Ti) by the observations of McWilliam and Rich (1994), (hereafter MR94), whereas for other α-elements (e.g. Ca, Si) the observed trend was different. Other discrepancies regarding the Mg overabundance came from Sadler et al. (1996). In order to better assess these points, Matteucci et al. (1999) studied a larger set of abundance ratios, by means of a detailed chemical evolution model whose parameters were calibrated so that the metallicity distribution observed by MR94 could be fitted. They concluded that an evolution much faster than that in the solar neighbourhood and even faster than that of the halo (see also Renzini 1993) is necessary for the MR94 metallicity distribution to be reproduced, and that an IMF index flatter ($x = 1.1 - 1.35$) than that of the solar neighbourhood is needed as well. They also made predictions about the evolution of several abundance ratios which were meant to be confirmed or disproved by subsequent observations, namely that α-elements should in general be overabundant with respect to Fe, but some (e.g. Si, Ca) less than others (e.g. O, Mg), and that the [^{12}C/Fe] ratio should be solar at all metallicities.

Samland et al. (1997) developed a self-consistent chemo-dynamical model for the evolution of the Milky Way components starting from a rotating protogalactic gas cloud in virial equilibrium, which collapses owing to dissipative cloud-cloud collisions. They found that self-regulation due to a bursting star formation and subsequent injection of energy from Type II supernovae led to the development of "contrary flows", i.e. alternate collapse and outflow episodes in the bulge. This caused a prolonged star formation episode lasting over $\sim 4 \times 10^9$ yr. They included stellar nucleosynthesis of O, N and Fe, but claimed that gas outflows prevent any clear correlation between local star formation rate and chemical enrichment. With their model, they could reproduce the oxygen gradient of HII regions in the equatorial plane of the Galactic disk and the metallicity distribution of K giants in the bulge (Rich 1988), field stars in the halo and G dwarfs in the disk, but they did not make predictions about abundance ratios in the bulge. In general, hierarchical clustering models of galaxy formation do not support the conclusion of a fast formation and evolution of the bulge. In Kauffmann (1996) the bulges form through violent relaxation and destruction of disks in major mergers. The stars of the destroyed disk build the bulge, and subsequently the bulge has to be rebuilt. This implies that late type spirals should have older bulges than early type ones, since the build-up of a large disk needs a long time during which the galaxy has to evolve undisturbed. This is not confirmed by observations, since the high metallicity and the the narrow age distribution observed in bulges of local spirals are not compatible with their merger origin (see Wyse 1999).

Mollá et al. (2000) proposed a multiphase model in the context of the dissipative collapse scenario of the Eggen et al. (1962) picture. They supposed that the bulge formation occurred in two main infall episodes, the first from the halo to the bulge, on a timescale $\tau_H = 0.7$ Gyr, and the second from the bulge to a so-called core population in the very nuclear region of the Galaxy, on a timescale $\tau_B \gg \tau_H$. The three zones (halo, bulge, core) interact via supernova winds and gas infall. They concluded that there is no need for accretion of external material to reproduce the main properties of bulges and that the analogy to ellipticals is not justified. Because of their rather long timescale for the bulge formation, these authors did not predict a noticeable difference in the trend of the [α/Fe] ratios but rather suggested that they behave more akin to that in the solar neighbourhood (contrary to several indications from abundance data, e.g. MR94).

Immeli et al. (2004) investigated the role of cloud dissipation in the formation and dynamical evolution of star forming gas rich disks by means of a 3D chemodynamical model. They found that the galaxy evolution proceeds very differently depending on whether the gas disk or the stellar disk first become unstable. This in turn depends on how efficiently the cold cloud medium can dissipate energy. If the gas cools efficiently, a starburst takes place which gives rise to enhanced [α/Fe] ratios, thus in agreement with a fast bulge formation.

A more recent model was proposed by Costa et al. (2005), in which the best fit to observations is achieved by means of a double infall model. An initial fast (0.1 Gyr) collapse of primordial gas is followed by a supernova-driven mass loss and then by a second slower (2 Gyr) infall episode, enriched by the material ejected by the bulge during the first collapse. Costa et al. (2005) claimed that the mass loss is necessary to reproduce the abundance distribution observed in planetary nebulae, and because the predicted abundances would otherwise be higher than observed. However, it should be noted again that the abundances derived from PNe can be affected by internal stellar processes and therefore are meaningless for studying galactic chemical evolution. With their model, they were able to reproduce the trend of [O/Fe] abundance ratio observed by Pompéia et al. (2003) and the data of nitrogen versus oxygen abundance observed by Escudero and Costa (2001) and Escudero et al. (2004). It must be noted however that Pompéia et al. (2003) obtained abundances for "bulge-like" dwarf stars. This "bulge-like" population consists of old ($\sim$10–11 Gyr) metal-rich nearby stars whose kinematics and metallicity suggest an inner disk or bulge origin and a mechanism of radial migration, perhaps caused by the action of a Galactic bar, but the birthplace of these stars is undoubtedly not certain.

2.3.6.3 Interpretation of Bulge Data and Other Galaxies

In summary, MR94 first measured the metallicity distribution and the [α/Fe] ratios in the Bulge and confirmed partly the predictions of Matteucci and Brocato (1990) that all of the α-elements should be enhanced relative to Fe for a large range of [Fe/H]. In fact, MR94 found that not all the α-elements were enhanced, in particular oxygen. In the following years, medium- and high-resolution spectroscopy of bulge stars was performed (Rich and McWilliam 2000; Fulbright et al. 2006, 2007; Zoccali et al. 2006; Lecureur et al. 2007; Alves-Brito et al. 2010; Bensby et al. 2010; Gonzalez et al. 2011; Hill et al. 2011), and it seems to indicate that also O is enhanced, thus supporting the suggestion of a fast formation of the bulge. The metallicity distribution of stars in the bulge and the [α/Fe] ratios greatly help in selecting the most probable scenario for the bulge formation. In Fig. 2.34 we present the predictions by Matteucci (2003) of the [α/Fe] ratios as functions of [Fe/H] in galaxies of different morphological type. In particular, for the Galactic bulge or an elliptical galaxy of the same mass, for the solar vicinity region and for an irregular magellanic galaxy (LMC and SMC).

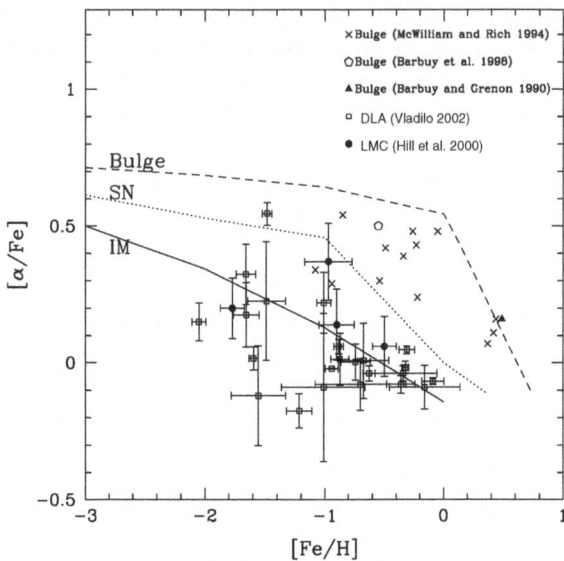

Fig. 2.34 The predicted [α/Fe] versus [Fe/H] relations for the Galactic bulge (*upper curve*), the solar vicinity (*median curve*) and irregular galaxies (*low curve*). Data for the bulge are reported for comparison. Data for the LMC and DLA systems are also shown for comparison, indicating that DLAs are probably irregular galaxies. Figure and references are from Matteucci (2003)

The underlying assumption is that different objects undergo different histories of star formation, being very fast in the spheroids (bulges and ellipticals), moderate in spiral disks and slow and perhaps gasping in irregular gas rich galaxies. The effect of different star formation histories is evident in Fig. 2.34 where the predicted [α/Fe] ratios in the bulge and ellipticals remain high and almost constant for a large interval of [Fe/H]. This is due to the fact that, since star formation is very intense, the bulge reaches very soon a solar metallicity thanks only to the SNe II; then, when SNe Ia start exploding and ejecting Fe into the ISM, the change in the slope occurs at larger [Fe/H] than in the solar vicinity. In the extreme case of irregular galaxies the situation is opposite: here the star formation is slow and when the SNe Ia start exploding the gas is still very metal poor. This scheme is quite useful since it can be used to identify galaxies only by looking at their abundance ratios. A model for the bulge behaving as shown in Fig. 2.34 is able to reproduce also the observed metallicity distribution of bulge stars, as first shown by Matteucci and Brocato (1990). The scenario suggested in that paper favors the formation of the bulge by means of a short and strong starburst, in agreement with Elmegreen and Bruce (1999) and Ferreras et al. (2003). A similar model, although updated with the inclusion of the development of a galactic wind and more recent stellar yields, has been presented by Ballero et al. (2007): it shows how a model with intense star formation (star formation efficiency $\sim 20\,\mathrm{Gyr}^{-1}$) and rapid assembly of gas (0.1 Gyr) can best reproduce the more recent accurate data on abundance ratios and metallicity distribution. This was confirmed later by

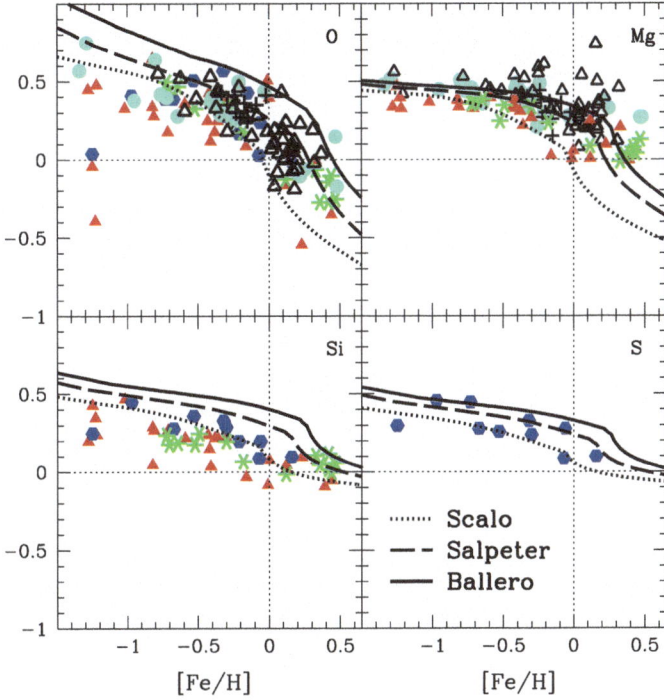

Fig. 2.35 Comparison between the predictions of the Cescutti and Matteucci (2011) model for the bulge of our Galaxy, using 3 different IMF for [O/Fe], [Mg/Fe], [Si/Fe] and [S/Fe] versus [Fe/H] and the observations in the bulge. The observational data for the bulge are: the *filled circles* from McWilliam (2009) and Fulbright et al. (2007); the *open triangles* from Lecureur et al. (2007); the *filled triangles* from Alves-Brito et al. (2010); the plus signs are the infrared results from Rich and Origlia (2005); the *filled hexagons* are the infrared results by Ryde et al. (2009); and the stars are the results for microlensed dwarf stars by Bensby et al. (2010)

Cescutti and Matteucci (2011) who considered the metallicity distribution function by Zoccali et al. (2008). The only difference with the Ballero et al. (2007) conclusions was that also a normal Salpeter (1955) IMF could be acceptable for reproducing the bulge data, while in Ballero et al. (2007) a very flat IMF was suggested as the only possibility. In Fig. 2.35 we show some of the predictions of Cescutti and Matteucci (2011) concerning the abundance ratios in bulge stars compared to data. As one can see in Fig. 2.35, the plateau in the [α/Fe] is longer than in the solar neighbourhood, since in the bulge the slope of the [α/Fe] ratio starts changing drastically only for [Fe/H] > 0.0 dex. The long plateau is well explained by the model of Cescutti and Matteucci (2011) assuming a very fast formation of the bulge (<0.5 Gyr). It is worth noting that the [O/Fe] ratio has a steeper slope than [Mg/Fe] and this could be due to differences in the nucleosynthesis of these elements (e.g. McWilliam et al. 2008).

The IMF assumed for the bulge is usually flatter than the IMF of the solar neighbourhood and this is generally dictated by the fit of the bulge metallicity distribution

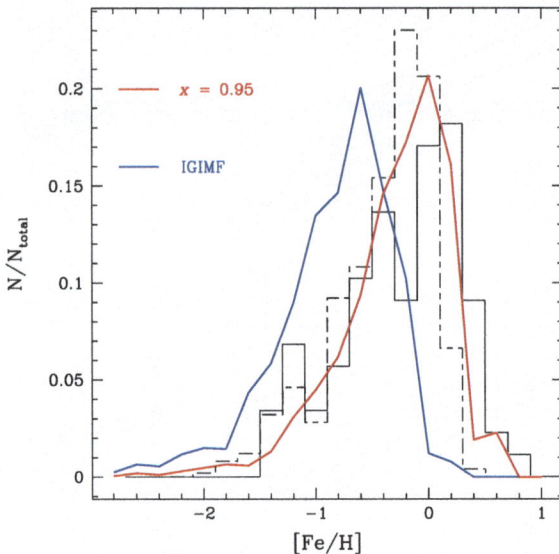

Fig. 2.36 The predicted and observed metallicity distribution in the Galactic bulge. The data are from Zoccali et al. (2003) (*dashed histogram*) and Fulbright et al. (2006) (*continuous histogram*). In particular, the model with the peak at the lower metallicity is computed with an IMF which is similar to that of the solar vicinity and indicated by IGIMF, whereas the distribution which best fits the data is computed with a flat IMF (x = 0.95 for $M > 1M_\odot$)

which peaks at a higher [Fe/H] than the G-dwarf metallicity distribution in the solar vicinity. In fact, as it is clear from Fig. 2.35, a Scalo IMF does not produce good agreement with the observations. Numerical calculations have indicated that the main parameter influencing the peak of the distribution is the IMF, as clearly shown in Figs. 2.36 and 2.37: these are the IMF and the efficiency of star formation.

In summary, the comparison between the models (Ballero et al. 2007; Cescutti and Matteucci 2011) on one side, and the metallicity distribution and the [α/Fe] ratios on the other, strongly indicates that the Galactic bulge is very old and must have formed very quickly during a strong starburst (with a SF efficiency much higher than in the disk). The metallicity distribution, in particular, seems to suggest an IMF flatter than in the disk with an exponent for massive stars in the range $x = 1.35 - 0.95$. However, to assess more precisely this point we need more data: in particular, a flatter IMF predicts that the overabundances of α-elements relative to Fe and to the Sun should be higher in the bulge than in the disk. This is not entirely clear from the available data, although Zoccali et al. (2006) conclude that the [O/Fe] ratios in bulge stars are higher than in thick and thin disk stars, Meléndez et al. (2008) suggest that these ratios are the same in bulge and thick disk stars. Finally, the timescale for the bulge formation by accretion of gas lost from the halo is 0.1–0.3 Gyr and certainly no longer than 0.5 Gyr.

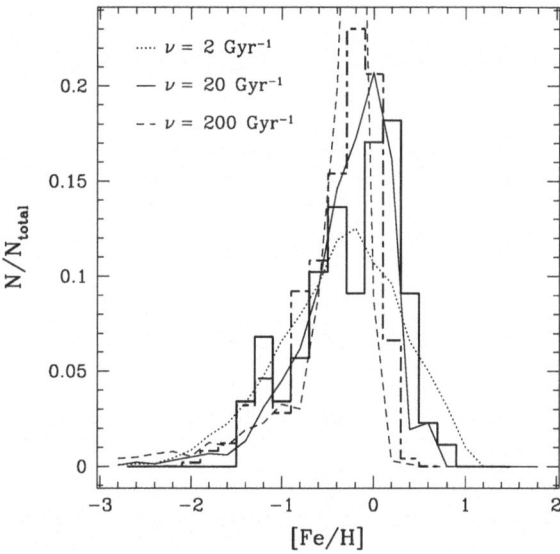

Fig. 2.37 The predicted and observed metallicity distribution in the Galactic bulge. The data are from Zoccali et al. (2003) (*dashed histogram*) and Fulbright et al. (2006) (*continuous histogram*). The lines are the predictions of models with the same IMF but different SF efficiencies, as indicated in the Figure

2.3.6.4 Recent Developments in the Galactic Bulge

In the last years a great deal of observational work has appeared for the Galactic bulge and has suggested a more complex scenario for the bulge formation than thought before. In particular, several papers (Babusiaux et al. 2010; Bensby et al. 2011; Gonzalez et al. 2011; Hill et al. 2011; Robin et al. 2012) have suggested the existence of two main stellar populations, one similar to the population described up to now and typical of a true bulge, and another with bar kinematics more typical of a pseudo-bulge. Grieco et al. (2012) have modelled these two populations by calling them "metal-poor" (MP, the one relative to the true bulge) and "metal-rich" (MR, the one related to the bar). The MP population should have formed very fast as in the previous models ($\sim$0.3 Gyr) whereas the metal-rich one should have formed more slowly ($\sim$3–4 Gyr) and from a pre-enriched gas coming either from the halo or the inner disk. In Fig. 2.38 we show the predicted and observed recent bulge stellar metallicity distribution.

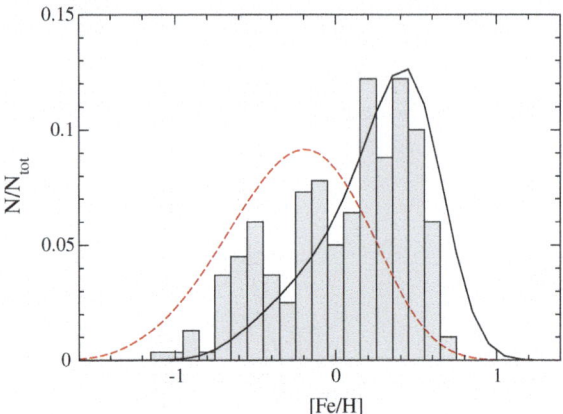

Fig. 2.38 Metallicity distribution function for the bulge stars. The histogram are the data from Hill et al. (2011), which is suggesting the existence of two stellar populations. The curves are the predictions from Grieco et al. (2012) relative to these populations. The dotted line refer to the HP population and the continuous line to the HR populatuion. The theoretical curves have been convolved with a gaussian to take into account an observational error of 0.25 dex

2.4 What We Have Learned About the Milky Way

From the discussions of the previous sections we can extract some important conclusions on the formation and evolution of the Milky Way, derived from chemical abundances. In particular:

- The inner halo formed on a timescale of 1–2 Gyr at maximum, the outer halo formed on longer timescales perhaps from accretion of satellites or gas.
- The disk at the solar ring formed on a timescale not shorter than 7 Gyr.
- The whole disk formed inside out with timescales of the order of 2 Gyr or less in the inner regions and 10 Gyr or more in the outermost regions.
- The abundance gradients arise naturally from the assumption of the inside-out formation of the disk. A threshold density for the star formation helps in steepening the gradients.
- The bulk of bulge stars is very old and formed very quickly on a timescale smaller than even the inner halo and not larger than 0.5 Gyr.
- The IMF seems to be different in the bulge and the disk, being flatter in the bulge, although more abundance data are necessary before drawing firm conclusions.

2.5 The Time-Delay Model and the Hubble Sequence

In this section we will discuss how different star formation histories affect the evolution of galaxies of different morphological type and in particular how the abundance patterns are expected to change with the star formation.

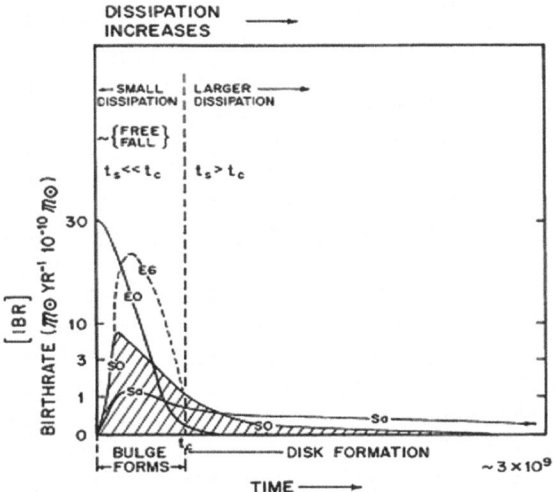

Fig. 2.39 Schematic representation of the change of the SFR with time for galaxies of types E, S0, and Sa. The *dashed vertical line* at the collapse time t_c separates regions of low energy dissipation (to the *left*) from those of high energy dissipation (to the *right*). Bulges form in the *left region*, disks in the *right*. The integral under the curves gives the total number of stars formed (per unit galaxy mass). The integral under the S0 curve is shaded. Figure from Sandage (1986)

2.5.1 Star Formation and Hubble Sequence

Sandage (1986), on the basis of a work of Gallagher et al. (1984) who measured SFR in galaxies, suggested a possible interpretation of the Hubble sequence in terms of different star formation histories. In this picture, ellipticals and bulges must have suffered an intense and strongly declining SFR, whereas late type galaxies must have undergone through a less intense, almost constant (spirals) and even increasing with time (irregulars) SFR. In Fig. 2.39 is illustrated such a behaviour of the SFR for E, S0 and Sa galaxies. In Fig. 2.39 there are two important timescales: t_c, the gas collapse time, and t_s, the star formation time, namely the timescale on which the gas in a galaxy is consumed by means of star formation (the inverse of the SF efficiency). The interplay between these two quantities can be crucial for the formation of the different galactic morphological types. In fact, if $t_s \ll t_c$, most of the stars form before the collapse is over and the gas does not have time to dissipate energy and settle into a disk. In this case, the resulting galaxy will be a spheroid, whereas if $t_s > t_c$ the gas has time to dissipate energy and form a spiral galaxy. This picture is certainly too simplistic to fully describe the reality of galaxy formation but it seems to work well when we are going to interpret galaxy formation by studying the stellar populations in galaxies.

Later on, Kennicutt (1998a, b) measured again the SFR in star forming galaxies and suggested similar behaviours for the different galactic morphological types, as shown in Fig. 2.40. In Fig. 2.40, the behaviour of the SFR for spirals is obtained by

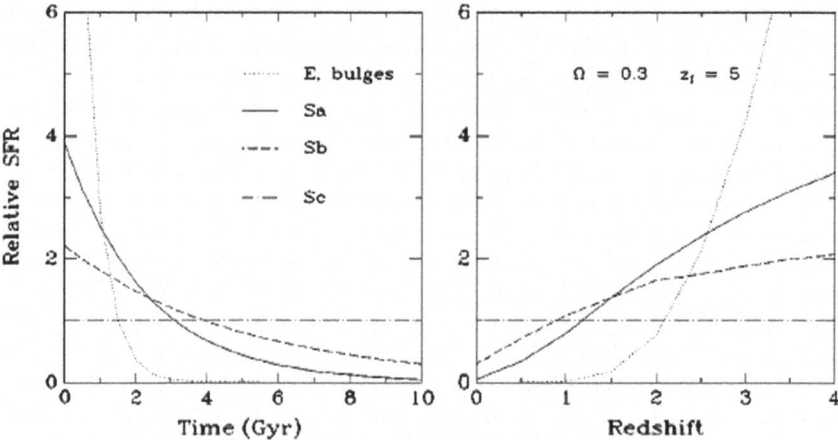

Fig. 2.40 Schematic illustration of the stellar birthrate for different Hubble types. The *left panel* shows the evolution of the relative SFR with time following Sandage (1986). The curves for spiral galaxies are exponentially declining SFRs which fit the mean values of the birthrate parameter b as measured by Kennicutt et al. (1994). The curve for elliptical galaxies and bulges is an arbitrary dependence for an e-folding time of 0.5 Gyr, for comparative purposes only. The right panel shows the corresponding evolution in SFR with redshift, for an assumed cosmological density parameter $\Omega = 0.3$ and formation redshift $z_f = 5$. Figure from Kennicutt (1998b)

fitting the mean value of the parameter b, measured by Kennicutt et al. (1994). The parameter b is the ratio between the present time SFR and the average SFR in the past, as defined in Eq. (2.16). Early type spiral galaxies are characterized by rapidly declining SFRs, with $b \sim 0.01$–0.1, whereas late type spirals have formed stars since a long time at an almost constant rate with $b = 1$. Finally, ellipticals and S0 have long ago ceased forming stars and have $b = 0$.

In Fig. 2.41 we show the SF histories which give rise to the [α/Fe] versus [Fe/H] relations of Fig. 2.34.

In Fig. 2.42 we show the predicted Type Ia SN rates according to the SFRs of Fig. 2.41. The assumed progenitor model for Type Ia SNe is the single degenerate with the delay time distribution as in Matteucci and Recchi (2001).

2.5.1.1 The Typical Timescale for SN Ia Enrichment

The predicted Type Ia SN rates for galaxies with different morphologies show a difference in the maximum SN Ia rate, which is reached quite early in ellipticals and it occurs later and later moving to late types. Matteucci and Recchi (2001) suggested to assume the time for the occurrence of the maximum SNIa rate as the typical timescale for the chemical enrichment from these SNe. It depends on the star formation history of each galaxy, on the IMF and on the stellar lifetimes. As we have already shown, the IMF together with stellar lifetimes represent the distribution of

2 Chemical Evolution of the Milky Way and Its Satellites

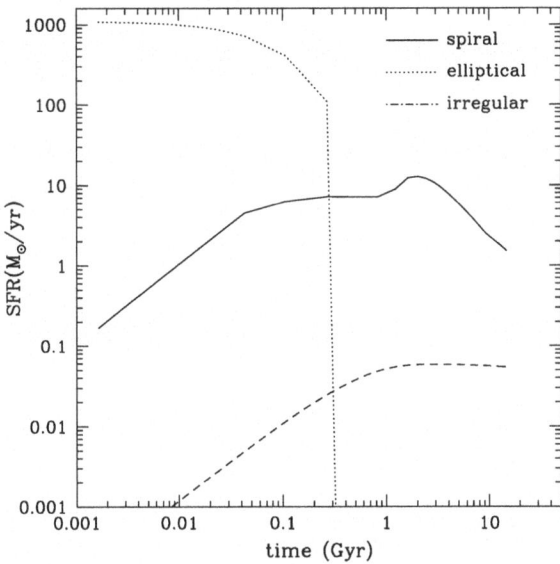

Fig. 2.41 The predicted histories of SF in galaxies of different morphological type, with decreasing efficiency of SFR from ellipticals to irregulars. Figure from Calura (2004)

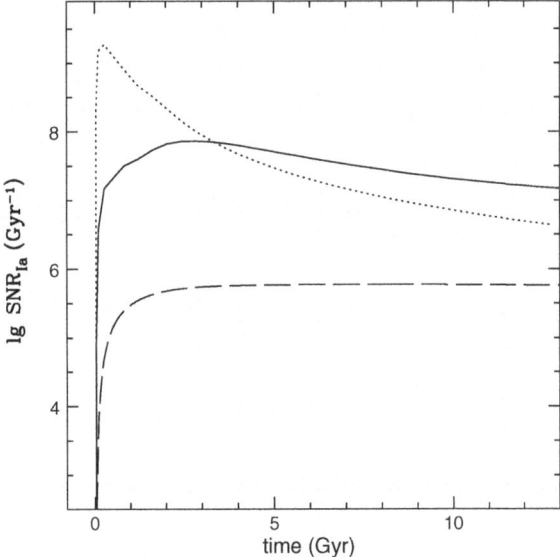

Fig. 2.42 Predicted Type Ia SN rates (expressed in Gyr^{-1}) obtained according to the SF histories of Fig. 2.41. As one can see the elliptical (*dotted line*) reaches a maximum at 0.3 Gyr, whereas the spiral (*continuous line*) at ~2 Gyr and the irregular (*dashed line*) has a rate increasing to its maximum at around 5 Gyr and then is roughly constant for the rest of the galactic lifetime. The minimum time delay for the Type Ia SNe to appear is 30 Myr, hardly visible in this plot

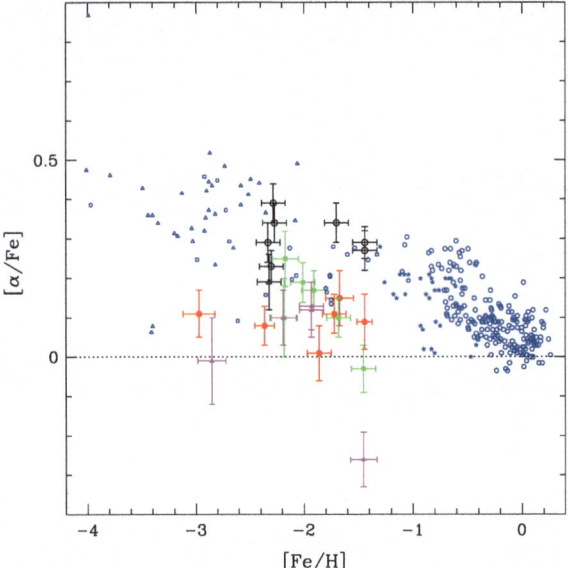

Fig. 2.43 Observed [α/Fe] versus [Fe/H] in the Milky Way (small points) and in dSphs (points with error bars). Figure from Shetrone et al. (2001)

the time-delays (DTD) with which SNe explode; therefore, when a DTD is specified, the Type Ia SN rate depends on the SF history only. In summary, in ellipticals and bulges this timescale is 0.3–0.5 Gyr since the beginning of star formation, in the solar vicinity there is a first peak at 1 Gyr then it decreases slightly and increases again till 3 Gyr (due to the two-infall episodes). In irregular galaxies the maximum is reached at ∼4 Gyr and then the rate remains constant.

2.6 Dwarf Spheroidals of the Local Group

A different pattern for the [α/Fe] versus [Fe/H] relation compared to the solar vicinity is observed in dwarf spheroidal galaxies (dSphs) of the Local Group, as shown in Fig. 2.43, and this can be easily interpreted in the framework of the time-delay model coupled with different star formation histories.

Before interpreting the [α/Fe] diagram, we recall the current ideas about the formation of the dSphs.

2.6.1 How do dSphs Form?

Cold dark matter (CDM) models for galaxy formation predict that the dSphs, systems with luminous masses of the order of $10^7 M_\odot$, are the first objects to form stars and

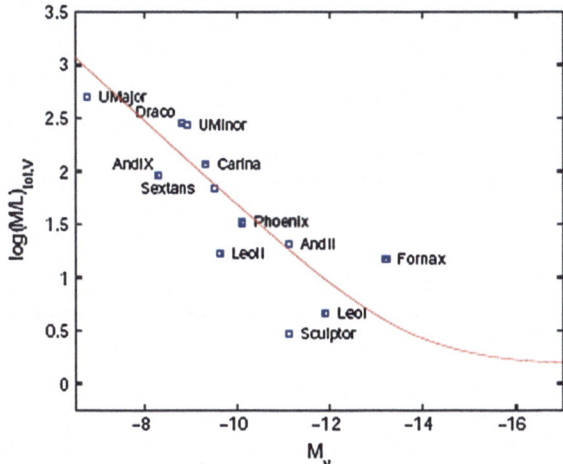

Fig. 2.44 Dark matter in dSphs: mass to light ratios versus absolute V magnitude for some Local Group dSphs. The *solid curve* shows the relation expected if all the dSphs contain about $4 \cdot 10^7 \, M_\odot$ of dark matter interior to their stellar distributions. Figure from Gilmore et al. (2007)

that all stars in these systems should form on a timescale <1 Gyr, since the heating and gas loss, due to reionization, must have halted the SF soon. However, observationally all dSph satellites of the Milky Way contain old stars indistinguishable from those of Galactic globular clusters and they seem to have experienced SF for long periods (>2 Gyr, Grebel and Gallagher (2004). The histories of SF for these galaxies are generally derived from the color-magnitude diagram (e. g. Mateo 1998). By looking at the [α/Fe] versus [Fe/H] relations for dSphs, as shown in Fig. 2.43, one can immediately suggest, on the basis of the time-delay model, that their evolution should have been characterized by a slow and protracted SF, at variance with the suggestion of a fast episode of SF truncated by the heating due to reionization.

2.6.1.1 Dark Matter in dSphs

The dSph satellites of the Milky Way are considered the smallest dark matter dominated systems in the universe. In the past years there have been a few attempts at deriving the amount of dark matter in dSphs, in particular by measuring the mass to light ratios versus magnitude for these galaxies (e.g. Mateo 1998; Gilmore et al. 2007). Gilmore et al. (2007) suggested that the dSphs have a shallow central dark matter distribution and no galaxy is found with a dark mass halo less massive than $5 \cdot 10^7 M_\odot$, as shown in Fig. 2.44.

2.6.2 Observations of dSphs

In recent years there has been a fast development in the field of chemical evolution of dSphs of the Local Group due to the increasing amount of data on chemical abundances derived from high resolution spectra (e.g. Smecker-Hane and William 1999; Bonifacio et al. 2000, 2004; Shetrone et al. 2001, 2003; Tolstoy et al. 2003; Bonifacio et al. 2004; Venn et al. 2004; Sadakane et al. 2004; Fulbright et al. 2004; McWilliam and Smecker-Hane 2005a, b; Monaco et al. 2005; Geisler et al. 2005, 2007). The abundances of α-elements (O, Mg, Ca, Si,) plus the abundances of s- and r- process elements (Ba, Y, Sr, La and Eu) were measured with unprecedented accuracy. Besides these high-resolution studies we recall also the measure of the metallicities of many red giant stars in several dSphs by Tolstoy et al. (2004); Koch et al. (2006); Helmi et al. (2006) and (Battaglia et al. 2006) obtained from the low-resolution Ca triplet. An interesting result of Helmi et al. (2006) is that they did not find stars with [Fe/H] < -3.0 dex and that the metallicity distribution of the stars in dSphs is different from that of halo stars in the Milky Way. Other important information comes from the photometry of dSphs of the Local Group and in particular from the color-magnitude diagrams. From these diagrams one can infer the history of SF of these galaxies. We recall the studies of Hernandez et al. (2000); Dolphin (2002); Bellazzini et al. (2002); Rizzi et al. (2003); Monelli et al. (2003); Dolphin et al. (2005). The color-magnitude diagrams seem to indicate that the majority of dSphs had one rather long episode of SF with the exception of Carina for which four episodes of SF have been suggested (Rizzi et al. 2003).

2.6.3 Chemical Evolution of dSphs

Several papers have appeared in the last years concerning the chemical evolution of dSphs. For example Carigi et al. (2002) computed models for the chemical evolution of four dSphs by adopting the SF histories derived, from color-magnitude diagrams, by Hernandez et al. (2000). In their model they assumed gas infall and computed the gas thermal energy heated by SNe in order to study galactic winds. In fact, the dSphs must have lost their gas in one way or another (galactic winds and/or ram pressure stripping) since they appear completely without gas. They assumed that the wind is sudden and devoids the galaxy of gas instantaneously. The adopted IMF is the Kroupa et al. (1993) IMF as in the solar vicinity. Carigi et al. predicted a too high metallicity for dSphs and did not match the correct slope for the observed [α/Fe] ratios, as shown in Fig. 2.45.

Then, Ikuta and Arimoto (2002) proposed a closed box model (no infall nor outflow) for dSphs. In this Simple Model they had to assume some external cause to stop star formation, such as ram pressure stripping. They tested different IMFs and suggested that these galaxies had suffered very low SFRs (1–5 % of that in the solar neighbourhood) and that the SF had a long duration (>3.9–6.5 Gyr). In Fig. 2.46 are shown their predictions for [Mg/Fe] in dSphs. Also here, the predicted slope of the [Mg/Fe] ratio is flatter than observed.

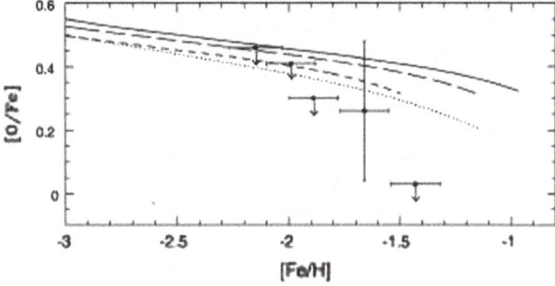

Fig. 2.45 Observed and predicted [O/Fe] versus [Fe/H] relation for the galaxy Ursa Minor. Models and Figure by Carigi et al. (2002)

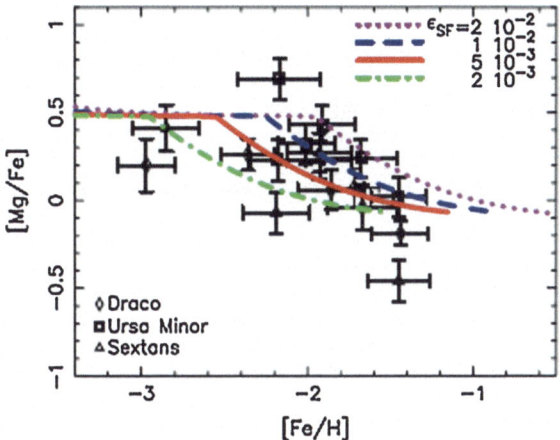

Fig. 2.46 Observed and predicted [Mg/Fe] versus [Fe/H] relation for the dSph Draco, Ursa Minor and Sextans. The different curves refer to different SF efficiencies (ϵ_{SF}) expressed in Gyr^{-1}, which are equivalent to the quantity ν. Figure from Ikuta and Arimoto (2002)

More recently, Fenner et al. (2006) suggested a model with a galactic wind for Sculptor: they indicated an efficiency of SF of $0.05\,Gyr^{-1}$. They concluded, from the study of the [Ba/Y] ratio, that chemical evolution in dSphs is inconsistent with the SF being truncated after reionization (at redshift z = 8). In fact, the high value of this ratio measured in stars indicates strong s-process production from low mass stars which have very long lifetimes.

2.6.3.1 The Results of Lanfranchi and Matteucci

(Lanfranchi and Matteucci 2003, 2004 hereafter LM04) developed models for dSphs of the Local Group. First they tested a "standard model" devised for describing an average dSph galaxy. This model was based on the following assumptions:

- one long star formation episode of duration ~ 8 Gyr,
- a small star formation efficiency, namely the star formation rate per unit mass is 1–10% of that in the solar vicinity,
- a strong galactic wind develops when the thermal energy of the gas equates the binding energy of the gas. The rate of gas loss is assumed to be several times the SFR, as in eq. (2.27), with typical values of λ between 5 and 15.

The condition for the onset of the wind is written as:

$$(E_{th})_{ISM} \geq E_{Bgas} \quad (2.37)$$

namely, that the thermal energy of the gas is larger or equal to its binding energy. The thermal energy of gas due to SN and stellar wind heating is:

$$(E_{th})_{ISM} = E_{th_{SN}} + E_{th_w} \quad (2.38)$$

with the contribution of SNe being:

$$E_{th_{SN}} = \int_0^t \epsilon_{SN} R_{SN}(t') dt', \quad (2.39)$$

while the contribution of stellar winds is:

$$E_{th_w} = \int_0^t \int_{12}^{100} \varphi(m) \psi(t') \epsilon_w dm dt' \quad (2.40)$$

with $\epsilon_{SN} = \eta_{SN} \epsilon_o$ and $\epsilon_o = 10^{51}$ erg (typical SN energy), and $\epsilon_w = \eta_w E_w$ with $E_w = 10^{49}$ erg (typical energy injected by the stellar wind of a $20 M_\odot$ star taken as representative). η_w and η_{SN} are two free parameters and indicate the efficiency of energy transfer from stellar winds and SNe into the ISM, respectively, quantities still largely unknown. It is assumed that $\epsilon_w = 0.03$ for the stellar winds, and that $\epsilon_{SN} = 0.03$ for Type II SNe and $\epsilon_{SN} = 1.0$ for Type Ia SNe, as suggested by Recchi et al. (2001). The total mass of the galaxy is expressed as $M_{tot}(t) = M_*(t) + M_{gas}(t) + M_{dark}(t)$ with $M_L(t) = M_*(t) + M_{gas}(t)$ and the binding energy of gas is:

$$E_{Bgas}(t) = W_L(t) + W_{LD}(t) \quad (2.41)$$

with:

$$W_L(t) = -0.5 G \frac{M_{gas}(t) M_L(t)}{r_L} \quad (2.42)$$

which is the potential well due to the luminous matter and with:

$$W_{LD}(t) = -G w_{LD} \frac{M_{gas}(t) M_{dark}}{r_L} \quad (2.43)$$

which represents the potential well due to the interaction between dark and luminous matter, where $w_{LD} \sim \frac{1}{2\pi} S(1 + 1.37S)$, with $S = r_L/r_D$, being the ratio between the galaxy effective radius and the radius of the dark matter core. The typical model for a dSph starts with an initial baryonic mass of $10^8 M_\odot$ and ends up, after the wind, with a luminous mass of $\sim 10^7 M_\odot$. The dark matter halo is assumed to be ten times larger than the luminous initial mass but diffuse ($S = 0.1$). The galactic wind in these galaxies develops after several Gyr from the start of SF, according to the different assumed SF efficiency. Soon after, but not immediately, the wind has started, the SF decreases very strongly until it halts completely.

- The IMF is that of Salpeter (1955) for all galaxies.
- Each galaxy is supposed to have formed by infall of gas clouds of primordial chemical composition, on a timescale not longer than 0.5 Gyr.

In Fig. 2.47 we show the [α/Fe] ratios for different α-elements and for different efficiencies of SF, as predicted by the standard model of Lanfranchi and Matteucci (2003).

As one can see, the [α/Fe] ratios show a clear change in slope followed by a steep decline, in agreement with the data. The change in slope corresponds to the occurrence of the galactic wind which starts emptying the galaxy of gas. In such a situation the SF starts to decrease as does therefore the production of the α-elements from massive stars, whereas Fe continues to be produced since its progenitors have long lifetimes. This produces the steep slope: the low SF efficiency and the wind, which decreases furtherly the SF. In this situation, the time-delay model predicts an earlier and steeper decline of the [α/Fe] ratios, as we have already discussed.

In LM04, the histories of star formation of specific galaxies were taken into account and they developed models for six dSphs: Carina, Ursa Minor, Sculptor, Draco, Sextans and Sagittarius. In Table 2.1 we show the assumed SF histories and the assumed model parameters. In particular, in column 1 are the galaxy names, in column 2 are the SF efficiencies, in column 3 the wind parameter, in column 4 the number of SF episodes, in column 5 the time at which the SF episodes start, in column 6 the duration, in Gyr, of the SF episodes, in column 7 the times for the occurrence of the galactic wind and in column 8 the assumed IMF. The histories of SF have been taken from: Rizzi et al. (2003) for Carina, Dolphin et al. (2005) for Draco, Sextans and Ursa Minor and Dolphin (2002) for Sagittarius and Sculptor.

In Figs. 2.48, 2.49, 2.50, 2.51 and 2.52 we show the predictions for specific dSphs by LM04. As one can see, the [α/Fe] data in the dSphs are well reproduced and in particular the steep decline of the [α/Fe] ratio is well reproduced. This steep decline is due again to the low efficiency SFR, a feature common also to the other models, coupled with a strong and continuous galactic wind which gradually empties the galaxies of gas. In the previous models either the galactic wind was not present or it was assumed instantaneous or not as strong as in LM04, thus predicting a flatter slope for the descent of the [α/Fe] versus [Fe/H].

Lanfranchi et al. (2006) computed also the expected abundances of s- and r-process elements in dSphs, by adopting the same nucleosynthesis prescriptions used for the chemical evolution of the Milky Way. In particular, they adopted the

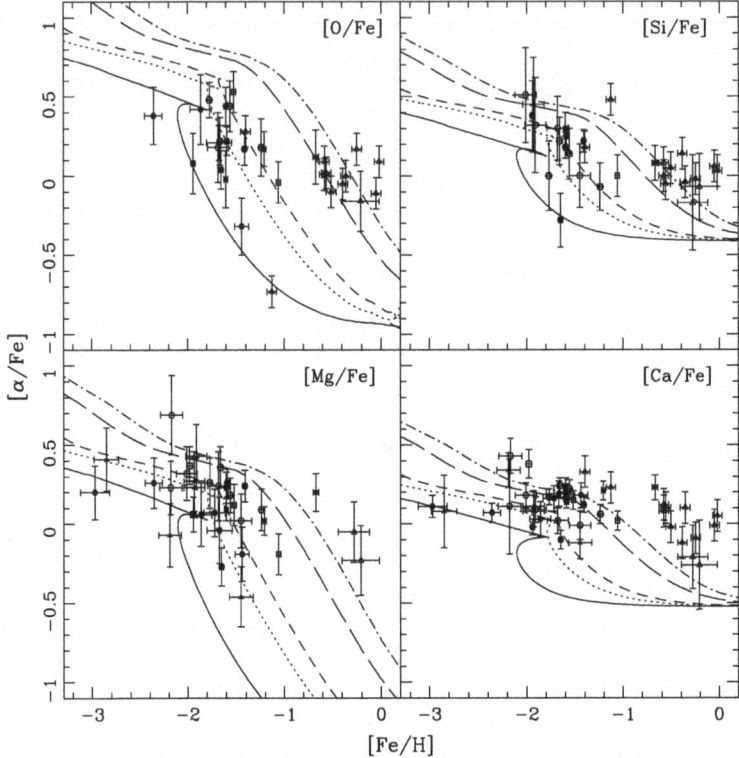

Fig. 2.47 Observed and predicted [α/Fe] versus [Fe/H]. The different lines refer to the "standard model" with different SF efficiency ν going from 1 (*dashed-dotted lines*) to 0.01 Gyr^{-1} (*continuous lines*). The points represent stars in different dSphs: Sagittarius (*open triangles*), Draco (*filled hexagons*), Carina (*filled circles*), Ursa Minor (*open hexagons*), Sculptor (*open circles*), Sextans (*filled triangles*), Leo I (*open squares*) and Fornax (*filled squares*). Figure and references from Lanfranchi and Matteucci (2003)

Table 2.1 Models for dSph galaxies. $M_{tot}^{initial}$ is the baryonic initial mass of the galaxy, ν is the star-formation efficiency, λ is the wind efficiency, and n, t and d are the number, time of occurrence and duration of the SF episodes, respectively

Galaxy	$M_{tot}^{initial}(M_\odot)$	ν (Gyr^{-1})	λ	n	t (Gyr)	d (Gyr)	IMF
Sextan	5×10^8	0.01–0.3	9–13	1	0	8	Salpeter
Sculptor	5×10^8	0.05–0.5	11–15	1	0	7	Salpeter
Sagittarius	5×10^8	1.0–5.0	9–13	1	0	13	Salpeter
Draco	5×10^8	0.005–0.1	6–10	1	6	4	Salpeter
Ursa Minor	5×10^8	0.05–0.5	8–12	1	0	3	Salpeter
Carina	5×10^8	0.02–0.4	7–11	2	6/10	3/3	Salpeter

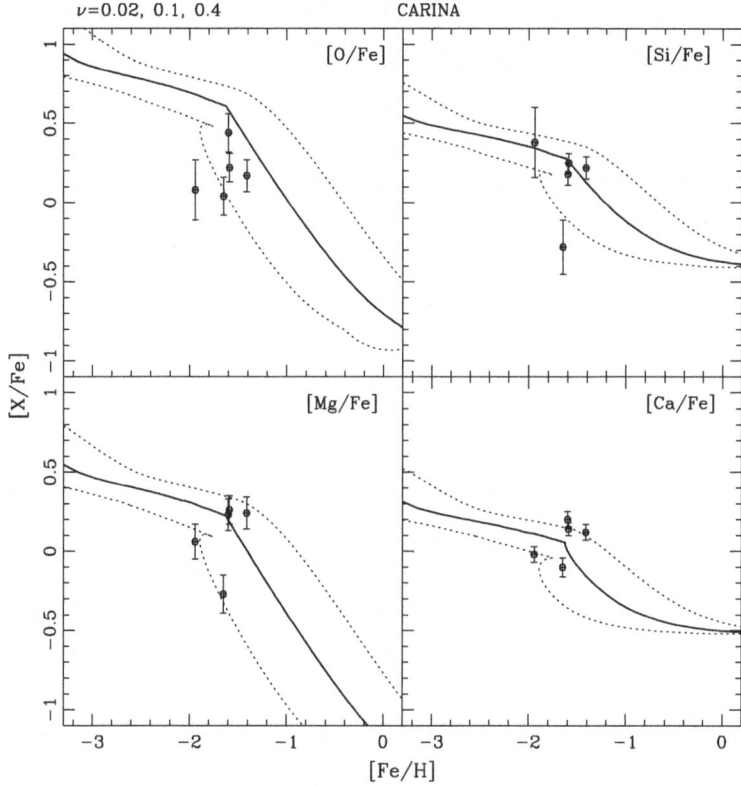

Fig. 2.48 Observed and predicted [α/Fe] versus [Fe/H] relation for the galaxy Carina. The *different lines* represent models with different SF efficiency. The *continuous line* represents the best model and corresponds to the efficiency $\nu = 0.1\,\mathrm{Gyr}^{-1}$. Figure from LM04

prescriptions of Cescutti et al. (2006) for Ba, Y, La, Sr and Eu: Ba, Sr, La and Y are mainly s-process elements produced on long timescales by low mass stars ($1-3\,M_\odot$), but they have also a small r-process component originating in stars in the mass range $12-30\,M_\odot$. The Eu instead is considered as a pure r-process element produced in the stellar mass range $12-30\,M_\odot$.

In Figs. 2.53 and 2.54 we show the predictions of Lanfranchi et al. (2006) for s- and r-process elements in Sculptor compared with the available data. Also in this case the agreement looks good, although more data are necessary before drawing firm conclusions. The general tendency for the α-elements in dSphs is to be less overabundant relative to Fe and the Sun than in the stars of the solar vicinity with the same [Fe/H]. This is due to the lower SFR in dSphs (the effect is increased by the galactic wind) which acts to shift the curve [α/Fe] versus [Fe/H] for the solar vicinity towards left in the diagram, whereas a stronger SF than in the solar neighbourhood moves the solar vicinity curve toward right in the diagram (see Fig. 2.34). The same holds for s- and r- process elements: in this case, since [s/Fe] versus [Fe/H] first

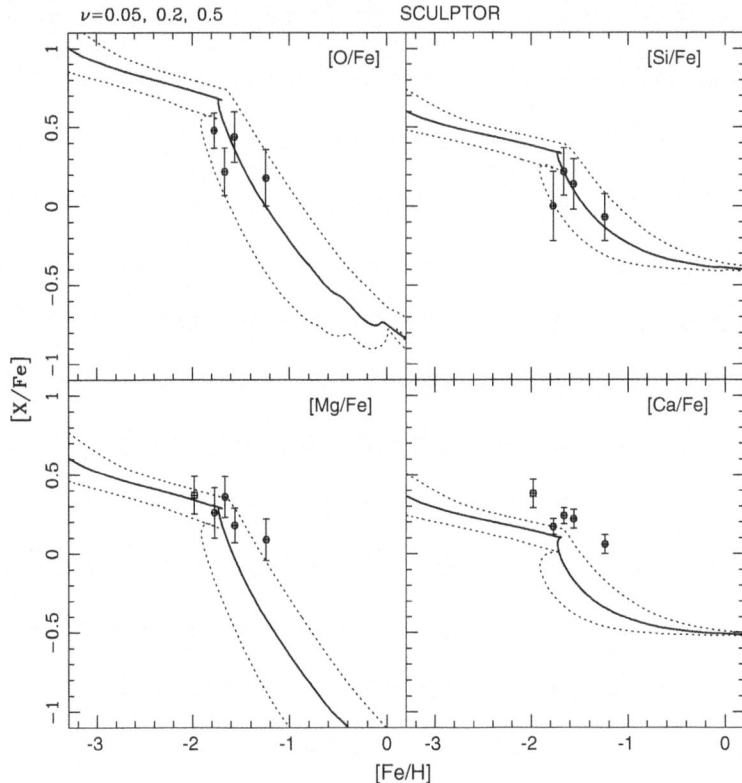

Fig. 2.49 Observed and predicted [α/Fe] versus [Fe/H] relation for the galaxy Sculptor. The *different lines* represent models with different SF efficiencies ($\nu = 0.05, 0.2, 0.5\,\mathrm{Gyr}^{-1}$). The *continuous line* represents the best model ($\nu = 0.2\,\mathrm{Gyr}^{-1}$). Figure from LM04

increases sharply at low metallicities and then it flattens at higher ones (the opposite of what happens for the α-elements), the dSphs show a higher [s/Fe] than the stars in the solar vicinity at the same [Fe/H]. This shows again the effect of the time-delay model but it still needs to be proven by observations.

Finally, another important constraint for model of galactic chemical evolution is represented by the stellar metallicity distribution. In Fig. 2.55 we show the predictions for the stellar metallicity distribution of Carina compared with the observed one and the agreement is very good. The observed distribution is from Koch et al. (2006) who measured the metallicity of 437 giants in Carina by means of Ca triplet and then transformed it into [Fe/H] through a suitable calibration. In Fig. 2.56 we also show the comparison between the stellar metallicity distribution in Carina and the G-dwarf metallicity distribution in the solar vicinity. As one can see, the Carina distribution lies in a range of smaller metallicities due to the lower efficiency of SF assumed for this galaxy.

2 Chemical Evolution of the Milky Way and Its Satellites

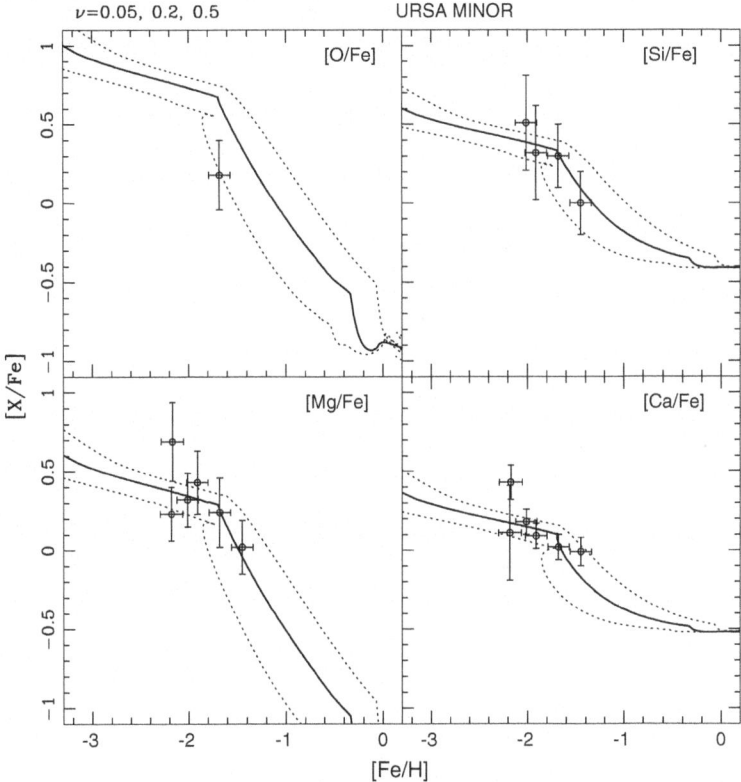

Fig. 2.50 Observed and predicted [α/Fe] versus [Fe/H] relation for the galaxy Ursa Minor. The *different lines* represent models with different SF efficiencies. The *continuous line* represents the best model ($\nu = 0.2\,\text{Gyr}^{-1}$). Figure from LM04

However, a word of caution is appropriate: in fact, the Ca triplet in principle traces the abundance of Ca and not that of Fe and we know that Ca and Fe evolve in a different way since Ca is mainly produced in Type II SNe, whereas Fe is produced mainly in Type Ia SNe. This different evolution of Ca and Fe leads, in the Koch paper, to an uncertainty of 0.2 dex. Besides that, the globular clusters which serve as calibrators for obtaining [Fe/H] lie in the range -2.0–1.0 dex, whereas Koch's data extend down to lower metallicities.

The good fit of the stellar metallicity distribution indicates that both the assumed history of SF and the IMF are close to reality.

LM04 predicted the stellar metallicity distribution for all the six dSphs and, while for Carina the agreement is quite good, for Sculptor they cannot reproduce the bimodal stellar distribution suggested by Tolstoy et al. (2004) and shown in Fig. 2.57, since their model is a one-zone model. In Fig. 2.58 are shown the predictions of LM04 for the Sculptor galaxy: it is clear from this Figure that to reproduce the two different stellar populations, one has to assume a multizone model

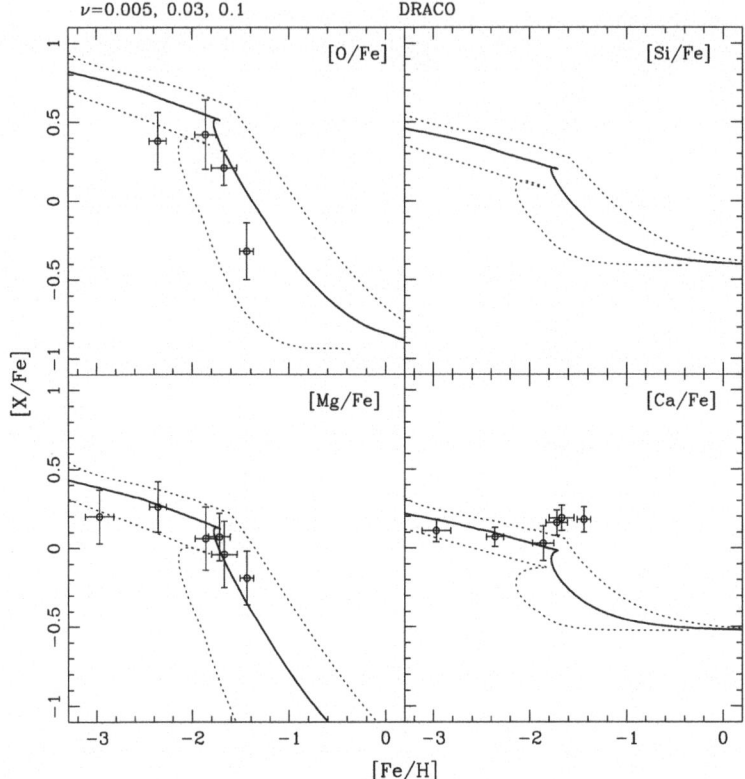

Fig. 2.51 Observed and predicted [α/Fe] versus [Fe/H] relation for the galaxy Draco. The *different lines* represent models with different SF efficiencies. The *continuous line* represents the best model ($\nu = 0.03\,\mathrm{Gyr}^{-1}$). Figure from LM04

possibly with different efficiencies of SF. Kawata et al. (2006) explained the bimodality of stellar populations in Sculptor as a consequence of dissipative collapse which produces higher metallicities at the center of the galaxy.

2.6.4 What Have we Learned About dSphs?

From the study of the chemical evolution of dSphs and the Milky Way we can derive the following conclusions:

- By comparing the [α/Fe] ratios in the Milky Way and in dSphs of the Local Group we can conclude that these systems had different histories of SF.

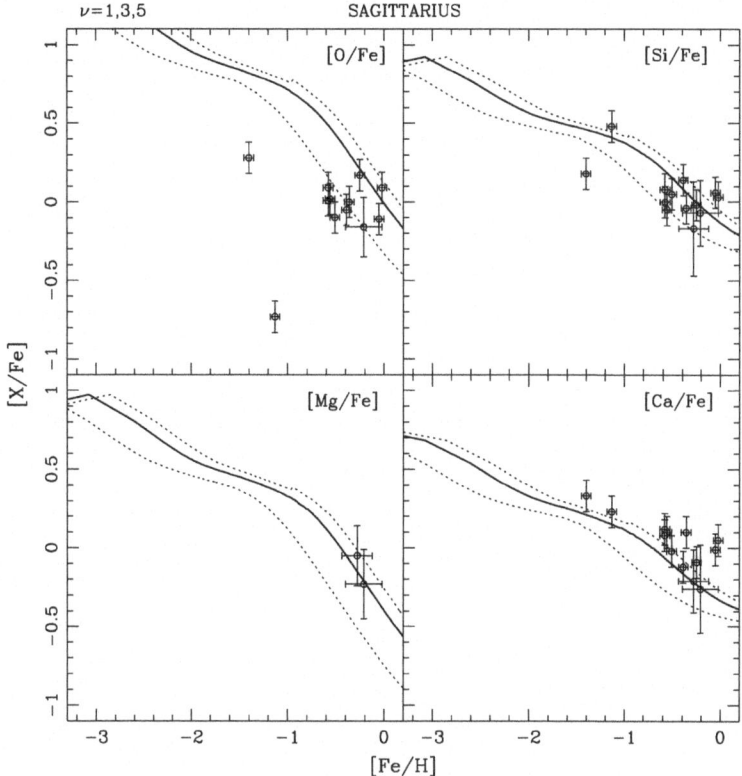

Fig. 2.52 Observed and predicted [α/Fe] versus [Fe/H] relation for the galaxy Sagittarius. The *different lines* represent models with different SF efficiencies. The *continuous line* represents the best model ($\nu = 3.0\,\text{Gyr}^{-1}$). Here the efficiency of SF is relatively high but the strong galactic wind makes it effectively much lower. Figure from LM04

- The [α/Fe] ratios in dSphs are always lower than in the Milky Way at the same [Fe/H]. This is a consequence of the time-delay model which predicts this behaviour for systems which suffered a lower star formation than the solar vicinity.
- The occurrence of strong galactic winds or gas loss in general is necessary to keep the SF low and it produces the steep decrease of the [α/Fe] ratio observed in dSphs (see Lanfranchi and Matteucci 2007).
- The [s/Fe] ratios are predicted to be higher than the same ratios in Milky Way stars with the same [Fe/H]. This is again a consequence of the time-delay model.
- The dSphs of the Local Group contain very old stars but they suffered extended periods of SF, far beyond the reionization epoch (Fenner et al. 2006). This is suggested both from the color-magnitude diagrams of these galaxies and from the level of the abundances of s-process elements such as Ba, which could not have been observed if the SF had stopped at the reionization epoch.

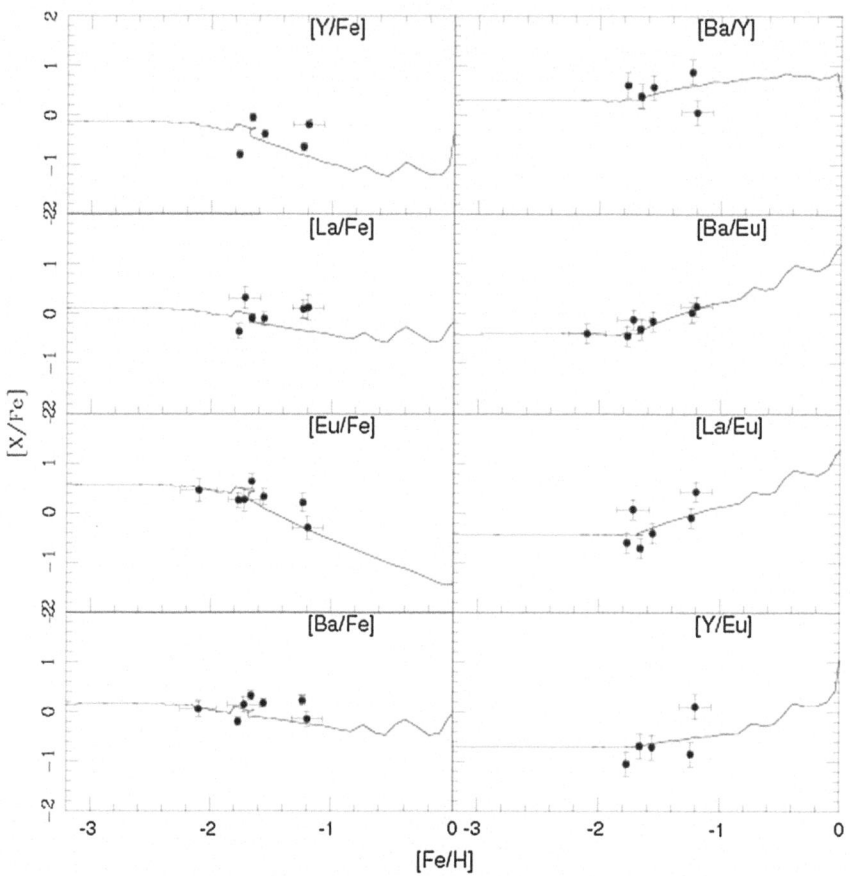

Fig. 2.53 Predicted and observed [s,r/Fe] versus [Fe/H] for the galaxy Sculptor. The model is from Lanfranchi et al. (2006), where the references for the the data can be found

- All the previous conclusions suggest that it is unlikely that the dSphs have been the building blocks of the Milky Way, as predicted by current CDM models (see review by Geisler et al. 2007 for a detailed discussion of this point). Robertson et al. (2005) studied the formation of the Galactic stellar halo by means of different accretion histories for the dark matter halo of the Milky Way in the framework of the λCDM model. They concluded, on the basis of the [α/Fe] ratios in Galactic halo stars, in dwarf irregulars and dSphs, that it is more likely that the Galactic dark matter halo was formed by an early accretion of dwarf irregular galaxies, which formed stars for a short time and then were destroyed. Concerning dSphs, they suggest that their chemical abundances should have been affected by galactic winds and that the dSphs should have been accreted and destroyed over the entire Milky Way lifetime.

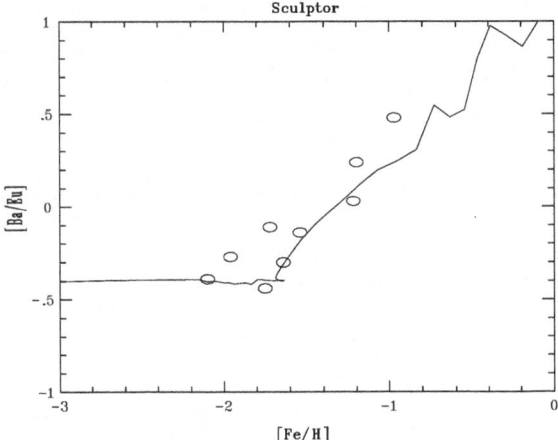

Fig. 2.54 Predicted and observed [Ba/Eu] versus [Fe/H] for the galaxy Sculptor. The model is from Lanfranchi et al. (2006), the data and the figure are from Geisler et al. (2007)

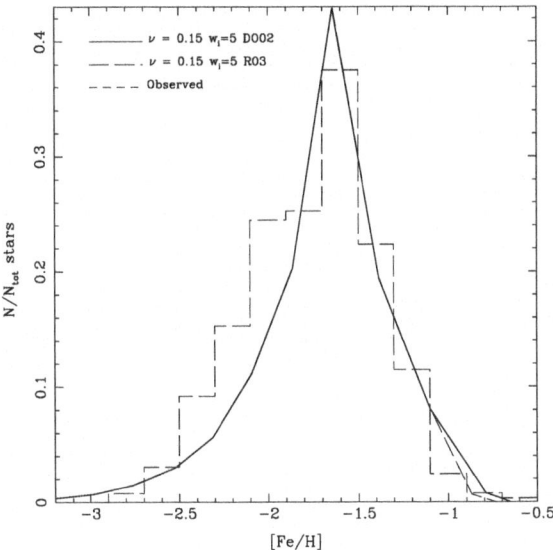

Fig. 2.55 Stellar metallicity distribution for Carina. Model from Lanfranchi et al. (2006). The assumed SF efficiency is $\nu = 0.15\,\text{Gyr}^{-1}$ and the wind efficiency is $\lambda = 5$. Two different histories of SF have been tested here: the one of Dolphin (2002) (*continuous line*) and that of Rizzi et al. (2003) (*long dashed line*), but this does not produce important differences in the results. The main difference between the two histories of SF is the number of bursts(3 in Dolphin and 4 in Rizzi et al). Data from Koch et al. 2006. Figure from Lanfranchi et al. (2006)

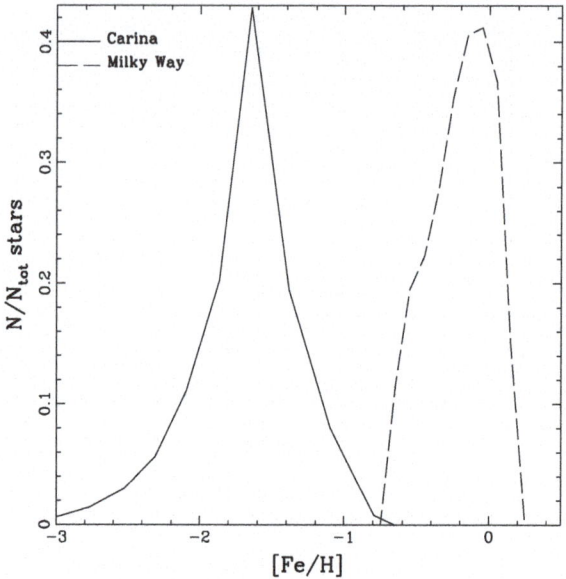

Fig. 2.56 Predicted stellar metallicity distribution for Carina compared to the predicted G-dwarf metallicity distribution in the solar neighbourhood (*dashed line*)

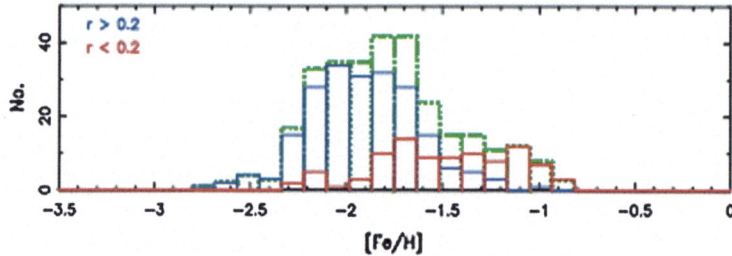

Fig. 2.57 Observed stellar metallicity distribution for Sculptor. Data and figure are from Tolstoy et al. (2004): all the stars of Sculptor are indicated by the *dotted line*. The central stars are those indicated by the lower histogram with *continuous line*, whereas the stars beyond $R > 0.2$ Kpc are indicated by the *upper* histogram with *continuous line*

2.7 Ultra-Faint Dwarfs in the Local Group

In the recent years even fainter systems than dwarf spheroidals have been found by SDSS around the Milky Way (Belokurov et al. 2007) and their individual stars have been resolved. These systems have been called Ultra Faint Dwarfs (UFDs). There are 8 systems with visual magnitude in the range $-7 < M_V < -4$ (Bootes I, UMa I, UMa II, Leo IV, Leo V, CVn II, Coma and Hercules) and most of the rest are at $M_V > -3$ (e.g. Wi I, Segue I, SegueII, Bootes II & III). These galaxies are predominantly metal poor and old systems and their masses are quite small: for

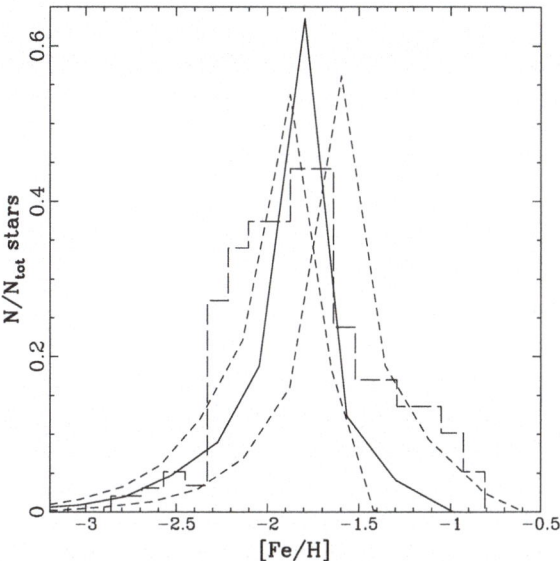

Fig. 2.58 Observed and predicted stellar metallicity distribution for Sculptor. The data (the *dotted line* of Fig. 2.59) are represented by the histogram (*long dashed*). The models are from LM04: the *solid line* represents the best model, whereas the *dotted lines* represent models with higher (the curve on the *right*) or lower (the curve on the *left*) SF efficiency

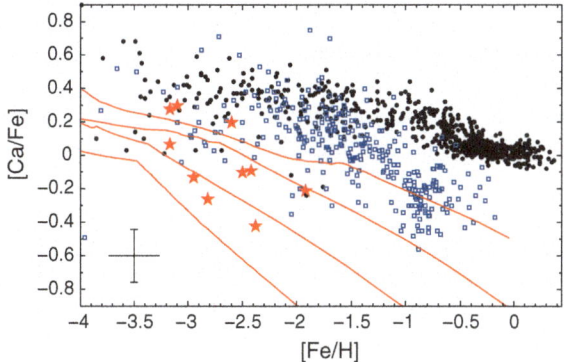

Fig. 2.59 [Ca/Fe] ratio in MW disk and halo stars (*black dots*), dSph stars (*blue squares*) and Hercules targets (*red symbols*). Overplotted are models of chemical evolution based on the Lanfranchi and Matteucci (2004) model, computed for star forming efficiencies of (*top to bottom*) $n = 0.1, 0.01, 0.001,$ and 0.0001 Gyr 1. Figure from Koch et al. (2012)

example, Hercules has an estimated stellar mass of $(4 - 7) \cdot 10^4 \, M_\odot$ (Martin et al. 2008). So far, Koch et al. (2008, 2012); Kirby et al. (2008); Frebel et al. (2010); Gilmore et al. (2013) and Vargas et al. (2013) have measured abundances in UFDs. The abundance ratios measured in these very faint and metal poor systems should suggest their history of star formation which should have been characterized by a very low efficiency. This seems to be true for Hercules, as shown in Fig. 2.59, where data from the Milky Way, dwarf spheroidals and Hercules are reported and compared to chemical evolution model results obtained by adopting very low star formation efficiencies, lower than those adopted for the models of dwarf spheroidals previously discussed.

2.8 Other Spirals

In this section we will briefly describe the properties of other spirals of the Local Group for which we have enough chemical information.

In external spirals we can measure:

- the SFR mainly from H_α emission and this suggests a correlation between the SFR and the total surface gas density, as discussed previously.
- Abundance gradients are also found in disks of local spirals (see Garnett et al. 1997): in particular, it is found that abundance gradients, expressed in dex/Kpc, are steeper in smaller disks but the correlation disappears if they are expressed in dex/R_d (where R_d is the disk scalelength). It is interesting to note that a universal gradient slope per unit scale length may be explained by viscous disk models (e.g. Lin and Pringle 1987; Sommer-Larsen and Yoshii 1989). Further information on gradients is that they are flatter in disks with bars: probably the bar induces radial flows which can wash-out the abundance gradients if their velocity is high enough (see Tinsley 1980).
- The gas distribution: one finds differences in the gas distribution along the disk between field and cluster galaxies, these latter being subject to ram pressure stripping.
- Integrated colors of galaxy disks give information on the distribution of stellar populations along the disks. Several authors (Josey and Arimoto 1992; Jimenez et al. 1998; Prantzos and Boissier 2000) have suggested that color gradients, as well as metallicity gradients, can be reproduced by assuming an inside-out formation for disks, as has been suggested for the disk of the Milky Way.
- In Fig. 2.60 we show the results of a paper by Boissier et al. (2001). They conclude that more massive disks are redder, more metal rich and more gas-poor than smaller ones. On the other hand, their estimated SF efficiency (defined as the SFR per unit mass of gas) seems to be similar among different spirals: this leads them to conclude that more massive disks are older than less massive ones. The various quantities in Fig. 2.60 are plotted as functions of the disk circular velocity which is a measure of the dark matter halo of each galaxy. The various curves are obtained by varying the spin parameter λ_s, expressed as a λ_s/λ_{MW}, where λ_{MW} refers to the Milky Way. In fact, in the framework of semi-analytical models of galaxy formation the evolution of galactic disks can be described by means of scaling laws calibrated on the Milky Way with V_c and λ_s as parameters (e.g. Mo et al. 1998).

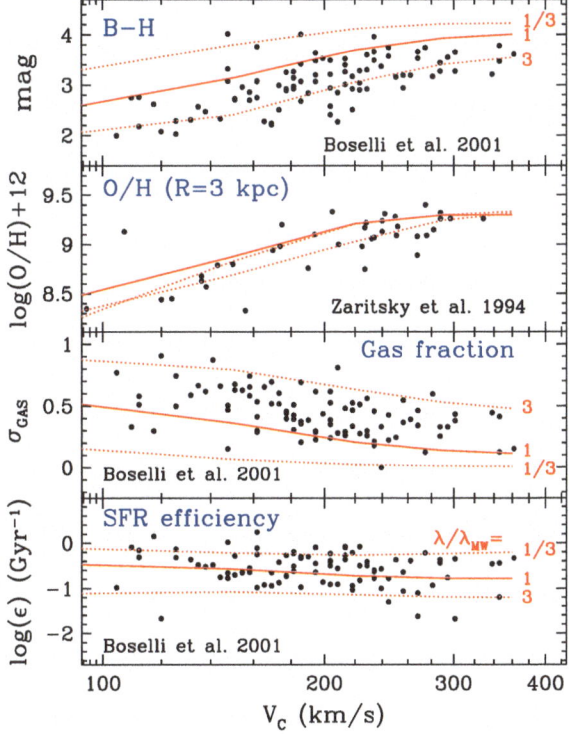

Fig. 2.60 Distribution of the (B-H) color, the $12 + \log$(O/H), the gas fraction and the SFR efficiency (the SFR per unit mass of gas) versus the circular velocity of disks of local spirals. The *different lines* refer to different values of the spin parameter λ. Figure from Boissier et al. (2001)

2.8.1 Chemical Models for External Spirals

Several models of chemical evolution of Local Group spirals have been developed in the past years (e.g. Diaz and Tosi 1984; Mollá et al. 1996; Chiappini et al. 2003). Diaz and Tosi (1984) first modeled the chemical evolution of M31, M33, M83 and M101. Mollá et al. (1996) modeled several spirals of the Local Group (M31, NGC300, M33, NGC628, NGC3198, NGC6946). In Fig. 2.61 we show the predictions, compared to observations, of the Mollá et al. model for M31.

As an another example of abundance gradients and gas distribution in a local spiral galaxy we show in Fig. 2.62 the observed and predicted gas distribution and abundance gradients for the disk of M101. In this case the gas distribution and the abundance gradients are reproduced with systematically smaller timescales for the disk formation relative to the MW (M101 formed faster), and the difference between the timescales of formation of the internal and external regions is smaller ($\tau_{M101} = 0.75 r(\text{Kpc}) - 0.5$ Gyr, Chiappini et al. (2003)) compared to Eq. (2.34).

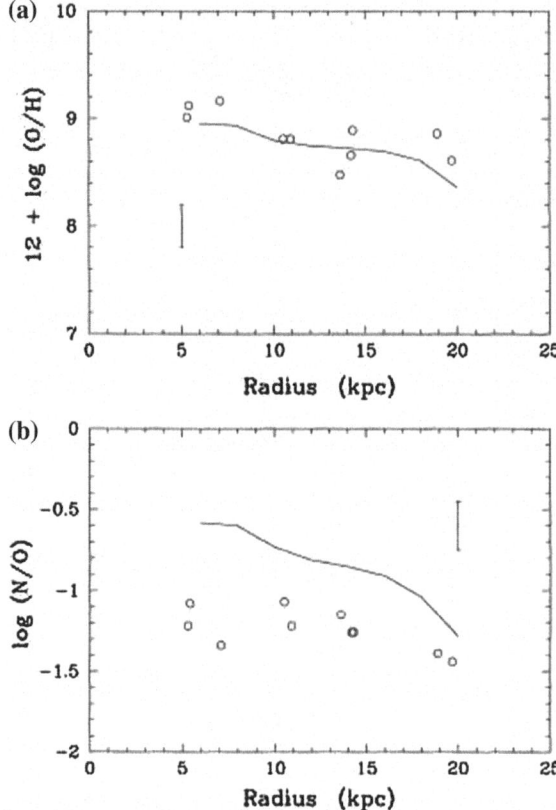

Fig. 2.61 Predicted and observed abundance gradients in M31. Figure and models are from Mollá et al. (1996)

Therefore, in the Chiappini et al. (2003) paper it is suggested that the fact that more luminous spirals (e.g. M101) tend to have shallower abundance gradients than less luminous ones can be interpreted as due to a faster formation (down-sizing) of large spirals, as also hinted at by the results of Boissier et al. (2001).

In summary, from the available studies of spirals in the Local Group we can suggest the following:

- The disks of spirals have all formed inside-out and the more massive disks have formed faster than the less massive one.
- This translates into a faster gas accretion rate and consequently into a faster SFR.
- In other words, the most massive disks are also the oldest, a conclusion which is not in line with classical predictions from CDM models.

Fig. 2.62 *Upper panel* predicted and observed gas distribution along the disk of M101. The observed HI, H_2 and total gas are indicated in the Figure. The *large open circles* indicate the models: in particular, the *open circles* connected by a *continuous line* refer to a model with central surface mass density of $1000 M_\odot \, \text{pc}^{-2}$, while the *dotted line* refers to a model with $800 M_\odot \, \text{pc}^{-2}$ and the *dashed line* to a model with $600 M_\odot \text{pc}^{-2}$. *Lower panel* predicted and observed abundance gradients of C,N,O elements along the disk of M101. The models are the lines and differ for a different threshold density for SF, being larger in the *dashed model*. All the models are by Chiappini et al. (2003)

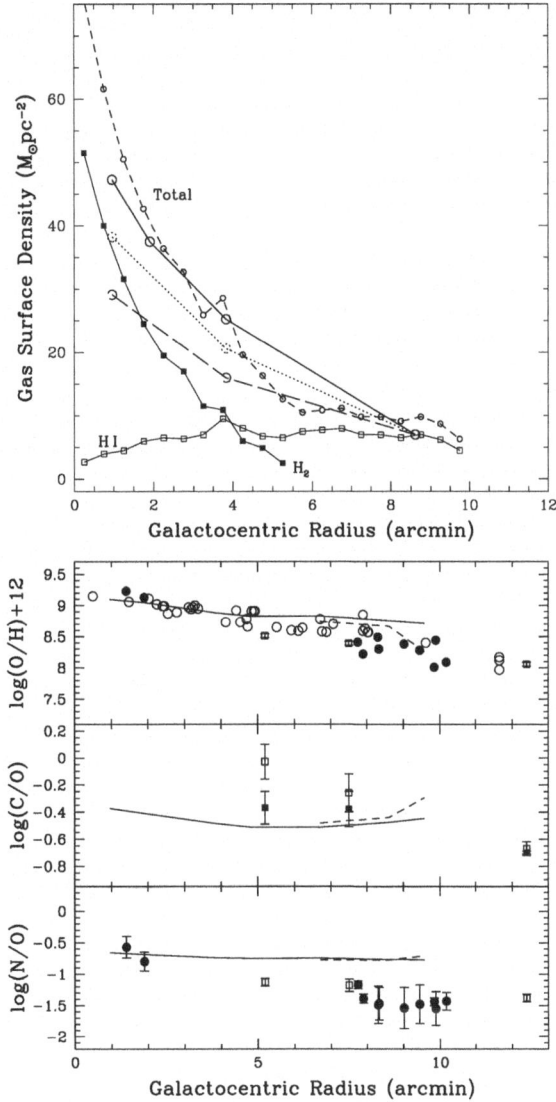

2.9 Cosmic Chemical Evolution

With the name cosmic chemical evolution we indicate the chemical evolution taking place in comoving volumes large enough to be representative of the whole universe (Pei and Fall 1995). The evolution can be described in terms of comoving densities of gas and stars Ω_{gas} and Ω_{stars}, both measured in units of the present critical density ($\rho_c = \frac{3H_o^2}{8\pi G}$) and the mean abundance of heavy elements in the ISM, Z, including

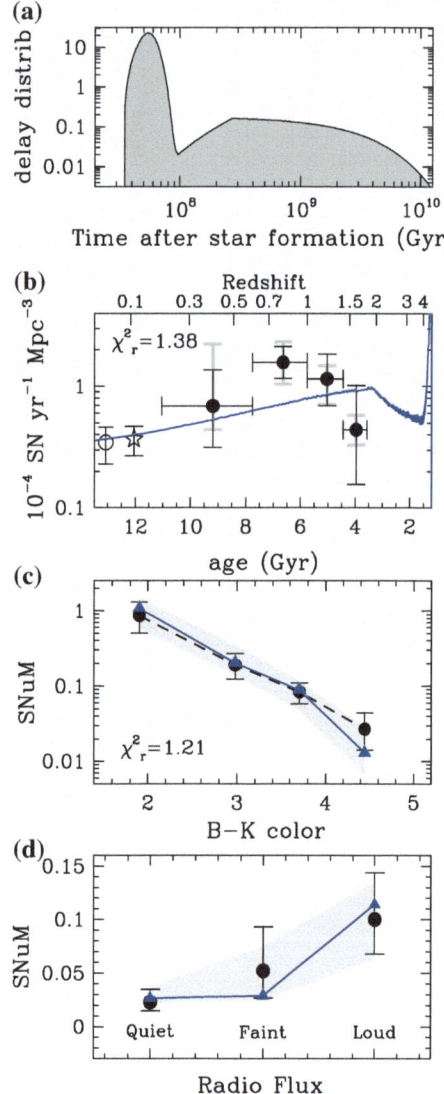

Fig. 2.63 Theoretical cosmic Type Ia SN rates compared with observational data from Mannucci et al. (2006). The progenitor model adopted assumes the delay-time distribution suggested by Mannucci et al. (2005, 2006) (panel **a**), whereas the cosmic SFR is the one of Calura and Matteucci (2003) (panel **b**). In panels **c** and **d** are shown predictions and data for the Type Ia SN rate per unit galactic mass (SNuM) versus color and radio flux in radio galaxies, respectively. For the references about the data see Mannucci et al. (2006). Figure adapted from Mannucci et al. (2006)

dust. Under IRA we can write, following Pei and Fall:

$$\dot{\Omega}_{gas} + \dot{\Omega}_{stars} = \dot{\Omega}_f \tag{2.44}$$

and

$$\Omega_{gas}\dot{Z} - y_Z \dot{\Omega}_{stars} = (Z_f - Z)\dot{\Omega}_f \tag{2.45}$$

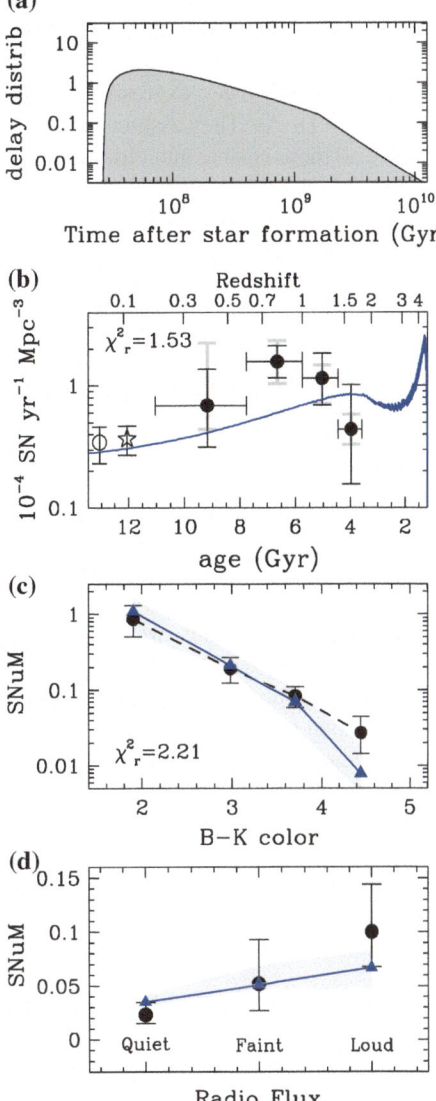

Fig. 2.64 Theoretical cosmic Type Ia SN rates compared with observational data from Mannucci et al. (2005, 2006). The progenitor model adopted assumes the delay-time distribution suggested by Matteucci and Recchi (2001) (panel **a**), whereas the cosmic SFR is the one of Calura and Matteucci (2003) (panel **b**). In panels **c** an **d** are shown predictions and data for the Type Ia SN rate per unit galactic mass (SNuM) versus color and radio flux in radio galaxies, respectively. For the references about the data see Mannucci et al. (2006). Figure adapted from Mannucci et al. (2006)

where the dots represent differentiation with respect to the cosmic time y_Z is the yield of metals per stellar generation and Z_f is either the metallicity of the inflowing or outflowing gas. We remind that these two metallicities are different. The term $\dot{\Omega}_f$ can represent either the infall or the outflow rate according to its sign. For the closed-box model $\dot{\Omega}_f = 0$ and its solution is:

$$Z = y_Z ln(\Omega_{gas}/\Omega_{gas_\infty}) \tag{2.46}$$

with Ω_{gas_∞} being the gas comoving density at some suitably high redshift when there are still no stars and heavy elements.

By means of these equations Pei and Fall followed the evolution of DLAs (the quasar absorbers). They expressed the quantity Ω_{gas} in terms of the observable properties of DLAs. They assumed IRA.

Since all these cosmic quantities refer to an unitary volume of the universe which contains galaxies of all morphological types, Calura and Matteucci (2004) proposed another approach to the cosmic chemical evolution, which takes into account galaxies of different morphological type. They computed the cosmic chemical enrichment of the universe by means of detailed models of chemical evolution of galaxies of all morphological types, relaxing IRA and assuming for each galaxy type a different history of SF, as discussed in the previous sections. They defined the comoving cosmic density of stars and gas for galaxies of different morphological type (ellipticals, spirals and irregulars) as:

$$\rho_{*,k} = \rho_{B,k} \cdot (M_*/L)_{B,k} \quad (2.47)$$

for the stars and

$$\rho_{g,k} = \rho_{B,k} \cdot (M_g/L)_{B,k} \quad (2.48)$$

for the gas. The quantities $(M_*/L)_{B,k}$ and $(M_g/L)_{B,k}$ are the predicted M/L ratios for stars and gas, respectively. $L_{B,k}$ is the blue luminosity for each galaxy type (k indicates the morphological type) and $\rho_{B,k}$ is the comoving blue luminosity for a given galaxian morphological type.

They computed the mean mass weighted metallicity of galaxies by summing the metallicities predicted for the different morphological types as:

$$<Z_{galaxies}> = \frac{\sum_k \rho_{g,k} Z_{g,k} + \sum_k \rho_{*,k} Z_{*,k}}{\sum_k (\rho_{g,k} + \rho_{*,k})} \quad (2.49)$$

and obtained:

$$<Z_{galaxies}> = 0.0175 = 0.9 Z_\odot \quad (2.50)$$

with 56, 42 and 2 % of metals produced in ellipticals, spirals and irregulars, respectively. Therefore, the conclusion is that the average metallicity in galaxies is almost solar and that most of the metals in the universe have been produced by elliptical galaxies. They also predicted the average [O/Fe] ratios for each galaxy type both in the gas and in stars.

In particular:

$[O/Fe]_{*,Ellipt} = 0.4$ dex, $[O/Fe]_{gas,Ellip} = -0.33$ dex, $[O/Fe]_{*,Spiral} = 0.1$ dex and $[O/Fe]_{gas,Spiral} = 0.01$ dex.

Then, they computed the metallicity in the intergalactic medium (IGM) by considering all the metals ejected by galaxies (mainly ellipticals) into the IGM:

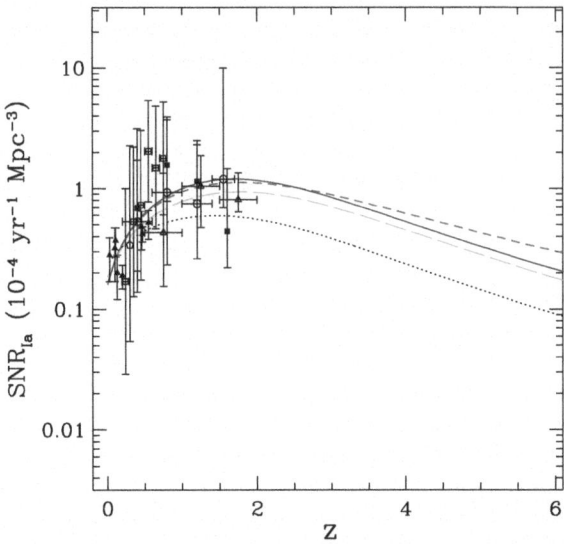

Fig. 2.65 Predicted cosmic SN Ia rates adopting the fit to the SFR density from Cole et al. (2001). The *short-dashed line* is for the Mannucci et al. (2006) DTD, the *solid line* is for the (Matteucci and Recchi 2001) DTD, while the *dashed* and *dotted lines* are for the wide DD and close DD DTDs, respectively (Greggio 2005). The Figure underlines the differences due to the choice of the DTDs. The data and the model are from Valiante et al. (2009). The redshift of galaxy formation is assumed to be $z_f = 6$

$$< Z_{IGM} > = \frac{\Omega_{Z,IGM}}{\Omega_{b,IGM}} \qquad (2.51)$$

with $\Omega_{b,IGM} = \Omega_b - \Omega_{b,*} - \Omega_{b,gas} = 0.0753$ being the baryonic density of the IGM and $\Omega_b = 0.02h^{-2}$ (from WMAP, Spergel et al. 2003, 2007) being the total baryonic content of the universe. Therefore, they obtained:

$$< Z_{IGM} > = 6.54 \times 10^{-4} = 0.03 Z_\odot. \qquad (2.52)$$

Finally, they computed the average metallicity of the universe by accounting for all the metals produced in galaxies over the lifetime of the universe:

$$< Z_{universe} > = \frac{\sum_k \Omega_{Z,k}}{\Omega_b} = \frac{\Omega_{Z,tot}}{\Omega_b} = 0.0017 = 0.09 Z_\odot \qquad (2.53)$$

where $\Omega_{Z,tot}$ represents the sum of all the metals produced in all galaxies and Ω_b represents the the total amount of baryons in the universe.

In summary, the mean metallicity inside galaxies of all morphological types is almost solar, whereas the mean metallicity of the universe is roughly 1/10 solar.

Calura and Matteucci (2003) computed also cosmic Type Ia SN rates (SNR_{cosm}), expressed in SNu (number of $SNe/10^{10}L_{B_\odot}$ per century). In particular, they took into account the contribution of all galaxy types in the following way:

$$SNR_{cosm}(z) = \frac{\sum_k SNR_k(z)}{\sum_k L_{B_k}} \qquad (2.54)$$

where SNR_{cosm} can represent the Type II, Ib/c, Ia SNe. The sums are over all the galactic morphological types, L_{B_k} is the total blue luminosity of the kth morphological type. In order to compute the SNR_{cosm} for each galaxy they assumed a SN model progenitor and a cosmic SFR (CSFR), calculated as:

$$\dot{\rho}_* = \sum_k \rho_{B,k}(z) \cdot (M_*/L)_{B,k}(z) \cdot \psi(z)_k, \qquad (2.55)$$

where $\rho_{B,k}(z)$ and $(M_*/L)_{B,k}(z)$ have been already defined and $\psi(z)_k$ represents the history of SF of a galaxy of kth morphological type. They assumed the SF histories of Fig. 2.41. In Figs. 2.63 and 2.64 we show some examples of predicted cosmic SN rates by adopting the same cosmic SFR (Eq. 2.55) but different assumptions about the Type Ia progenitor model.

As one can see from these Figures, the best DTD appears to be the one of Mannucci et al. (2006), although this conclusion is based only on the fit of the rates in radio galaxies. Concerning the cosmic Type Ia SN rate (panel b) the agreement is good for both DTDs except for the point at the highest redshift which is highly uncertain. We need more detections of SNe Ia especially at high redshift before drawing any conclusion on the high redshift cosmic Type Ia SN rate. Moreover, it is difficult to infer a particular DTD from the cosmic SN Ia rate since it depends also on the assumed cosmic star formation rate. If one assumes a fit to the observed CSFR, such as that of Cole et al. (2001) which fits also more recent data, then it obtains the results shown in Fig. 2.65. Here one can notice that it is very hard to choose a particular DTD, given the large error bars in the data.

Acknowledgments I warmly thank Eva Grebel and Ben Moore for inviting me to deliver these lectures in the beautiful village of Murren. I also thank my collaborators, Silvia Kuna Ballero, Francesco Calura, Gabriele Cescutti, Cristina Chiappini, Valentina Grieco, Gustavo Lanfranchi, Antonio Pipino, Simone Recchi and Emanuele Spitoni, whose precious help has allowed me to put together most of the material presented in these lectures. Finally, I am grateful to I.J. Danziger for his patience in reading the manuscript.

References

Alibés, A., Labay, J. & Canal, R., 2001, A&A, 370, 1103
Alves-Brito, A., Melndez, J., Asplund, M., Ramrez, I., Yong, D., 2010, A&A, 513, 35
Anders, E. & Grevesse, N. 1989, in Geochimica et Cosmochimica Acta, vol. 53, p. 197

Andrievsky S.M., Bersier D., Kovtyukh V.V. et al., 2002a, A&A, 384, 140
Andrievsky S.M., Kovtyukh V.V., Luck R.E. et al., 2002b, A&A, 381, 32
Andrievsky S.M., Kovtyukh V.V., Luck R.E. et al., 2002c, A&A, 392, 491
Andrievsky S.M., Luck R.E., Martin P. et al., 2004, A&A, 413, 159
Argast, D., Samland, M., Gerhard, O.E. & Thielemann, F.-K., 2000, A&A 356, 873
Asplund, M., Grevesse, N. & Sauval, A.J., 2005, ASP (Astronomical Society of the Pacific) Conf. Series, Vol. 336, p. 55
Babusiaux, C., Gómez, A., Hill, V., Royer, F., Zoccali, M., Arenou, F., Fux, R., Lecureur, A. et al. 2010, A&A, 519, 77
Ballero, S., Matteucci, F., Origlia, L. & Rich, R.M., 2007, A&A 467, 123
Barbuy, B. & Grenon, M., 1990. in: Bulges of Galaxies, eds. B.J. Jarvis & D.M. Terndrup, ESO/CTO Workshop, p. 83
Barbuy, B., Ortolani, S.& Bica, E., 1998, A&AS, 132, 333
Battaglia, G., Tolstoy, E., Helmi, A., Irwin, M. J., Letarte, B., Jablonka, P., Hill, V., Venn, K. A., Shetrone, M. D., Arimoto, N. & al., 2006, A&A, 459, 423
Bellazzini, M., Ferraro, F. R., Origlia, L., Pancino, E., Monaco, L. & Oliva, E., 2002, AJ, 124, 3222
Belokurov, V., Zucker, D. B., Evans, N. W., Kleyna, J. T., Koposov, S., Hodgkin, S. T., Irwin, M. J., Gilmore, G., Wilkinson, M. I., Fellhauer, M. et al., 2007, ApJ, 654, 897
Bensby, T., Asplund, M., Johnson, J. A., Feltzing, S., Melndez, J., Dong, S., Gould, A., Han, C., Adn, D., Lucatello, S., Gal-Yam, A., 2010, A&A, 521, L57
Bensby, T., Adén, D., Melendez, J., Gould, A., Feltzing, S., Asplund, M., Johnson, J. A., Lucatello, S., Yee, J. C., Ramrez, I., et al., 2011, A&A, 533, 134
Boissier, S., Prantzos, N., 1999, MNRAS, 307, 857
Boissier, S., Boselli, A., Prantzos, N. & Gavazzi, G., 2001 MNRAS, 321, 733
Bonifacio, P., Hill, V., Molaro, P., Pasquini, L., Di Marcantonio, P., Santin, P., 2000, A&A, 359, 663
Bonifacio P., Sbordone L., Marconi G., Pasquini L., Hill V., 2004, A&A, 414, 503
Calura, F. 2004 Ph.D., Thesis, Trieste University
Calura, F.& Matteucci, F., 2003, ApJ, 596, 734
Calura, F.& Matteucci, F., 2004, MNRAS 350, 351
Carigi, L., Hernandez, X. & Gilmore, G., 2002 MNRAS 334, 117
Carney B.W., Yong D., Teixera de Almeida, M.L., Seitzer P., 2005, AJ, 130, 1111
Carollo, D., Beers, T. C., Lee, Y. S., Chiba, M., Norris, J. E., Wilhelm, R., Sivarani, T., Marsteller, B., Munn, J. A. & Bailer-Jones, C. A. L., 2007, Nature, 450, 1020
Carraro G., Bresolin F., Villanova S. & al., 2004, AJ, 128, 1676
Casagrande, L., Schoenrich, R., Asplund, M., Cassisi, S., Ramirez, I., Melendez, J., Bensby, T., Feltzing, S., 2011, A&A, 530, 138
Cayrel, R., Depagne, E., Spite, M., Hill, V., Spite, F., François, P., Plez, B., Beers, T., & al., 2004, A&A, 416, 117
Cescutti, G., FranÇois, P., Matteucci, F., Cayrel, R. & Spite, M., 2006, A&A 448, 557
Cescutti, G., Matteucci, F., François, P. & Chiappini, C., 2007, A&A, 462, 943
Cescutti, G., Matteucci, F., 2011, A&A, 525, 126
Chabrier, G., 2003, PASP 115, 763
Chang, R.X., Hou, J.L., Shu, C.G. & Fu, C.Q., 1999, A&A, 350, 38
Chiappini, C, Hirschi, R., Meynet, G., Ekstroem, S., Maeder, A. & Matteucci, F., 2006, A&A, 449, L27
Chiappini, C., Matteucci, F. & Ballero, S., 2005, A&A, 437, 429
Chiappini, C., Matteucci F. & Gratton R. 1997, ApJ 477, 765
Chiappini, C., Matteucci, F., & Romano, D., 2001, ApJ, 554, 1044
Chiappini, C., Romano, D & Matteucci, F., 2003, MNRAS, 339, 63
Chieffi, A. & Limongi, M., 2004, ApJ, 608, 405
Chieffi, A. & Limongi, M., 2002, ApJ, 577, 281
Chiosi, C., 1980, A&A, 83, 206

Cole, S., Norberg, P., Baugh, C.M., Frenk, C.S., Bland-Hawthorn, J., Bridges, T., Cannon, R., Colless, M. et al., 2001, MNRAS, 326, 255
Costa, R.D.D., Escudero, A.V., Maciel, W.J., 2005, in the proceedings of the conference "Planetary Nebulae as Astronomical Tools". AIP Conference Proceedings, Volume 804, pp. 252
Daflon, S.& Cunha K., 2004, ApJ, 617, 1115
Diaz, A.I. & Tosi, M., 1984, MNRAS, 208, 365
Dolphin A.E., 2002, MNRAS, 332, 91
Dolphin, A. E., Weisz, D. R., Skillman, E. D.& Holtzman, J. A., 2005, astro-ph/0506430, Invited Review at the meeting "Resolved Stellar Populations", held in Cancun, Mexico, 18–22 April 2005; paper with full resolution figures available at http://purcell.as.arizona.edu/pubs/cancun.ps.gz
Dopita, M.A.& Ryder, S.D., 1994, ApJ, 430, 163
Eggen, O.J., Lynden-Bell, D. & Sandage, A.R., 1962, ApJ, 136, 748
Elmegreen, Bruce G., 1999, ApJ, 517, 103
Escudero, A.V., Costa, R.D.D., 2001, A&A, 380, 300
Escudero, A.V., Costa, R.D.D., Maciel, W.J., 2004, A&A 414, 211
Fenner, Y., Gibson, Brad K., Gallino, R. & Lugaro, M., 2006, ApJ, 646, 184
Ferreras, I., Wyse, F.G. & Silk, J., 2003, MNRAS, 345, 1381 521, 81
François, P., Matteucci, F. Cayrel, R., Spite, M., Spite, F. & Chiappini, C., 2004, A&A, 421, 613
François, P., Depagne, E., Hill, V., Spite, M., Spite, F., Plez, B., Beers, T. C., Barbuy, B., Cayrel, R. Andersen, J. & al., 2006, AIP Conference Proceedings, Vol. 847, p. 205
Frebel, A., Simon, J. D., Geha, M., Willman, B., 2010, ApJ, 708, 560
Fulbright, J. P., Rich, R. M. & Castro, S., 2004, ApJ, 612, 447
Fulbright, J.P., McWilliam, A. & Rich, R.M., 2006, ApJ, 636, 831
Fulbright, J.P., McWilliam, A. & Rich, R.M., 2007, ApJ, 661, 1152
Forestini, M. & Charbonnel, C., 1997, A&A 123, 241
Gallagher, J. S., III, Hunter, D. A. & Tutukov, A.V., 1984, ApJ 284, 544
Garnett, D.R., Skillman, E.D., Dufour, R.J.& Shields, G.A., 1997, ApJ, 481, 174
Geisler, D., Wallerstein, G., Smith, V. V. & Casetti-Dinescu, D. I., 2007, PASP, 119, 939
Geisler, D., Smith, V. V., Wallerstein, G., Gonzalez, G. & Charbonnel, C., 2005, AJ, 129, 1428
Gilmore, G., Wilkinson, M. I., Wyse, R. F. G., Kleyna, J. T., Koch, A., Evans, N. W. & Grebel, E. K., 2007, ApJ, 663, 948
Gilmore, G., Norris, J. E., Monaco, L., Yong, D., Wyse, R. F. G., Geisler, D., 2013, ApJ, 763, 61
Gonzalez, O.A., Rejkuba, M., Zoccali, M., Hill, V., Battaglia, G., Babusiaux, C., Minniti, D., Barbuy, B., Alves-Brito, A., Renzini, A. et al., 2011, A&A, 530, 54
Grebel, E. K. & Gallagher, J. S., III, 2004, ApJ, 610, L89
Greggio, L., 2005, A&A, 441, 1055
Greggio, L. & Renzini, A., 1983a, A&A, 118, 217
Greggio, L. & Renzini, A., 1983b, MemSaIt, Vol. 54, 311
Grieco, V., Matteucci, F., Pipino, A., Cescutti, G., 2012, A&A, 548, 60
Han, Z. & Podsiadlowski, Ph., 2004 MNRAS 350, 1301
Hartwick, F.D.A., 1976, ApJ, 209, 418
Hernandez X., Gilmore G., Valls-Gabaud D., 2000, MNRAS, 317, 831
Helmi, A., Irwin, M. J., Tolstoy, E., Battaglia, G., Hill, V., Jablonka, P., Venn, K., Shetrone, M., Letarte, B., Arimoto, N. et al., 2006, ApJ, 651, L121
Hill, V., François, P., Spite, M., Primas, F. & Spite, F., 2000, A&A, 364, L19
Hill, V., Lecureur, A., Gómez, A., Zoccali, M., Schultheis, M., Babusiaux, C., Royer, F., Barbuy, B., Arenou, F., Minniti, D., Ortolani, S., 2011, A&A, 534, 80
Hirschi, R. 2007, A&A 461, 571
Iben, I. Jr. & Tutukov, A., 1984, ApJ 284, 719
Ikuta, C. & Arimoto, N., 2002, A&A 391, 55
Immeli, A., Samland, M., Gerhard, O. & Westera, P., 2004, A&A 413, 547

Israelian, G., Ecuvillon, A., Rebolo, R., García-López, R., Bonifacio, P., Molaro, P., 2004, A&A, 421, 649
Iwamoto, K., Brachwitz, F., Nomoto, K., Kishimoto, N., Umeda, H., Hix, W. R. & Thielemann, F-K., 1999, ApJS, 125, 439
Jimenez, R., Padoan, P., Matteucci, F. & Heavens, A.F., 1998, MNRAS, 299, 123
Josey, S. A. & Arimoto, N., 1992, A&A, 255, 105
Kauffmann, G., 1996, MNRAS 281, 487
Kawata D., Arimoto N., Cen R. & Gibson B.K., 2006, ApJ, 641, 785
Karakas, A. & Lattanzio, J.C., 2007, PASA, 24, 103
Karakas, A., 2010, MNRAS, 403, 1413
Kennicutt, R.C. Jr., 1998a, ApJ, 498, 541
Kennicutt, R.C. Jr., 1998b, ARA&A, 36, 189
Kennicutt, R. C., Jr., Tamblyn, P. & Congdon, C. E., 1994, ApJ, 435, 22
Kirby, E. N., Simon, J. D., Geha, M., Guhathakurta, P., Frebel, A., 2008, ApJ, 685, L43
Kobayashi, C., Tsujimoto, T., Nomoto, K., Hachisu, I. & Kato, M., 1998, ApJ 503, L155
Koch, A., Grebel, E. K., Wyse, R. F. G., Kleyna, J. T., Wilkinson, Mark I., Harbeck, D. R., Gilmore, G. F. & Evans, N. W., 2006, AJ, 131, 895
Koch, A., Koch, McWilliam, A., Grebel, E. K., Zucker, D. B., Belokurov, V., 2008, ApJ, 688, L13
Koch, A., Matteucci, F., Feltzing, S., 2012, in FIRST STARS IV - FROM HAYASHI TO THE FUTURE -. AIP Conference Proceedings, Volume 1480, pp. 190–193 (2012)
Kotoneva, E., Flynn, C., Chiappini, C., Matteucci, F., 2002, MNRAS, 336, 879
Kroupa, P., 2001, MNRAS, 322, 231
Kroupa, P., Tout, C.A. & Gilmore, G., 1993, MNRAS 262, 545
Lanfranchi, G. & Matteucci, F., 2003, MNRAS, 345, 71
Lanfranchi, G. & Matteucci, F., 2004, MNRAS, 351, 1338 (LM04)
Lanfranchi, G. & Matteucci, F., 2007, A&A 468, 927
Lanfranchi, G. & Matteucci, F. & Cescutti, G., 2006, MNRAS, 365, 477
Langer, N. & Henkel, C., 1995, Space Science Reviews, Vol. 74, p. 343
Larson, R.B., 1972, Nature 236, 21
Lecureur, A., Hill, V., Zoccali, M., Barbuy, B., Gomez, A., Minniti, D., Ortolani, S. & Renzini, A., 2007, A&A, 465, 799
Limongi, M. & Chieffi, A., 2003, ApJ, 592, 404
Lin, D. N. C. & Pringle, J. E., 1987, ApJ, 320, L87
Luck R.E., Gieren W.P., Andrievsky S.M. et al. 2003, A&A, 401, 939
Luck, R. E., Lambert, D. L., 2011, AJ, 142, 136
Maeder, A., 1992, A&A, 264, 105
Maeder, A., Meynet, G., 2001, A&A, 373, 555
Mannucci, F., Della Valle, M., Panagia, N., Cappellaro, E., Cresci, G., Maiolino, R., Petrosian, A. & Turatto, M., 2005, A&A, 433, 807
Mannucci, F., Della Valle, M.& Panagia, N., 2006, MNRAS 370, 773
Marigo, P., 2001, A&A, 370, 194
Marigo, P., Bressan, A. & Chiosi, C., 1996, A&A 313, 545
Martin, C.L., Kennicutt, R.C., Jr, 2001, ApJ, 555, 301
Martin, N. F., de Jong, J. T. A., Rix, H.-W., 2008, ApJ, 684, 1075
Mateo M.L., 1998, ARA&A, 36, 435
Matteucci, F., 2001, The Chemical Evolution of the Galaxy, ASSL, Kluwer Academic Publisher
Matteucci, F., 2003, Ap&SS, 284, 539
Matteucci, F., Brocato, E., 1990, ApJ, 365, 539
Matteucci, F. & Chiosi, C., 1983, A&A, 123, 121
Matteucci, F. & François, P., 1989, MNRAS 239, 885
Matteucci, F. & Greggio, L., 1986, A&A, 154, 279
Matteucci, F. & Recchi, S., 2001, ApJ, 558, 351
Matteucci, F., Romano, D., Molaro, P., 1999, A&A, 341, 458

Martinelli, A. & Matteucci, F., 2000, A&A, 353, 269
McWilliam, A. & Rich, R. M., 1994, ApJS, 91, 749
McWilliam, A. & Smecker-Hane, T. A., 2005a, ApJ 622, L29
McWilliam, A. & Smecker-Hane, T. A., 2005b, in Cosmic Abundances as Records of Stellar Evolution and Nucleosynthesis in honor of David L. Lambert, ASP Conference Series, Vol. 336, Proceedings of a symposium held 17–19 June, 2004 in Austin, Texas. Edited by Thomas G. Barnes III and Frank N. Bash. San Francisco: Astronomical Society of the Pacific, p. 221
McWilliam, A., Matteucci, F., Ballero, S., Rich, R. M., Fulbright, J. P., Cescutti, G., 2008, AJ, 136, 367
Meléndez, J. & Barbuy, B., 2002, ApJ 575, 474
Meléndez, J., Asplund, M., Alves-Brito, A., Cunha, K. Barbuy, B., Bessell, M. S., Chiappini, C., Freeman, K. C., Ramrez, I., Smith, V. V., Yong, D., 2008, A&A, 484, L21
Meynet, G. & Maeder, A., 2002a, A&A, 390, 561
Meynet, G., Maeder, A., 2002b, A&A, 381, L25
Meynet, G.& Maeder, A., 2003, A&A, 404, 975
Meynet, G.& Maeder, A., 2005, A&A, 429, 581
Meynet, G, Ekstroem, S. & Maeder, A., 2006, A&A, 447, 623
Mo, H. J., Mao, S. & White, S. D. M., 1998, MNRAS, 295, L71
Mollá, M., Ferrini, F. & Gozzi, G., 2000, MNRAS, 316, 345
Mollá, M., Ferrini, F. & Diaz, A. I., 1996, ApJ, 466, 668
Monaco L., Bellazzini M., Bonifacio P. et al., 2005, A&A, 441, 141
Monelli, M., Pulone, L., Corsi, C. E., Castellani, M., Bono, G., Walker, A. R., Brocato, E., Buonanno, R., Caputo, F., Castellani, V. et al., 2003, AJ 126, 218
Mott, A., Matteucci, F., Spitoni, E., 2013, MNRAS, 435, 2918
Navarro, J. F., Frenk, C. S.& White, S. D. M., 1997, ApJ 490, 493
Nissen, P. E. 2003, in CNO in the Universe, ed. C. Charbonnel, D. Schaerer, & G. Meynet, ASP Conf. Ser., 304, 60
Ness, M., Freeman, K., 2012, Assembling the Puzzle of the Milky Way, Le Grand-Bornand, France, Edited by C. Reyl; A. Robin; M. Schultheis; EPJ Web of Conferences, Volume 19, id.06003
Nomoto, K., Hashimoto, M., Tsujimoto, T., Thielemann, F.-K. & al., 1997, Nucl. Phys. A., 616, 79
Nomoto, K., Tominaga, N., Umeda, H., Kobayashi, C. & Maeda, K., 2006, Nucl. Phys. A777, 424
Oey, M. S., 2000, ApJ, 542, L25
Pagel, B.E.J., 1997, "Nucleosynthesis and Chemical Evolution of Galaxies", Cambridge University Press
Pardi, M.C., Ferrini, F. & Matteucci, F., 1995, ApJ, 444, 207
Pei, Y.C. & Fall S.M., 1995, ApJ, 454, 69
Pettini, M., Ellison, S. L., Bergeron, J.& Petitjean, P., 2002, A&A 391, 21
Pompéia, L., Barbuy, B., Grenon, M., 2003, ApJ, 592, 1173
Portinari, L., Chiosi, C. & Bressan, A., 1998, A&A, 334, 505
Portinari, L., Chiosi, C., 2000, A&A, 355, 929
Prantzos, N., 2003, A&A, 404, 211
Prantzos, N. & Boissier, S., 2000, MNRAS, 313, 338
Recchi, S., Matteucci, F. & D'Ercole, A., 2001, MNRAS 322, 800
Reid, M.J., 1993, ARA&A, 31, 345
Renzini, A., 2006, ARA&A, 44, 141
Renzini, A., 1993, in "Galactic Bulges", IAU Symp. 153, eds. H. DeJonghe & H.J. Habing (Kluwer Academic Publishers, Dordrecht), p. 151
Renzini, A., Voli, M., 1981, A&A, 94, 175 (RV81)
Rauscher, T., Heger, A., Hoffman, R. D. & Woosley, S. E., 2002, ApJ 576, 323
Rich, R.M., 1988, AJ, 95, 828
Rich, R.M. & McWilliam, A., 2000, in Discoveries and Research Prospects from 8- to 10-Meter-Class Telescopes, Ed. Jacqueline Bergeron, Proc. SPIE Vol. 4005, p. 150
Rich, R.M., Origlia, L., 2005, ApJ, 634, 1293

Rizzi, L., Held, E. V., Bertelli, G. & Saviane, I., 2003, ApJ, 589, L85
Robertson, B., Bullock, J. S., Font, A. S., Johnston, K. V.& Hernquist, L., 2005, ApJ, 632, 872
Robin, A.C., Luri, X., Reylé, C., Isasi, Y., Grux, E., Blanco-Cuaresma, S., Arenou, F., Babusiaux, C. et al., 2012, A&A, 543, 100
Romano, D., Karakas, A. I., Tosi, M., Matteucci, F., 2010, A&A, 522, 32
Ryde, N., Edvardsson, B., Gustafsson, B., Eriksson, K., Kufl, H. U., Siebenmorgen, R., Smette, A., 2009, A&A, 496, 701
Sadakane K., Arimoto N., Ikuta C. et al., 2004, PASJ, 56, 1041
Sadler, E.M., Rich, R.M.& Terndrup, D.M., 1996, AJ, 112, 171
Salpeter, E.E., 1955, ApJ, 121, 161
Samland, M., Hensler, G., Theis, C., 1997, ApJ 476, 544
Sandage, A., 1986, A&A 161, 89
Scalo, J.M., 1986, Fund. Cosmic Phys. 11, 1
Scalo, J.M., 1998, The Stellar Initial Mass Function, A.S.P. Conf. Ser., Vol. 142 p. 201
Schmidt, M., 1959, ApJ 129, 243
Searle, L. & Zinn, R., 1978, ApJ, 225, 357
Sellwood, J. A., Binney, J. J., 2002, MNRAS, 336, 785
Shetrone, M. D., Coté, P.& Sargent, W. L. W., 2001, ApJ, 548, 592
Shetrone M., Venn K.A., Tolstoy E., Primas F., 2003, AJ, 125, 684
Schoenrich, R., Binney, J., 2009, A&A, 396, 203
Siess, L., Livio, M. & Lattanzio, J., 2002, ApJ, 570, 329
Smecker-Hane T. & Mc William A., 1999, in Spectrophotometric Dating of Stars and Galaxies, Eds. Hubeny I. et al., ASP Conference Proceedings, Vol. 192, p. 150
Sommer-Larsen, J. & Yoshii, Y., 1989, MNRAS, 238, 133
Spergel, D. N., Verde, L., Peiris, H. V., Komatsu, E., Nolta, M. R., Bennett, C. L., Halpern, M., Hinshaw, G., Jarosik, N., Kogut, A. & al., 2003, ApJS, 148, 175
Spergel, D. N., Bean, R., Dor, O., Nolta, M. R., Bennett, C. L., Dunkley, J., Hinshaw, G., Jarosik, N., Komatsu, E., Page, L. & al., 2007, ApJS, 170, 335
Spite, M., Cayrel, R., Plez, B., Hill, V., Spite, F., Depagne, E., François, P., Bonifacio, P., Barbuy, B., Beers, T.& al., 2005, A&A, 430, 655
Spitoni, E., Matteucci, F., 2011, A&A, 531, 72
Talbot, R. J., Jr. & Arnett, W. D., 1975, ApJ 197, 551
Thielemann, F.K., Nomoto, K. & Hashimoto, M., 1996, ApJ 460, 408
Tinsley, B.M., 1980, Fund. Cosmic Phys. Vol. 5, 287
Tinsley, B.M., 1979, ApJ, 229, 1046
Tolstoy, E., Venn, K. A.. Shetrone, M., Primas, F., Hill, V., Kaufer, A., Szeifert, T., 2003, AJ, 125, 707
Tolstoy, E., Irwin, M. J., Helmi, A., Battaglia, G., Jablonka, P., Hill, V., Venn, K. A., Shetrone, M. D., Letarte, B., Cole, A. A. & al., 2004, ApJ 617, 119
Tornambé, A. & Matteucci, F., 1986, MNRAS, 223, 69
Tornambé, A. & Matteucci, F., 1986, MNRAS, 223, 69
Umeda, H. & Nomoto, K., 2001, in The Physics of Galaxy Formation, ASP Conference Proceed. Vol. 222, p. 45
Valiante, R., Matteucci, F., Recchi, S., Calura, F., 2009, NewA, 14, 638
van den Hoek, L.B. & Groenewegen, M.A.T., 1997, A&AS, 123, 305 (HG97)
Vargas, L. C., Geha, M., Kirby, E. N., Simon, J. D., 2013, arXiv1302.6594
Venn K. A., Irwin M., Shetrone M.D. & al., 2004, AJ, 128, 1177
Ventura, P., D'Antona, F. & Mazzitelli, I., 2002, A&A, 393, 215
Vladilo, G., 2002, A&A, 391, 407
Yong D., Carney B.W.& Teixera de Almeida M.L., 2005, AJ, 130, 597
Yong D., Carney B.W.& Teixera de Almeida M.L., Pohl B.L., 2006, AJ, 131, 2256
Weidner, C. & Kroupa, P., 2005 ApJ, 625, 754
Whelan, J. & Iben, I. Jr., 1973, ApJ, 186, 1007

White, S.D.M., & Rees, M.J., 1978, MNRAS, 183, 341
Woosley, S.E. & Weaver, T.A., 1995, ApJS, 101, 181 (WW95)
Woosley, S. E. & Weaver, T. A., 1994, ApJ, 423, 371
Wyse, R.F.G., 1999, Ap&SS, 267, 145
Wyse, R.F.G. & Gilmore, G., 1992, AJ, 104, 144
Wyse, R. F. G. & Silk, J., 1989, ApJ, 339, 700
Zoccali, M., Lecureur, A., Barbuy, B., Hill, V., Renzini, A., Minniti, D., Momany, Y., Gómez, A., Ortolani, S., 2006, A&A, 457, L1
Zoccali, M., Renzini, A., Ortolani, S., Greggio, L., Saviane, I., Cassisi, S., Rejkuba, M., Barbuy, B., Rich, R.M. & Bica, E., 2003, A&A, 399, 931
Zoccali, M., Hill, V., Lecureur, A., Barbuy, B., Renzini, A., Minniti, D., Gómez, A., Ortolani, S., 2008, A&A, 486, 177

Index

Symbols
21cm radiation, 21, 29

A
Abundance gradients, 69, 184, 214
Accretion shocks, 32
Acoustic peaks, 6
Age gradients, 69
Age metallicity relation, 82
Age-metallicity relation, 71
AGN feedback, 22
Asteroseismology, 51

B
Baryon fraction, 22, 27, 32, 49, 63, 67
Black holes, 10, 11
Bondi accretion, 30
Bondis limit, 33
Brown dwarf stars, 150

C
Chandrasekhar mass, 157
Chemical abundances, 121
Chemical evolution models, 159
Chemical species, 146
Chemical tagging, 78, 86, 92, 93, 100
COBE, 56
Cold dark matter haloes, 22, 26, 29, 62
Cold flows, 35
Color-magnitude relation, 73
Cooling flows, 34
Cosmic microwave background (CMB), 4
Cosmochronemtry, 51
Cosmochronology, 101

D
Dark ages, 28
Dark matter
   discovery, 109
   dwarf spheroidals, 199
   haloes, 111
   LCDM, 104
   rotation curves, 111
Dark matter haloes, 12
   mass function, 22
Dark matter substructure, 66
Disk heating, 69
Disk surface density, 169
Downsizing, 19
Dwarf galaxies, 39, 82
Dwarf spheroidals, 198

E
Eddington accretion, 11
Eddington limit, 11

F
Feedback, 21
First stars, 59

G
G dwarf problem, 72, 161, 176
G dwarf stars, 188
GAIA, 54, 106
Galactic archaeology, 46
Galactic halo, 40
Galactic potential, 40
GAMA, 25
Gamma-ray bursts, 151

Gas accretion, 21, 27, 31, 36, 48, 192, 219
    cold flows, 35
    hot flows, 34
    warm flows, 38
Globular clusters, 15, 46, 53, 80, 82, 93

H
Hermes, 97
High velocity clouds (HVC), 20, 32, 39, 67
Hipparcos, 53, 119
Hubble flow, 20
Hubble sequence, 195
Hubble's law, 3

I
Initial mass function (IMF), 8, 148, 192
Instantaneous recycling approximation, 158

J
James Webb Space Telescope (JWST), 47
Jeans mass, 7, 30

K
Kelvin Helmholz instability, 42
Kennicutt law, 168
Kroupa IMF, 148

L
Local group satellites, 64, 66, 203, 208
Local supernovae rate, 169

M
M31, 27, 61, 111
M81, 20
Magellanic Clouds, 77
Magellanic Stream, 41, 77
Main sequence stars, 149
Mean free path, 31
Merger rates, 38
Mergers, 38, 58
Metal poor stars, 94
Metallicity, 155, 221
Metals, 94, 146
Milk Way
    metal poor halo, 60
Milky Way
    abundance gradient, 183
    age gradient, 66

    bar, 59
    bulge, 55, 186, 193
    chemical evolution, 164
    dark halo, 62
    disk, 53, 59
    formation, 48, 194
    formation timescale, 49, 194
    globular clusters, 80
    halo, 40
    halo stars, 48
    mass, 63
    star formation history, 74
    star formation rate, 31, 49, 169
    stellar halo, 60, 84
    thick disk, 49, 70, 165
    thin disk, 165
    think disk, 174
Missing baryons, 25
Missing satellites, 67
Molecular clouds, 90
Monolithic collapse, 46, 164
Moving groups, 84, 100

O
Open clusters, 70, 78

P
Photoinisation, 15
Planetary nebulae, 56, 189
Population III stars, 7, 8, 10, 59, 101, 155
Protostellar accretion, 8
Pseudobulges, 58

R
R-process, 94, 151, 179
RAVE, 120
Rayleigh-Taylor instability, 42
Reionization, 15, 25, 28, 29, 47, 67
RR Lyrae stars, 56

S
S-process, 152, 179
Sagittarius, 54, 85
Satellite accretion, 61, 64, 71, 75, 86
Scalo initial mass function, 167
Schmidt law, 147
SEGUE, 120
Specific angular momentum, 65
Square kilometer array (SKA), 28
Star clusters, 53, 69, 77, 84, 89, 93

Star formation, 90
Star formation rate, 25, 49, 147
Stellar accretion, 30
Stellar ages, 52, 118, 123
Stellar halo, 46
Stellar migration, 69, 86, 104, 177, 189
Stellar nucleosynthesis, 152, 191
Stellar photomet, 113
Stellar populations, 82
Stellar streams, 60, 75, 100
Stellar yields, 152
Supernovae, 9, 91, 94, 156, 166
Surface brightness profiles, 68

T
Tidal streams, 75, 84
Top hat collapse, 14
Tully-Fisher law, 65

Turbulence, 89
Type Ia supernovae, 151
Type II supernovae, 179, 188
TypeIa supernovae, 156

U
Ultra faint dwarf galaxies, 212

V
Virial radius, 13, 30
Virial temperature, 32
Virialisation, 13, 14

W
White dwarf stars, 53

The manufacturer's authorised representative in the EU is Springer Nature Customer Service Centre GmbH, Europaplatz 3, 69115 Heidelberg, Germany. If you have any concerns regarding our products, please contact ProductSafety@springernature.com

Printed and bound by CPI Group (UK) Ltd, Croydon, CR0 4YY
23/03/2026
02076681-0001